Galileo Studies

EUROPEAN PHILOSOPHY AND THE HUMAN SCIENCES
General Editor: John Mepham

This is a new series of translations of important and influential works of European thought in the fields of philosophy and the human sciences.

Other titles in the series:

Pietro Chiodo
Sartre and Marxism

Marcel Detienne
The Gardens of Adonis:
Spices in Greek Mythology

Serge Moscovici
Society Against Nature:
The Emergence of Human Societies

Marcel Detienne & Jean-Pierre Vernant
Cunning Intelligence in Greek Culture and Society

Jean-Pierre Vernant
Myth and Society in Ancient Greece

Galileo Studies

ALEXANDRE KOYRÉ

TRANSLATED FROM THE FRENCH BY
JOHN MEPHAM

HUMANITIES PRESS · NEW JERSEY

Published in the USA, 1978 by
HUMANITIES PRESS INC.
Atlantic Highlands, New Jersey 07716

This translation and arrangement
© The Harvester Press Limited 1978

First published in France by Hermann & Cie., Editeurs as *Etudes Galiléennes*.
© Librarie Scientifique Hermann et Cie, 1939

Library of Congress Cataloguing in Publication Data

Koyré, Alexandre, 1892-1964.
   Galileo studies.

   (European philosophy and the human sciences)
   Translation of Études galiléennes.
   Includes bibliographical references.
   1. Galilei, Galileo, 1564-1642. 2. Descartes, Rene, 1596-1650.
3. Dynamics. 4. Science—Philosophy.   I.   Title.
QB36.G2K6813   1978   520'.92'4   77-18003
ISBN 0-391-00760-2

Typesetting by
Cold Composition Ltd,
220 Vale Road, Tonbridge, Kent

Printed in England by
Redwood Burn Limited, Trowbridge, Wiltshire and
Esher, Surrey

TO THE MEMORY OF EMILE MYERSON

ACKNOWLEDGEMENT

The publishers acknowledge with gratitude the financial assistance from the French Government which has made this publication in the English language possible.

CONTENTS

TRANSLATOR'S NOTE

English translations of primary sources are my own except in the following cases. All quotations from Galileo's *Dialogue Concerning the Two Chief World Systems* are from the translation by Stillman Drake (University of California Press, Berkeley and Los Angeles, 1962; referred to in the notes as *Dialogue*). I have also used Professor Drake's translation of Galileo's *Il Saggiatore* (*The Assayer*, in *Discoveries and Opinions of Galileo*, Double day Anchor Books, Garden City, New York, 1957). As for Galileo's *Discorsi e Dimostrazioni Matematiche, Intorno à Due Nuove Scienze* I have usually made my own translations, but there are some passages which I have taken directly from the overall rather unsatisfactory available English translation by Henry Crew and Alfonso de Salvio (Dover Publications, New York, no date) which is unfortunately called *Dialogues Concerning Two New Sciences*; to avoid confusion, this work by Galileo is referred to throughout as *Discorsi* or *Discourses*. There are also two passages from this text of which I have used the English translation given in Marshall Clagett's extremely useful *The Science of Mechanics in the Middle Ages* (University of Wisconsin Press, Madison, 1961). Translations of Benedetti are from Stillman Drake and I. E. Drabkin, *Mechanics in Sixteenth-Century Italy: Selections from Tartaglia, Benedetti, Guido Ubaldo and Galileo* (University of Wisconsin Press, Madison, 1969). In all the above cases I have made modifications to the English translations so as to bring them into conformity with the text as a whole.

References to Galileo, *Opere*, are to *Le Opere di Galileo Galilei*, Edizione Nationale, ed. Antonio Favaro. References to Descartes, *Oeuvres*, are to *Oeuvres de Descartes*, eds. Charles Adam and Paul Tannery, Paris, 1897-1913.

I have throughout translated *vitesse* (Italian: *velocità*) as 'speed' so as to avoid the more specific post-Newtonian connotations of the word 'velocity'. *Pesanteur* has been translated as 'gravity' except where it is used to refer specifically to a property of bodies contrasted with lightness, in which case it is translated as 'heaviness'. But in any case the context should always make it clear that the word retains this ambiguity and does not have the meaning which it has come to have since Newton. The French *moment* (Italian *momento*, Latin *momentum*) is usually translated as 'moment' although in fact neither this nor 'momentum' could do justice to the great variety of meanings which this word had in scientific texts in the period under discussion. Notes giving some indication as to which meanings are relevant are provided. In general pre-Newtonian ambiguities survive in modern non-technical French rather more than in English (egs. *pesanteur* = gravity or heaviness; *espace* = space or distance; *vitesse* = speed or velocity) and so make the translator's task a difficult one.

PART I

AT THE DAWN OF
CLASSICAL SCIENCE

Veniet tempus quo posteri nostri tam aperta nos nescisse mirentur.
(Seneca, *Nat. quaes.*, VII.25.2.)

INTRODUCTION

FORTUNATELY it is no longer necessary nowadays to insist on the interest of the historical study of science. It is no longer even necessary—after the magisterial work of those such as Duhem and Emile Meyerson, Cassirer and Brunschvicg—to insist on the *philosophical* interest and fruitfulness of this study.[1] For the study of the evolution (and the revolutions) of scientific ideas—the only history (together with the related history of technology) which can give a meaning to the much glorified and much decried idea of progress—shows us the human mind at grips with reality, reveals to us its defeats and victories; shows us what superhuman effort each step on the way to knowledge of reality has cost, effort which has sometimes led to a veritable 'mutation' in human thought,[2] that is to a transformation as a result of which ideas which were 'invented' with such effort by the greatest of minds become accessible and even simple, seemingly obvious, to every schoolboy.

The scientific revolution of the seventeenth century was without doubt such a mutation; one of the most important, perhaps even *the* most important, since the invention of the Cosmos by Greek thought, it was a profound intellectual transformation of which modern physics (or more precisely classical physics[3]) was both the expression and the fruit.

Some have wanted to describe and explain this transformation as a kind of complete inversion of spiritual attitude. Henceforth the active life takes precedence over the contemplative; modern man strives for domination over nature whereas mediaeval or ancient man sought only contemplation of it. That classical physics is mechanistic—the physics of Galileo, Descartes and Hobbes, the view of science as active, operative, making of man 'the master and possessor of nature'—is to be explained on this view by this desire for active domination. The application to nature of the categories of thought of *homo faber*[4] would be then simply one version of this attitude. Cartesian science—and *a fortiori* that of Galileo—would be seen on this view as 'engineer's science'.[5] While this way of viewing the matter is no doubt correct in general, and sometimes even in detail (we need only think of the inversion in value and ontological status between contemplation and action which has occurred in modern philosophy; or of some of the images and explanations in Cartesian physics, with its pullies, strings and levers) it seems nevertheless to suffer from all the faults of a global explanation. Moreover it neglects the technological effort of the Middle Ages and the spiritual attitude of alchemy. In fact the activist attitude which it describes was that of Bacon (whose role in the history of the scientific revolution was completely negligible[6]) and not that of

Descartes or Galileo. The mechanistic basis of classical physics far from being an artisanal[7] or engineering conception is in fact precisely the negation of this.[8]

There has also been much discussion of the role of observation and experiment, of the birth of an 'experimental attitude'.[9] It is certainly true that the experimental character of classical science is one of its most typical features. In fact, however, it is easy to misunderstand just what is involved here. The only role in the birth of classical physics played by observation, in the sense of *simple* observation, the observation of common sense, was that of an obstacle. The physics of the Parisian nominalists—even that of Aristotle—was often much more akin to such observation than was that of Galileo.[10] As for experimentation—the methodical interrogation of nature—it presupposes both the language in which its questions are to be posed and a terminology which makes it possible to interpret nature's replies. But if it is in a mathematical (or more precisely, geometrical) language that nature is interrogated by classical physics then this language, or to put it more accurately the decision to use it—a decision which corresponds to a change of metaphysical attitude[11]—cannot itself be the product of the experiment which is conditioned by it.

An alternative, and more modest, attempt to characterise classical physics is based on an identification of what distinguishes it as *physics*. One can, for example, point to the role in Galilean physics of the related ideas of speed and force[12], of 'moment', interpreting them as the expression of a profound insight, an intuition of the *intensity* of physical processes, and even of their intensity at an instant.[13] This is certainly correct—one has only to think of the instantaneism of Cartesian physics,[14] of the idea of element or moment of speed, i.e. of speed at an instant—but it is in fact much more a characteristic of Newtonian physics which is based on the idea of force, than of Cartesian or Galilean physics which tried to avoid it. In fact it was even more characteristic of the 'Parisian' physics of Buridan and Nicole Oresme. Classical physics is certainly a dynamics. But it was not as such that it was born. It appeared in the first instance as a kinetics.[15]

Finally there are those who try to characterise classical physics by reference to the role in it of the principle of inertia.[16] Again this is certainly correct. One only has to think of the fundamental role of the idea of inertia in all of classical science, and of the fact that this principle, which was unknown to the ancients, implicitly underlies Galilean physics and is explicitly basic in that of Descartes. Nevertheless, this characterisation seems somewhat superficial. It is not enough simply to point to these facts. It must be explained why modern physics has been able to adopt the principle of inertia i.e., It must be explained why and how this idea, which seems to us so very obvious, has been able to acquire this *status* as an *a priori*, self-evident truth, whereas for the Greeks and for mediaeval thought it seemed, on the contrary, to suffer from a self-evident and irremediable absurdity.[17]

I believe that the intellectual attitude of classical science can be characterised by the following two changes, which are moreover intimately related: geometrisation of space and dissolution of the Cosmos, that is to say the disappearance from within scientific reasoning of the Cosmos as a presupposition[18] and the substitution for the concrete space of pre-Galilean physics of the abstract space of Euclidean geometry. It was this substitution that made the invention of the law of inertia possible.

We have already said that this intellectual attitude seems to have been the outcome of a decisive mutation. It is this which explains why the discovery of things which seem to us nowadays childishly simple required such prolonged efforts—and not always crowned with success—by the greatest of geniuses, by Galileo and Descartes. This is because it was not a matter of battling against theories which were simply inadequate or erroneous, but of changing the very intellectual framework itself, of overthrowing an intellectual attitude, one which was when all is said and done a perfectly natural one,[19] and substituting for it another, one which was not natural at all. And it is this which explains why—in spite of appearances to the contrary, appearances of historical continuity to which Caverni[20] and Duhem[21] give such emphasis—classical physics, issuing from the thought of Bruno, Galileo and Descartes, was not in fact continuous with the mediaeval physics of "the Parisian precursors of Galileo": it was from the very beginning located on a different terrain, a terrain that we would like to define as Archimedean. For the precursor and inspirer of classical physics was not Buridan or Nicole Oresme but Archimedes.[22]

The history of mediaeval and Renaissance scientific thought (physics)—with which we are beginning to become more familiar thanks above all to the admirable works of Duhem—can be divided into three periods. Or more accurately, since the chronological order does not correspond at all well to this division, the history of scientific thought can be seen *grosso modo* as having three stages, each corresponding to a different type of thought. First there was Aristotelian physics: secondly there was *impetus* physics, inaugurated like everything else by the Greeks, but elaborated above all during the fourteenth century by the Parisian school of Buridan and later of Nicole Oresme:[23] finally there was mathematical, experimental, Archimedean or Galilean physics.

We can in fact see precisely these three stages in Galileo's earliest works which can therefore not only teach us something about the history—or the prehistory—of his thought, about the motives and influences which conditioned his work, but which also provide us with a fascinating historical sketch, as it were distilled and clarified by the marvellous mind of the author, of the whole development of pre-Galilean physics. Thus the interest, for the history of scientific thought, of a careful study of these early writings, cannot be overestimated.[24]

1
ARISTOTLE

First the Aristotelian stage. We can find in Galileo's *Juvenilia*[25] a sizable fragment of a physics course, or more accurately a course in cosmology, more or less such as was taught in the sixteenth century in most European universities. Unfortunately this fragment is incomplete. It contains only part of a commentary on *De Caelo*. We could supplement it with *De Motu* by F. Bonamico[26] who was Professor of Philosophy at Pisa at the very period when Galileo was a student there. There is no doubt at all that Galileo attended his courses. But it is not strictly necessary to fall back on Bonamico's enormous compilation. Galileo's fragment, even though it is incomplete, provides us with a very lucid—remarkably lucid—account of Aristotelian cosmological physics, at least as it was understood in the Middle Ages.

This cosmological physics is so widely known that we have no need to give an account of it here, not even in following Galileo. But we ought to recall the basic principles. This will allow us at the same time to warn against a certain misunderstanding and underestimation of Aristotle's works which is all too common nowadays.

We know perfectly well that Aristotelian physics is false, that it is irremediably superseded.[27] None the less it is a physics, that is to say a highly (though not mathematically) developed theory.[28] It is neither a simple verbal extension of common sense nor an infantile fantasy. It is a theory, i.e. a doctrine which, while it does of course have its starting point in the data of common sense, nevertheless submits these to a systematic, extremely coherent and rigorous elaboration.

The common sense facts on which the Aristotelian elaboration is based are very simple, and we would accept them just as much as did Aristotle. It seems to all of us perfectly 'natural' that heavy bodies fall to the earth.[29] We would be just as astonished as Aristotle or St. Thomas to see a heavy object, a stone or an ox, rise freely into the air. We would scarcely think of this as 'natural' but would look for an explanation of this phenomenon in the action of some hidden mechanism.

Similarly we find it quite 'natural' to see the flame of a match pointing 'upwards' and to put our saucepans 'on' the fire. We would be very surprised—and we would look for an explanation—if, for example, we saw the flame turn itself around to point 'downwards'. But this, it might be said, is a simplistic, childish argument, for it is only here that science begins, where we look for an explanation of what seems 'natural'. Of course. But when thermodynamics postulates as a basic principle that heat does not pass from a cold to a hot body is it doing anything else but repeating in its own words a common sense intuition to the effect that something hot will cool down 'naturally' whereas something cold does not 'naturally' heat itself up? Again, when we say that a system's centre of gravity tends to the lowest position and does not get higher again of its own accord is this not once more simply a repetition in new words of a fundamental common sense intuition, the intuition that is expressed in Aristotelian physics by the distinction between *natural* and *violent* motion?[30]

But Aristotelian physics does not confine itself to expressing in its own words this common sense phenomenon which we have just mentioned. The phenomenon is transposed into a different framework. The distinction between 'natural' and 'violent' motion is located within a general conception of physical reality,[31] a conception of which the key aspects appear to be: (a) the belief in the existence of specific 'natures' and (b) the belief in the existence of a Cosmos,[32] i.e. the belief in the existence of principles of order by virtue of which the totality of real beings form a (naturally) well-ordered whole.

Whole, cosmic order; these ideas imply that everything in the Universe is, or ought to be, distributed and arranged in a highly determinate manner; that things are not indifferent as to whether they be here or there, but that, on the contrary, everything in the Universe has its proper place in conformity with its nature.[33] A place for each thing and everything in its place; the idea of 'natural place' is the theoretical expression of this necessity in Aristotelian physics.[34]

The idea of 'natural place' represents a purely static conception of order. For if everything were 'in order' everything would be at rest in its natural place, would remain there, and would not move away from it.[35]

Indeed why would anything leave 'its' place? In fact quite the opposite; it would exert a resistance to anything that tried to remove it from 'its' place—something that could only be accomplished by *violence*—and would attempt to return there whenever, as a result of such *violence*, it found itself located elsewhere.

Thus all motion implies a cosmic disorder, a disequilibrium, whether it is itself the direct effect of such a disequilibrium resulting from the application of an external force (violence), or is on the contrary the effect of a compensatory effort of a thing to rediscover its lost and violated equilibrium, to bring things back to their natural, appropriate places, where they will be able once again to find rest. It is precisely this return to order which constitutes what we have called natural motion.[36]

Upsetting of equilibrium, return to order: it is understood that order takes the form of a stable state, a state which tends to persist indefinitely. There is, therefore, no need for an explanation of rest, at least of the natural rest of a body in its proper place. Its own nature is itself the explanation; it is this which explains, for example, the earth's being at rest at the centre of the world. It is also understood that motion is necessarily a transitory state. Natural motion naturally ceases when its goal has been reached. As for violent motion, Aristotle is too much of an optimist to believe that this abnormal state could be an enduring one. Moreover, since violent motion is disorder which creates disorder, to allow that it could last indefinitely would be, in fact, to give up the very idea of the Cosmos: hence the reassuring formula for violent motion—nothing which is *contra naturam potest esse perpetuum*.

Thus what we have just said comes to this, that motion in Aristotelian physics is essentially a transitory state. But this proposition is strictly speaking inaccurate in two respects. On the one hand, although in fact the motion *of each moving thing*, as least as far as 'sublunary' things, the things of our perceptual experience, are concerned, is essentially a transitory and finite state, nevertheless, if we consider the world as a whole then motion is necessarily an eternal phenomenon.[37] And it follows from this that it is also eternally necessary. It is a phenomenon which can only be explained by reference to its origin in the very structure of the Cosmos, i.e. by postulating as the cause of the transitory and variable motions of sublunary beings a constant, perpetual, and therefore 'natural' motion of the bodies and spheres of the heavens.[38] On the other hand, motion is not strictly speaking a state: it is a process, a becoming in and by which beings are formed, actualised and completed.[39] Certainly being is the goal and end-point of becoming; and the goal of motion is rest. But this unchangeable rest of the fully realised being is something quite different from the impotent and ponderous immobility of a being which is not capable of motion. The former is actuality, the latter is simply a lack or privation. Therefore motion—process, becoming, change—is ontologically intermediate between these. It is the being of all that changes, and which *is* only in being changed and modified.[40] The famous Aristotelian definition of motion— the fulfilment or actualisation of what exists potentially, in so far as it exists potentially—(a definition which Descartes found utterly incomprehensible) expresses marvellously well the fact that motion is the being—the actualisation[41]—of that which is not God.

Therefore to move is to change, *aliud et aliud se habere*; to be, says Aristotle, 'alteration of what is alterable *qua* alterable'. On the one hand this implies a fixed point of reference in relation to which the moving thing alters and alters:[42] and this means, in the case of local motion, that there must be a fixed point in relation to which the moving things moves, an absolute coordinate centre, the centre of the Universe. On the other hand, since all change, all process, must have a cause which is also its explanation, this implies that all motion must be caused by a motor which sustains the motion as long as it endures. In other words motion, unlike rest, does not persist of its own accord. Rest, which is a state or a privation, has no need of a cause to explain its persistence. Motion, which is a process, an actuality, even a continuous actualisation, is impossible without one. If the cause is removed the motion ceases; *cessante causa cessat effectus*.[43]

In the case of 'natural' motion this cause, this motor, is the nature of the body itself, its form, which seeks to bring it back to its place. This is what sustains the motion. In contrast to this a non-natural motion requires, throughout its duration, the *continuous* action of an *external* motor, contiguous with the moving body. If the motor is taken away the motion will cease. Equally, if the motor is separated from the moving body the motion will cease. For Aristotle does not accept action at a distance.[44] According to him all transmission of motion implies contact. There are, therefore, only two ways in which motion can be transmitted: pushing and pulling.[45] It will be appreciated, then, that Aristotelian physics is a marvellous theory, marvellously coherent, and which really has (apart from being false) only one single defect: it is contradicted every day by the perfectly commonplace act of throwing things. But no theorist worthy of the name is put off by an objection deriving from common sense. When he comes across a phenomenon which is not in agreement with his theory, then he denies the phenomenon. And when he cannot deny it then he explains it instead. It is in the explanation of this phenomenon—that of the projectile which keeps moving in spite of the absence of a motor—which is seemingly incompatible with his theory, that Aristotle demonstrates his genius.[46] His theory of the projectile—a systematic elaboration of a remark by Plato[47]—in fact consists in explaining the apparently motorless motion of the projectile by the reaction of the surrounding medium.[48]

An inspired explanation, but one which is quite implausible to common sense. So all attacks on Aristotelian dynamics made reference to this much disputed question: *A quo moveantur projecta?*[49]

We will come back to this question, but first we must pause to consider another aspect of Aristotelian dynamics: the denial of the vacuum and of motion in a vacuum.[50] For in this dynamics not only does a vacuum not make motion easier, it makes it altogether impossible. And this for reasons which are fundamental.

In Aristotelian dynamics every body is conceived as having a tendency to be in its natural place and to return to it whenever it is violently displaced therefrom. This tendency explains its (natural) motion, motion which takes it back to its (natural) place by the shortest and fastest route. It follows that all natural motion is in a straight line and that every body goes to its natural place as fast as it can, i.e. as fast as the surrounding medium allows it to. If, on the other hand, there were nothing to stop it, if the medium in which it moved offered no resistance to its motion (as would be the case in a vacuum) then it would travel towards its place at infinite speed. But an instantaneous motion seemed to Aristotle, not without reason, completely impossible.[51] Therefore natural motion cannot take place in a vacuum. As for violent motion, for example that of a projectile, motion in a vacuum would be equivalent to motion without a motor: a vacuum is not a medium and cannot itself be moved nor transmit or sustain motion. Moreover, in a vacuum (i.e. in the space of Euclidean geometry) there are no preferred places or directions. In a vacuum there could be no natural places. A body in a vacuum would not know where to go, would have no reason to move in one direction rather than another, and therefore would have no reason to move at all.

Once again Aristotle is right. Empty space (Euclidean space) is incompatible with the idea of a cosmic order[52] because not only would there be no natural places, there would in fact be no *places* at all. Thus the idea of a vacuum is incompatible with that of motion-process. Perhaps it is not even compatible with the idea of a real, bodily motion. The void is nothing, and the idea of locating something in this nothing is absurd. One can only locate geometrical objects, not real bodies, in geometrical space. Therefore, Aristotle would say, we must not confuse geometry with physics. The physicist is concerned with the real (the qualitative), the geometer only with abstractions.[53]

2

MEDIAEVAL DISCUSSIONS:
FRANCESCO BONAMICO

As we have pointed out above the adversaries of Aristotelian dynamics had throughout the ages confronted it with the phenomenon of the persistence of the motion of a moving body separated from the motor. The classical examples of such motion, the wheel (or sometimes the sphere), the stone thrown in the air, the arrow, can be found in critiques of Aristotle from Hipparchus and John Philoponus[54] to Buridan, Nicole Oresme and Albert of Saxony, and right up to Leonardo da Vinci, Benedetti and Galileo.

We will not retrace here the history of this problem.[55] To get an idea of the state of the discussion it will be sufficient to look at Galileo's own teacher, Bonamico.[56] Here is his exposition of the problem of the projectile.[57]

'Of contraries the method and the science is the same. To natural motion we can oppose motion which is contrary to nature: thus, having discussed natural motion now we must, following the rule which we have established concerning motion, discuss that which is against nature, that which is produced by violence. This is two-fold: it is either simply against nature, or it is against nature in some particular way only. It is said that something is moved by force if that which is moved does not itself exert the force, that is when it does not itself have the propensity by which it is moved because it is not perfected by this motion by reaching that place in which it is conserved, i.e. that which is appropriate to its form; rather the motion is to a place in which its motion may be corrupted. Now anything whatever will resist its death as much as it can; the movable body is thus so unwilling to go there that unless the virtue [*virtus*] of the motor overcomes its resistance it will never move, and if the violating faculty does not prevail the body will always retreat to its original place. Thus it does not aid the motive *conatus* at all, as does a stone thrown downwards with great *impetus*: in the latter case its own virtue is added to the action of the mover to produce a much faster motion. Thus the principle of such [simply violent] motion is completely external and alien, and it is helped in its work only by the medium which receives the mover's *impetus* and imparts it to the movable body. The fact is that that which moves absolutely against nature has no force whatsoever conferred on it: . . . but it is overcome by the mover such that it follows the same line that it would travel along if its motion were natural [only in the opposite direction]. Therefore it moves more rapidly at the beginning than at the end. On the other hand that which moves against nature only in some particular way does not fully resist. It has no tendency to move as it does, and being subjected to violence it does not move along the same line as it would if it were moving naturally, but is deflected sideways away from it. This is why the medium helps it and comes to the aid of its motion, so that a given stone is thrown further and with a faster motion when thrown laterally than when thrown directly, vertically upwards. However, no body [moved in this way] has a simple tendency to go to the place towards which it is impelled, and nor does it stay there by nature, but after the moving force is used up it reverts to its natural motion and place, describing a line appropriate to its nature, a vertical line between the highest point of its motion and the centre of the world, and it moves faster as it progresses. The principles responsible for violent motion can be various and those which affect the matter are usually contraries, as is seen in the case of lightning which, being fire, is expelled by the surrounding water and acts on bodies with its own force, and also in cases such as that of heavy bodies being lifted up by the wind, and the *raptus* of certain movable bodies such as the upper air, the *impetus* of water, and the action of the air swirling around in a whirlwind, and generally in the pushing, dragging, lifting and turning which are mostly done by living things.

'But as the causes of violent motion in general have been sufficiently discussed above we will now discuss particular cases and to start with we will investigate the cause of that other motion which is generally called projection. This cause is much more difficult to discover and the most varied opinions have been put forward concerning it since antiquity. Thus Plato attributed the cause of this motion to antiperistasis, to use his own terminology. But how this cause is to be understood is not sufficiently explained by Plato, and Aristotle's propositions on the subject do not add much. The term is ambiguous since strictly speaking it designates the circuits of contraries, when one of the contraries surrounds the other and circulates, somehow drawing the other to the centre. Thus in the summer heat forces cold to the centre, and thus are born many fruits which are cold by nature; on the other hand in winter cold forces heat to the centre so that in winter the cores are warmer. Secondly, and more commonly, the term designates only local motion, as when the medium produces local motion in a body which it pushes along while at the same time, according to Plato, the medium's own motion derives from that of the body. For anything which moves, in as much as it moves, is at the same time moved. It transmits no force to the moving body, and none is transferred to anything but itself, and this is why it moves with the same motion as the moving body itself by one and the same motion.

'Thus when something is projected the parts of the surrounding medium go in turn to occupy the place of the rear end of the moving body, so that if A moves B it takes over its place and if B propels C it occupies its place, and so on. But it can be asked whether this results from an expansion of the body which is effecting the circuit, or rather by means of a succession which occurs as a result of a vacuum. Simplicius takes [Plato's] words to mean the latter, but Aristotle refutes the theory thus interpreted with the following arguments. According to this theory the medium collects together behind the moving body (thus the medium must be liquid and easily moved) so as to prevent the formation of a vacuum. When this is achieved the body can proceed. But whether it is argued that the medium which comes behind the moving body simply fills in the space which the latter relinquishes, or that it pushes forward whatever it comes into contact with, there are many difficulties which lead us to reject this opinion.

'As for the second of these alternatives, which Simplicius himself took Plato's words to mean, the following reasons are enough to show that it is mistaken. First, it cannot be explained why, once the first [motion of the medium] ceases, other motions continue. For if motion takes place only by contact, as it does on this hypothesis, all [the bodies] must move with one motion, and when this ceases they must all stop, for they must occupy each other's places in succession . . . For if it does not happen like this then everything must remain at rest. Such is antiperistatic motion, if we are to believe Aristotle; no movable thing is moved unless the thing which moves it occupies its place, so that the moving thing and the mover move together and the speed of the later parts of the motion is no greater than at the beginning. But it is the opposite of this that is observed. If we are prepared to doubt what can be observed it could also be claimed that the slowing down of the motion, an indubitable phenomenon, would equally be impossible. . . .

'For the motion cannot occur unless the mover follows along. Thus the moment of following and of the motion must be one and the same. Moreover, the thrust of a vacuum is always the same, and thus so is the motion.' This implies that all motions will take place at the same speed. 'Moreover, nature seeks only contact, i.e. to eliminate the vacuum. So it is not clear why the air, having occupied the place of the stone and thus established contact, should continue to work. So if the air makes contact after the stone's first motion why does the motion proceed any further?

'As for the first of the two interpretations of antiperistasis, according to which it takes place by extrusion, this is similarly contradicted by many observations. First, the cause of the stone's projection would be enough to send it as far as the sky. For if the air takes over its place and thereby propels the stone in such a way that this succession is continuous, then the propulsion of the stone would continue as far as the air extends, or that body of the air which gives it the power of contact and which is thus as effective as the air itself. If this were so then a straw could be more easily projected than a stone, because a straw is lighter and tends upwards much more than does a stone. . . . For the same reason, if a thread were attached to the stone it would travel ahead of it, but we can observe that it actually hangs behind and is drawn along by the stone rather than being propelled by the air. Thus we can say that extrusion is not the cause of this motion, and hence Plato's opinion on the subject appears quite ludicrous.

'Having rejected Plato's views,[58] Aristotle decided that the moving body impresses a force on the air or the medium, of which the nature is equivocal, being neither simply heavy nor light. Consequently the air can receive *impetus* in any direction. But because *impetus* is against its nature (although as we have said elsewhere, a horizontal motion is less opposed to it than a motion which is simply either straight up or straight down, since the air is not only light but also heavy) it resists it, and where it is somewhat distant from the original mover it little by little loses the force impressed on it so that it is dissipated and finally used up; thus the projectile being no longer subjected to its violence reverts to its former condition and consequently hurries to return to the place which it had been forced to leave, just as iron when separated from the fire reverts to its own coldness. Philoponus and other Latin authors have very strongly criticised Aristotle on this matter, and have gone so far as to reject his authority.

'In the first place his position in no way avoids the difficulty which was our objection against Plato, namely that if the stone is carried by the air its motion will never end, for air which has received *impetus* no longer has any reason to return to rest. For this *impetus* is in conformity with its nature and its motion is thus no different from that of the descent of a stone which occurs in conformity with nature. So not only would the stone be moved through the whole extent of the air, but, if the air were infinite then its motion would have infinite duration. For it is not at all plausible that the air could move by itself, so that it could both set itself in motion and stop itself; for this is what characterises living things. Also it is not enough to say, as does Averroës, that the medium moves by its natural form but that this motion takes place only when occasioned by something external to it. For even if this were so, how then would the medium return to rest? For that which is the occasion for the motion is still present, and the medium is moving in conformity with its nature.

'Next, if this motion comes from the *impetus* which is introduced and impressed by the original mover, then the *impetus* of the projected stone would be greatest when the stone is nearest to the mover, and its motion would be faster then. But this is false, for the motion of projectiles increases at first for some distance, as can be observed from the fact that the effects produced by a sling or a ballista or by any kind of cannon are more violent at some distance than close to. To which it can be added that the stone would not be able to move against the wind [if it were moved by the air]. . . . The *impetus* of the air against the stone, deriving from the wind, is greater than that deriving from the projector. Moreover, he adds that a stone would be projected an equal distance by a motor in contact with it as by one separated from it, since they can both impress the same *impetus* on the air. Finally, a long and a short spear would be projected with the same speed since an equal *impetus* could be impressed on them by the projection.

'This is why Philoponus, and after him Albert and St. Thomas and many others, were of the opinion that the force is impressed by the original motor not at all on the air but on the movable body, thus on the stone. And depending on whether a greater or smaller force is impressed on it, the moving body is carried further and faster. Similarly, the force is sometimes received more easily and more rapidly, and sometimes with more difficulty and more slowly: this depends on those aspects [of the body] that influence the motion, such as its shape, size, quantity of matter etc., aspects which we have called above the concomitant causes of the motion. Thus a spear is carried further than a square body: and a tensed string vibrates longer and has greater power than one which is slack, because the former receives *impetus* better and retains it longer. If now they are asked why, in projection, the air does not move indefinitely, they reply that the motion is transmitted by the stone to those parts of the air which are closest to it, and by them to other, contiguous parts; and that this motion, as Aristotle himself says, . . . is not unchanging, for the moving body does not simply remain the same; furthermore, the motion is not natural, neither for the stone nor for the air, but is derived from something external to them, and is propagated outwards towards the circumference, as can be seen when a stone is thrown into water; it first makes faster but smaller rings [in the water] because of the greater ratio which exists at this point between the moving body and that which is moved; for the smaller the distance the more rapidly it is covered; later the stone makes bigger but slower rings, because the distance increases and the ratio of the mover to the moved body decreases.

'It is the same when a stone is thrown in the air; the motion becomes slower and finally ceases, then, after an interval of rest, the stone begins a natural motion, because these motions are either contraries or correspond to contraries. Thus when the hindrance is removed the body moves in conformity with its nature. Similarly it can be explained why a ball bounces more easily than a stone, for during its motion before the rebound it is strongly compressed, and after the rebound it expands again. Seeking in this way its innate size (in the same way that the elements seek their natural place as soon as any hindrance is removed) it receives a greater impulse from the repulsion.

'It has been concluded from this that this theory possesses all the characteristics of a good explanation of the problem, i.e. that it is in agreement with reason and does not contradict the senses: it provides a solution to all the problems that can be raised on this matter, and it gives an account of the inherent causes. Therefore it is strongly defended by the Latin authors against Aristotle himself.

'And such is the power of experiment in the method of natural science that we should follow it, putting aside all other devices of the understanding and of reason; so let us perform the following experiment. . . . Take a highly polished board and with a lathe or a sharp compass carve out a disc of wood, so that the disc can rotate in the cavity without touching the wood: fix the board somewhere, make an axle for the disc, and support this axle in little forks or slots. Then it will be seen clearly that the disc turning round in the circular cavity of the board is moved by the motion of the mover without any air pushing it. For although there is some air between the board and the disc there is so little of it that it would not have the force to produce this motion, and this all the more so in that the very smooth surface of the disc could not receive any impulsion from the surrounding air; for the smoother something is the less it can be gripped. . . .'

The interest of these passages is obvious. They demonstrate clearly the essential features of mediaeval science: the combination of a finalist metaphysics with common sense 'observation'. These same features—both of which were to be rejected by Galilean science—are equally to be found in the analysis of the problem of falling bodies.

The problem of the projectile was not the only crucial one for ancient and mediaeval commentaries on Aristotle's physics. Another, scarcely less formidable, was that of falling bodies, or more accurately that of *accelerated* fall.

For why is it that bodies fall with an ever increasing speed? In fact for Aristotle himself this was hardly a problem at all. Since the motion of a heavy body as it falls (or correspondingly, the motion of a light body as it rises) is the result of the object's natural tendency to go to its 'proper' place, what could be more 'natural' than that the motion should accelerate as it gets progressively nearer the goal?

But for the commentators, particularly for the mediaeval commentators, there is indeed a problem here, and a difficult one at that. Taking the Aristotelian idea of 'tendency' to be the same as 'force' they quite rightly asked how it comes about that a constant cause (the weight), acting in a natural manner, can produce a variable effect? Where does the acceleration come from?

Commentators suggested, roughly speaking, two kinds of answers to this question.[59] The Aristotelians tried to find a solution either in a variation (decrease) in the resistance of the medium (the air), or applying to the motion of falling bodies the theory developed for the projectile, in the reaction of the medium; the effect of this reaction, brought about by the motion itself, was thought to supplement that which is strictly speaking attributable to the body's own gravity (i.e. its heaviness).[60]

On the other hand the partisans of *impetus* physics tried to find a solution in a variation of the motive force—the *impetus*—impelling the body, in a kind of aggregation of the impulse to motion. Lacking a concept of inertia, however, this solution in fact depends on an equivocation, a shift in meaning from *impetus* (i.e. motive force) to *impetuousity* (i.e. a quality or property of the motion). Thus it was thought that the body, as it falls, acquires a certain *impetuousity*, and that this impetuousity of its motion, being added to the natural *impetus* of gravity (heaviness), would explain the increase in speed. Let us again consult Bonamico.[61]

'. . . Why do things which move in accordance with nature move faster at the end than at the beginning of the motion? Much has been said on this problem both in Aristotle's time and since then, right up to our own day. Many different causes have been proposed; on the one hand causes *per se*, such as the nature or the place [of the body] and on the other hand causes *per accidens* such as the removal of obstacles, rarifying heat, a certain adventitious gravity, and these either singly or in conjunction one with another. Such explanations can indeed seem plausible, and without the eyes of Argos it is easy to be mistaken, so we must investigate each of these causes very carefully. . . .

'In antiquity (for we will start by recounting the views and doctrines of the Greeks) Timaeus, Strato of Lampsacenus and Epicurus were of the opinion that everything is in fact heavy (*gravia*) and that nothing is in itself light. Now there are two termini to the motion, the one at the highest and the other at the opposite, lowest, point; but one of these, namely the lowest, is the place towards which everything naturally tends; the other, on the contrary, is that towards which things are carried by force. Thus, all things being heavy, they are carried downwards by their nature, and if one among them is higher or lower this only arises from the fact that the heaviest bodies exert a pressure on those which are less heavy, and consequently are located below them. It is not that anything is actually light and is carried upwards by a spontaneous tendency, but that both [the heavier and the lighter] belong to the genus of heavy things. If one of them seems to be light this is because the other is heavier and the former less so. Now because one of them is very heavy it exerts a pressure on that which is less heavy and goes below it, the less heavy one going above. Thus [upwards] motion occurs by a kind of extrusion because the heavier something is the more it expels and oppresses that which is less heavy and the more rapidly it does this. Thus the speed of this [upwards] motion does not in fact derive from an internal cause, but from an external one, and it is a violent and not a natural motion.

'Aristotle criticised these doctrines on the basis of facts perceived by the senses in every kind of motion. He concluded that there is a natural motion in every body, even upwards, for whatever is moved by some force moves faster if it is smaller than if it is bigger; moreover, everything moved by force moves faster at the beginning of its motion, for as the *impetus* which moves it fades away so equally does its motion cease, to be followed by a natural motion. This latter, on the contrary, is slower at the beginning but progressively increases and is fastest towards the end: for that which is carried somewhere by force moves away from that place in accordance with nature. In the motion of the elements we see, for example, that that of earth descending is faster as the mass is greater. Moreover, we see that earth moves more slowly at the beginning than later, and that it moves fastest as it arrives at the end of the motion and finally that when it arrives at the centre it does not move any more as long as it is not forced to do so. Similarly for things which are carried upwards. Therefore we say that such bodies as these are moved not by oppression or extrusion, nor indeed by any other force, but by nature.

'It could be said, however, that Aristotle does show most clearly that this motion is natural and also fastest towards the end, contrary to what was said by these ancient philosophers, but that this in no way provides us with the cause of the phenomena; this remains to be discovered. This is again a question on which much effort has been spent, and there are seven different theories about it. As for the cause proposed by Aristotle, it has been rejected as unacceptable.

'In fact Hipparchus (according to Simplicius in a small work devoted especially to this problem) thought that natural motion is faster at the end because at the beginning of its motion the body is hindered by an external force: as a result of this it cannot exercise its native force and consequently it moves idly; later, as this extrinsic and external force gradually fades away the natural force builds up again and, as it were, freed from obstacles, acts more effectively. This is how bodies progressively accelerate their speed. It is a process comparable to that of the cooling of water which has been heated up and then removed from the fire. For at first it cools down imperceptibly and appears to make almost no progress but, as the heat is exhausted, it recovers its original faculty, cools down more rapidly and finally goes so far that it becomes much colder than it had been before it was heated up. Apparently Aristotle did not reject this account: in fact it was to hypotheses such as these that he looked for support in his search for the causes of hail, justifying them by reference to the observations of fishermen.

'What Alexander did object to in Hipparchus was as follows: there are two reasons why the elements are carried to their natural places: first, there is produced at one and the same time both the place and the form, i.e. the place is theirs constitutionally: secondly, the fact that they are not at their proper place, that they are held elsewhere (fire, for example, in the region of the earth) and are prevented from moving; so that the removal of the fetters which retain them away from their proper places constitutes the second cause of the motion. The acceleration is explained by the fact that since they are produced in a place other than their own they are imperfect and cannot exercise their native power, whereas when the *impedimentum* is removed what then remains to prevent them from rushing there with the whole of their nature?

'This might be a good argument against Hipparchus, but it is not damaging for our own position. This is that the hindrance is always present until the elements reach their proper places, and once there it is completely eliminated; then they do not move but remain at rest in their natural places. I do not know what everybody else will think of this, but there are many who accept this view.

'Simplicius himself believes that the speed increases as a result of the fact that the medium's resistance is lower towards the end of the motion than at the beginning; the motion of the moving body, as it approaches its end, has only a smaller part of the medium to traverse, and this therefore resists less. The condition of the virtues which are in matter is such that, other things being equal, they are more vigorous in larger bodies. The medium resists the motion, and that is precisely the reason why local motion, i.e. change of place, takes time. We have explained above why the speed is greater where the medium is thinner, and why there would be no motion in a vacuum. However, this cause [postulated by Simplicius] is not the same as that given by Aristotle, who says that the speed increases at the end of the motion as a result of an addition of gravity, and not from the fact that there only remains a smaller part of the medium to be traversed. But since this is disputed we will not use it, for to do so would be to commit a *petitio principii*. Instead we will argue as follows, that other things being equal more air obstructs a larger body . . . than a smaller one. . . .

'Thus the air obstructs a larger body more than a smaller one, and yet the larger body falls faster than the smaller one. Therefore the medium's resistance cannot be the cause of the motion's being weaker at the beginning. Also, since this same cause, i.e. the decrease in the medium remaining to be traversed, is involved in violent as in natural motion, it should therefore produce the same effect. But since experience does not confirm this, but rather shows us the opposite, it is not credible that this is the cause of natural motion's increasing towards its end.

'We can read in the Latin interpreters[62] the view that the air is heated up by the motion; being heated it becomes rarer and consequently yields more easily to things moving through it; it follows from this that the longer something moves the more it heats the medium and renders it more rare and, moreover, the rarer it is the greater is the ease with which it is rarified. As a result of this the motion takes place with increasing ease, and therefore with greater speed. Thus an arrow would move much faster as it progresses, particularly if it is heated by the motion. Now according to Aristotle it does in fact heat up to such a degree that if it were made of lead it would melt. Nevertheless, as it moves it progressively slows down.

'All this seems to me to get the order of nature completely upside down. For the motion occurs *before* the heating up of the medium. So those who hold to the above view and place the rarification *before* the motion, postulate an effect which by nature precedes its cause. Nothing could be more foolish.

'Several interpreters attribute the cause of such effects to forces in the place itself, but they do not all conceive this in the same fashion: we can see that there are two ways of explaining the forces of the place. As we recounted above, some believe that the place has a force to conserve the moving body. Now all things have a natural desire to seek their own conservation, and thus bodies seek their natural place. . .

'Others say that there is in the place a force which attracts the movable body, as there is a force in a magnet which attracts iron. But against this view we would say: is it not the case that the greater the body the more it resists attractive forces? Certainly. Consequently larger bodies would fall more slowly than smaller ones. Moreover, a portion of earth would not be moved at any distance whatsoever, just as iron cannot be moved at any distance whatsoever by a magnet, because the vigour of a natural faculty is finite. This view would, moreover, destroy the force of the Aristotelian arguments by virtue of which we believe that earth would be carried to the centre of our world even from the centre of another world however distant. For it would not move here unless the attractive faculty, which is in the centre of our world, could reach it. . . Even though this argument is not of great force, nevertheless it is of value against those who attribute an attractive force to the place.

'And if, in addition, you postulate a propensity, then you contradict yourself.

'Against Averroës some object, though by a fallacious argument, that that which is most lacking is also that which is desired most. But the place is lacking more when something is further from it than when it is close to it. For the further a thing is from its place and from its form, the faster it moves towards it and reaches it. But those who argue thus certainly do not see that the desire, which is the cause of the motion, is greater in matter which is closer than in that which is distant from its goal. Thus plants do not desire sight, and moles do not desire light, while man, if he is blind, desires these things above all, because he is close to vision. Similarly, matter does not desire something good unless it has affection for it, and among those things for which it can have affection it desires those which are closest to it. In my opinion this happens in the same way that a lover who is waiting for his loved one desires her more and more as the hour approaches, so that one hour seems a very long time to him . . .

'I cannot see how the authors of this theory can avoid the conclusion that the [motive] power is greater at the beginning; then the speed would be so too. They have carelessly made the mistake of conflating degrees of privation and degrees of power, as if they were one and the same. It is perfectly true that at the beginning there is more privation, but there is less power; the power increases progressively whereas the privation diminishes. So, as we will show elsewhere, they confuse the latitude of the power with its degree. Now the latitude or distance of the power is greater at the beginning of the motion, just as is that between the greatest heat and eight degrees of cold: but the degree of power is greater later, for something at five degrees of cold more easily becomes extremely cold than does something extremely hot: thus the power and the propensity increase not in latitude but in degree. . .

'Now we come to other arguments.

'Some think that in general the efficacity is to be attributed to the degree of form and not to the quantity of matter (which we do not ourselves believe), because the degree of appetence is the same in bigger as in smaller portions [of matter]. It would necessarily follow from this that they must both move with the same degree [of speed], for the same degree of appetence can be placed in each, as if they were in the same degree of power or perfection. Now in our view this is of the greatest importance. However, this view is imperfect even though the cause postulated by these authors does seem to be the true one. But the speed is not determined by this cause alone, but many others contribute to it over and above the goal; namely, the efficient cause, and other principles *per accidens*; for example, the removal of the hindrance, and the nature of the moving thing; such causes can in fact be the causes of the motion.

'St. Thomas, and after him Albert of Saxony, came to the conclusion that the heaviness and lightness in the elements are double; according to them there is one which is *per se* and natural and another which is adventitious, the former deriving from the generative power and conserving the object in its proper place, the latter being acquired in the process of motion; and it is because of the latter that natural bodies move with a progressively increasing *impetus*. They show that this is so by those observations which we have described above, in particular when we pointed out that even in the absence of the motor a force is conserved in the moving body, by which the body is impelled just as if the original motor were still present. Therefore when the impulse of the original motor is no longer there, the motion only takes place because a certain force remains in the moving thing, and thus it moves with the same motion as before. It is true, however, that this force is extrinsic and adventitious, and that it is continuously used up; but in things which are moving in conformity with nature it increases, and this is why their speed increases. . .

'If the authors of this opinion are asked where this *impetus* comes from and what it is, to the latter question they reply that it is a certain quality which is a power, a power to move. To the former question they reply that this power derives from the form by way of the motion. It seems, however, that in dealing with this question the cause has once again been confused with the effect. For what is in question is the cause of the motion's speed: and they say that it is a faculty or an aptitude. But if the question is put the other way round and we ask what this aptitude derives from they say, from the motion! Now this is taken either as speed or simply [as motion]: if it is taken *simpliciter* then the motion would be the cause of its own speed: similarly if it is taken as speed. So once again the proposed cause is that which is assumed in the problem to be the effect.

'Among recent authors Ludovico Buccafiga proposes that the moving body agitates and as it were impels the medium in the same way that it disturbs and impels the first part of the medium. The latter then transmits its motion to contiguous parts, and these others carry the body itself along. Since they are in front of the moving body this makes the motion easier. But since at the end of the motion the *impetus* impelling the body is greater, so the air is also more apt to receive motion. Consequently the motion becomes faster at the end.

'To this others add the propulsion of the air which, as it follows the body, thrusts it forward, which has the effect that its motion becomes faster. This is generally supported by reference to numerous passages from Aristotle, and in particular from the eighth book of his *Physics* and the fourth of *De Caelo*, in which mention is made of this propulsion. This, however, goes against Aristotle's meaning taken in context, for he says that the motion becomes faster at the end as a result of the addition of gravity: to which they reply that this is not what Aristotle really thought, but that he said it only to satisfy vulgar opinion, so that on this point they do not accept the authority of Aristotle. We will discuss the truth of these passages elsewhere. Meanwhile we will show that what they propose is false. First, it can be seen to commit the same absurdity as did St. Thomas and Albert when they asserted that the adventitious *impetus* is the cause of the speed; this mixes up the cause and the effect. For they take the thrust by the air to be the cause of the speed even though this thrust itself comes from the moving body. Moreover, it can be asked just where do moving bodies get the force to impel the air from, and to impel it more the longer the motion lasts? Now since this greater impulsion derives from the increased speed, this phenomenon cannot be caused by the impulsion, as they would have it, but by the speed. And when all is said and done this cause is the heaviness [*gravitas*], which they reject: for what travels faster is heavier, and when the medium is more oppressed this is as a result of the heaviness, for this is operative to a greater extent in that which is heavy or light *simpliciter* than in that which is so only in some particular way. If this heaviness or the speed is only adventitious why then does it not in fact progressively diminish? They add to this that the parts of the medium impel more than they are impelled, and this the less so as they are progressively more distant from the moving virtue. But a natural mover would get progressively weaker, except when it is taking the object towards its form, and this is not the case with this adventitious virtue.'

Bonamico goes on to explain why the phenomenon of wind does not contradict this view.[63] This is because wind is a very complex phenomenon, composed of the motion of both the air and of exhalations. It is the latter which is the real cause of the wind's speed. Consequently he believes that generally speaking the adventitious *impetus* cannot explain acceleration, because it is caused by it, and that, on the contrary, we must propose an *impetus* which is in the body before motion begins. 'Moreover does not Aristotle reject the position of those who propose that it is the impulsion which makes the motion faster, because if that were so then the motion would slow down towards the end and would not increase in speed, and because a smaller body would be carried along more easily than a large one. So it can be seen that the heaviness is the cause of the speed because that which is heavier falls faster. If this impulsion is often located in the air by Aristotle this is because it is this which is used by nature in the motion of projectiles: but we are concerned here with natural motion. I conclude then that while they set out to support the view that motion inheres in the elements *per se*, they in fact attribute to motion a cause *per accidens*. . . Thus when they decide to depart from Aristotle they fall into error.'

Bonamico gives us here a fine and very instructive account of the criticisms and difficulties with which Aristotelian physics was confronted. However, it is not entirely accurate,[64] nor complete[65] as regards either mediaeval or even modern authors. For example, although he mentions Buccafiga and borrows from Scaliger[66] he does not say anything about either Tartaglia or Cardan, or even about Benedetti. Now while it could be argued that, strictly speaking, Cardan (who in different works adopted two contradictory positions) and even Tartaglia, did not contribute a great deal to *impetus* physics, the same could not be said of Benedetti. We must, therefore, now turn our attention to him.

3

IMPETUS PHYSICS : BENEDETTI

Giovanni Battista Benedetti[67] was a firm supporter of 'Parisian' physics. Like his precursors he held that the Aristotelian theory of the projectile was worthless. Thus he says:[68] 'Aristotle at the end of *Physics* Book VIII expresses the opinion that a body moved by force and separated from its original mover is or has been moved over a period of time by air or water, which follows it. But this cannot be so, since the air which enters the space left by the body, in order to eliminate the void, not only does not drive the body forward but actually hinders its motion. For air is forcibly drawn by the body, and though this air is separate from the air in front of the body, it similarly exerts resistance. To the extent that the air in the [forward] part is condensed, the air in the rear is rarefied. Hence, since the air is rarefied under compulsion [*per vim*], it prevents the body from continuing its flight with the same speed with which it started, since every agent suffers in acting. Therefore, while the air is forcibly acted on by the body, the body too is forcibly acted on by the air. Now this rarefying of the air is not natural, but violent. For this reason the air resists and tries to draw [the body] to itself, but since nature does not permit a void to exist between one and another of the bodies [i.e. between the projectile and the air], these bodies are always in contact. And since the moving body cannot leave the air behind, its speed is impeded.'

Thus, the increased speed of the projectile is not explained by the reaction of the medium; on the contrary this reaction can only impede it. As for the motion itself, whether it be natural or violent, it is always to be explained in terms of a motive force which is immanent in the moving body.[69] 'Every heavy body when moved either naturally or by force, receives on itself an impression and *impetus* of motion, so that, even if separated from the motive force, it moves by itself for some length of time. Indeed, if it is set in natural motion it will always increase its velocity: for then the *impetus* and impression are always being increased, since the motive force is always conjoined to the body. Thus, if we move the wheel with our hand and then remove the hand from it, the wheel will not immediately come to rest but will turn for some length of time.'

What is this *impetus*, this motive force, cause of the motion immanent in the moving body? It is difficult to say. It is some kind of quality, power or virtue which is impressed on the moving body, or rather which impregnates it, following and as a result of its contact with the motor (which possesses the *impetus*), by virtue of its coming to take part in the motion. It is also a kind of *habitus* which is acquired by the moving body, and this increasingly *the longer the time* during which it is acted on by the motor. Thus, for example, if a stone is thrown further by a sling than by hand, this is because it goes round in the sling *many times*, and is thereby more 'impressed'.[70]

'The true explanation why a heavy body is projected much further from a sling than from the hand is this:[71] in the revolution of the sling a greater impressing of the *impetus* of motion in the heavy body takes place than would be the case if the hand alone were used. For when the projectile is released from the sling, it takes its path, with nature as its guide, from the point where it leaves it, on a line tangent to the circle which it last made. And there is no doubt that a greater *impetus* of motion can be impressed on the body by the sling, since from repeated revolutions an ever greater *impetus* is received by the body. Now the hand is not the centre of the motion of that projectile while it is being revolved (*pace* Aristotle), nor is the sling the radius.' This means that the motion's circularity (according to Aristotle) is nothing to do with the matter under discussion. Moreover, the circular motion produces in the body an *impetus* to move *in a straight line*. 'Now it is true that the impressed *impetus* gradually and continuously decreases. Hence the downward tendency of the body, caused by its weight, enters at once and, mingling itself with the impressed force, does not permit line *ab* to remain straight for long, but causes it quickly to become curved. For the body, *a*, is moved by two forces, one impressed by violence, the other by nature—contrary to the view of Tartaglia who says that a body cannot at one and the same time be moved by violent and by natural motion.'

The explanation given by Benedetti might well seem rather muddled, but this is hardly very surprising for the idea of *impetus* is in fact a very muddled idea. Fundamentally it does no more than express in 'scientific' terms a conception based on everyday experience, on common sense phenomena.

For what is *impetus, forza, virtus motiva* if not, as it were, a condensation of our experiences of *elan* (as when running to take a jump) and muscular *effort*? So it is in excellent agreement with the 'facts'— whether real or not—which constituted the experimental basis of mediaeval dynamics: and in particular with the 'fact' of the initial acceleration of projectiles. In fact it even explains this phenomenon, for is not some lapse of time necessary before the *impetus* can take hold of the moving body? Furthermore, everybody knows that it is necessary to take a good run in order to jump over an obstacle, and that a cart which is pushed or pulled only gets under way slowly but then progressively increases its speed: for it, also, needs to acquire *elan*. And everybody—even children playing ball—knows that to hit a target hard one should stand back a bit, not get too close to the target, so that the ball has time to pick up *elan*.[72]

Impetus, impression, quality or motive virtue: these are all something which passes from the mover to the moved and which, having entered the moving body or having impregnated it or been impressed on it, has an effect on it. Therefore it is different from other qualities or virtues (thus different *impetuses* impede each other and cannot easily coexist in the moving body), even from natural ones. Thus, the *impetus* of violent motion, as Benedetti explains in a very curious passage, makes the object in which it is located lighter.[73]

'As a result of this tendency to rectilinear motion possessed by the parts of any round body, a top while spinning around with great violence remains for some time quite erect over its iron point, not inclining toward the centre of the world on one side any more than on another. For the inclination of each of its parts during such motion is hardly at all toward the centre of the world, but is directed far more transversely, at right angles to the line of direction [of free fall], i.e. at right angles to the vertical or the axis of the horizon. Thus such a body must necessarily remain erect. Now in saying that the line of inclination of those parts is 'hardly at all' toward the centre of the world, my reason is that those parts are never completely deprived of this tendency, a tendency which makes the spinning top rest on its point. However, it is true that the more swiftly it spins, the less does it press upon that point; indeed, the body becomes that much lighter. This is obvious if we consider the example of the missile shot by a bow or any other instrument or machine for hurling missiles. The swifter the missile is in its violent motion, the stronger is the tendency it has to traverse a straighter path and, therefore, the weaker is its tendency toward the centre of the world, and, for this reason, the lighter does it become. But if you wish to see this truth more clearly, imagine that while the body, i.e., the top, is spinning around very rapidly, it is cut up or divided into many parts. You will observe not that those parts immediately fall toward the centre of the world, but that they move in a straight line and, so to speak, horizontally. No one, so far as I know, has previously made this observation on the subject of the top. From such motion of the top or of a body of this kind it may be clearly seen how mistaken are the Peripatetics on the subject of the violent motion of a body. They hold that the body is driven forward by the air which enters [behind it] to occupy the space left by the body. But actually the opposite effect [i.e., resistance] is produced by the air.'

In Aristotelian physics the medium has a dual role to play; it is simultaneously resistance and motor. *Impetus* physics denies the motive action of the medium. Benedetti adds that even its retardatory action was not properly understood and that, in particular, Aristotle was quite incorrect in his estimation of it. The reason for this was that Aristotle did not properly understand, or rather did not understand at all, the role of mathematics in physical science. Consequently he almost everywhere got things wrong. Now, it is only on the basis of the 'unshakable foundations' of mathematical philosophy—which in fact means, on the basis of Archimedes—that Aristotle's physics can be replaced by a superior physics.

So Benedetti is fully aware of the significance of what he is attempting. He even adopts a heroic posture:[74] 'Such is the influence and authority of Aristotle that is is difficult and hazardous to write anything contrary to his teachings, and especially so for me, since I have always had the highest admiration for his wisdom. But impelled by the pursuit of truth—and if he were alive, he would himself be motivated by this love for truth—I have not hesitated to set forth certain matters in which I have been constrained to disagree with him. And I base my disagreement on the unshakable foundation of mathematical philosophy on which I always take my stand. . .

'Since we have undertaken the task of showing that Aristotle was mistaken on the subject of local motion, certain preliminary statements should be made that are completely true and known by themselves as objects of understanding. To begin with, any two bodies, heavy or light, equal in volume, of similar shape, and similarly situated, but made of different material, will maintain the same ratio between the speeds of their natural local motions as the ratio between their weights or lightnesses in one and the same medium. And this is obvious from their nature if we consider that greater speed or slowness can arise from no other source than from four causes (so long as the medium is uniform and quiet), i.e., from greater or lesser weight or lightness, from differences of shape, or, in the case of bodies of the same shape, from differences in the position of the body with respect to the line of inclination (which is a straight line drawn between the centre of the world and the circumference thereof), and, finally, from unequal size. Hence it will be clear that, in the absence of variation in the form of the body, either in size or in kind, or in the position of the body having that form, the motion will be in proportion to the moving force [*virtus*], i.e., to the weight or lightness. And what I say about the kind, the size, and the position of that form, I say with respect to the resistance of the medium itself. For the differences and inequalities of shapes and variation in position have no small effect on the motions of these bodies. For a small figure more easily divides the continuity of the medium than does a large one, and so also an acute figure divides the medium more swiftly than a blunt figure, and one with an angle moves more swiftly if the angle is in front than otherwise. Therefore, whenever two bodies are subjected to or receive one and the same resistance to [the motion of] their surfaces, [the speed of] their motions will turn out to be to each other in precisely the same proportion as their motive forces. And, conversely, whenever two bodies have one and the same heaviness or lightness, but are subject to differing resistances, [the speeds of] their motions will have the same ratio to each other as the inverse ratio of the resistances: . . . the body which, when compared to the other, is of equal weight or lightness, but is subject to smaller resistance, moves [in natural motion] more swiftly than the other in the same proportion as its surface is subject to a smaller resistance than is that of the other body. . . For example, if the surface of the larger body to that of the smaller body were in the ratio of 4 to 3, the speed of the larger body to that of the smaller would be in the ratio of 3 to 4. That is, the speed of the smaller body would be greater than that of the larger in the ratio of 4 to 3.'

An Aristotelian could, and even must, accept all that. But, Benedetti adds, there is something else that must also be accepted, namely:[75] 'that the speeds of the natural motion of a heavy body in different media are proportional to the weight of the body in those media. For example, if the total weight of a heavy body is indicated by *ai*, then when the body is placed in a medium less dense than itself (for if it were placed in a medium denser than itself it would not be heavy, but light, as Archimedes showed), that medium subtracts part *ei*, so that part *ae* of that weight remains free. And if the same body were then placed in some other medium denser [than the first] but still less dense than the body itself is, this medium would subtract part *ui* of the weight, so that part *au* of that weight would remain.

'I say that the ratio of the speed of the body through the less dense medium to its speed through the denser medium will be as *ae* to *au*. And this is more reasonable than if we were to say that the ratio of the speeds was as *ui* to *ei*: for the speeds attain their ratio solely from the motive forces (since the form of the body remains one and the same in quality, size and position). What I have just now said is quite in keeping with what I wrote above. For to say that the ratio of the speeds of [natural motion of] two heterogeneous bodies of the same shape and size and in the same medium is equal to the ratio of their weights is the same as saying that the ratio of the speeds of a single body moving [naturally] through different media is equal to the ratio of the body's weights in those media.'

Benedetti is certainly quite right, from his own point of view. If speeds are proportional to motive forces, and if a part of the motive force (the weight) is neutralised by the action of the medium, then only the remaining part is to be taken into account; so, in increasingly dense media the speed of a heavy body will decrease in an arithmetical progression, and not in a geometrical progression as Aristotle had thought. But Benedetti's argument, which is founded on Archimedes' hydrostatics, does not rest on the same assumptions as does that of Aristotle. For Aristotle a body's weight is one of its constant and absolute properties, and not a relative property as it was for Benedetti and the 'ancients'.[76] Therefore, for Aristotle, the whole weight acts in the various resisting media.[77] So Benedetti's view is that Aristotle's physics shows quite clearly that he did not 'know the cause of the heaviness or lightness of natural bodies, which is that the density (or rareness) of the heavy (or light) body is greater than the density (or rareness) of the fluid medium in which it is placed.'[78] Density and rareness, these are the absolute properties of bodies. Weight, i.e., heaviness or lightness, is only an effect. So as to avoid an error into which it would be easy for us to fall, Benedetti warns[79] 'that the ratio of the weights of the same body in different media is not the same as the [inverse] ratio of the densities of those media. Hence unequal ratios of speeds necessarily result. [And that] if heavy or light bodies of the same shape and material but of unequal size move naturally in the same medium, the ratio of their speeds is quite different from what Aristotle thought'. In particular, the smaller of two bodies of equal weight will move with the higher speed, because the medium's resistance will be less.[80]

According to Benedetti, in fact, Aristotle never understood anything about motion. He understood neither natural motion, since he did not even see 'that upwards and downwards rectilinear motion of natural bodies is not primarily and *per se* natural';[81] nor violent motion, since he did not see either that rectilinear motion, going and returning on the same line, is continuous and occurs without interruption,[82] or that motion on a straight line can go on for ever even though it is spatially finite.[83]

Clearly Aristotle's main error was to have neglected, or even to have excluded altogether from physics, the unshakable foundations of mathematical philosophy.

But we have not yet finished taking stock of the errors of Aristotle's physics.[84] We come now to the most serious of them: the denial of the vacuum. On this Benedetti speaks out quite plainly: Aristotle's proof of the nonexistence of the vacuum is entirely worthless.

As is well known Aristotle shows the impossibility of the vacuum by an argument *ad absurdum*. In a vacuum, i.e., in the absence of any resistance, motion would take place with an infinite speed.[85] Now, according to Benedetti, nothing could be more wrong. Given that speed is proportional to the relative weight of the body, i.e., to its absolute weight diminished but not divided by the medium's resistance, it follows directly that the speed does not increase indefinitely and that in the absence of resistance the speed does not beome infinite.[86] 'In order to show this more easily, let us imagine an infinity of corporeal media, one rarer than another in whatever proportion we please, beginning at unity . Let us also imagine a body Q denser than the first medium.' This body's speed, in the first medium, will obviously be finite. Now if it is placed in the various media that we have imagined its speed will certainly increase, but can never go beyond a certain limit. Thus motion in a vacuum is perfectly possible.

But what motion will it be, i.e., what speed will it have? On Aristotle's view, if motion in a vacuum were possible, the ratios of the speeds of different bodies in it would be the same as in a plenum. Once again, this is incorrect. 'This statement is . . . completely false. For in a plenum the ratio of the external resistances in the case of these bodies is subtracted from the ratio of the weights, so that the ratio of the speeds remains. And this last ratio would be annulled if the ratio of these resistances were equal to the ratio of the weights. For this reason the ratio of the speeds of the bodies in a void would be different from what it is in a plenum.'[87] The speeds of different bodies (i.e., of bodies composed of different material) would be proportional to their specific absolute weights, i.e., to their densities. As for bodies composed of the same material, in a vacuum they would have the same natural speed.[88] This is proved by the following argument.[89] 'Suppose that there are two bodies of the same material, o and g, and that g is half the size of o. And suppose further that there are two other bodies, a and e, of the same material as the first two, with each of them equal [in size] to g. Now let us conceive of both these latter bodies as placed at the ends of a line whose midpoint is i. Clearly, point i will have as much weight as the centre of *o*. And *i*, by virtue of bodies *a* and *e*, would move in a void with the same speed as the centre of *o*. And the fact that those bodies *a* and *e* are separated by the length of the line does not in any way change their speed [of natural motion]. Each would have the same speed as *g*. Therefore *g* would have the same speed as *o*.'

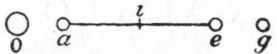

Motion in a vacuum,[90] the simultaneous fall of homogeneous bodies; we are already a long way from Aristotle's physics. But the unshakable foundations of mathematical philosophy, the example of Archimedean science which Benedetti always has in mind, make it impossible for him to call a halt at this point.[91] For Aristotle's error was not simply in having rejected the possibility of a void in the world; it was in having invented for himself a false picture of the world, and in having adapted his physics to it. It was the finitist foundations of his false cosmology—Benedetti was a Copernican[92]—which were the basis for his theory of 'natural places'. Actually 'there is no body, whether in the world or outside it (regardless of what Aristotle says about it) which does not have its place'.[93] Extramundane places? Why not? Would there be any objection to there being an infinite body beyond the sky?[94] Certainly Aristotle would deny this: but his reasons for doing so are not at all obvious.

'For he believes, without proof and even without providing any reasons, that the infinite parts of the continuum are not *in actu* but only *in potentia*. But this cannot be accepted; since the whole actually existing continuum is *in actu* then each of its parts must be so also; for it would be silly to believe that things which are *in actu* are composed of things existing only *in potentia*. And it must also not be said that the continuity of these parts makes them *in potentia* and deprived of all actuality. For example, let there be a continuous line *au*; divide it into equal parts at point *e*; it cannot be doubted that before the division the half-length *ae* (even though conjoined with the other, *eu*) is as much *in actu* as the whole line, *au*, though it is not differentiated as such by the senses. I assert the same of the half-length of *ae*, i.e., of the quarter part of the whole line *au*, and similarly of the eighth part, the thousandth, or whatever.' Therefore infinite multiplicity is no less real than the finite. The infinite is actual and not merely potential in nature. And the actual infinite can be understood just as much as can the potential.[95]

4

GALILEO

Now we come to Galileo.

In the treatises and essays on motion which he wrote in Pisa[96]—which were never completed—Galileo struggled to develop the dynamics of 'impressed force', of *impetus*, about which we have spoken at length above, to make it more coherent and comprehensive. At the same time he tried to take as far as possible the mathematisation (or rather the 'Archimedisation') of physics, of which we have seen the beginnings in the work of Benedetti. So we will find once again, in Galileo, the traditional arguments of his Parisian precursors, but now systematised, condensed and clarified.

In his Pisan works Galileo can be seen to be firmly, even passionately, anti-Aristotelian.[97] Aristotle, he tells us, never understood anything in physics.[98] In particular he was almost always wide of the mark on the subject of local motion. For he was unable to show that the motor must necessarily be contiguous with a body in motion without at the same time postulating that projectiles are moved by the surrounding air.[99]

Galileo can cite counter-instances, phenomena which Aristotelian theory cannot explain. Can it explain, for example, how it is that a heavy body—a piece of lead—can be thrown further than a light body (of the same size)? Or that an elongated body—a lance—flies better than a short one, and moreover can fly with its heavy tip at the front? How could it claim that an arrow fired into the wind could be carried forward by the reaction of the air? How can the reaction of the medium explain the persistent motion of a wheel or of a spinning-top, or of a marble sphere, mounted and rotating on an axis?[100]

Moreover, the Aristotelian way of looking at things is self-contradictory. For if a displacement of the air gives rise to a further displacement then the same effect will be produced over again, and the motion, once started, would go on indefinitely; there would even be an acceleration. But it is one of the fundamental principles of Aristotelian dynamics that all motion is limited and finite. Finally there is a logical argument: all that Aristotle achieves in transfering the role of motor to the air is a change of site of the problem, and a further self-contradiction, because this shift implicitly postulates that there is a *virtus motiva impressa* in the air. But why is the air thus privileged? And if one has to have a *virtus motiva*, why not make the simpler assumption that it is there in the moving body in all these cases under discussion?[101]

Consider, for example, the case of a stone thrown in the air. The stone rises; it has, therefore, acquired a certain quality or virtue which makes it rise. Since upwards motion is characteristic of light bodies therefore it must be a kind of lightness that the stone has acquired. It is this (non-natural) lightness which explains the upwards motion of the moving body: it is a *virtus impressa*, a *virtus motiva*.

Now this motive virtue, in other words the lightness, is conserved in the stone when it is no longer in contact with the mover, just as heat is conserved in a piece of heated iron after it is removed from the fire. This virtue, impressed when the stone is thrown, progressively diminishes once the stone is separated from the thrower, just as the heat in the iron diminishes once it is out of the fire. Thus the stone returns to rest just as the iron returns to its natural coldness, and just as there is a natural—and specific—capacity of bodies to absorb heat, similarly there is such a capacity for motion. The same force is impressed more in a body with higher resistance, i.e., a heavier body, than in one which is less resistant, (thus more in iron than in a feather), in the same way that heat is impressed more in iron than in air (and is thus conserved longer in the former).[102]

It is clear that Galileo, faithful to the inspiration of those who came before him,[103] is developing here the physics of 'impressed force'. He conceives this force, of which the motion of the moving body is an effect, on the model of heat and cold, the forces or qualities of Aristotelian physics. These are substantial qualities, at least in the sense that they can be separated from their source and transfered to another body. And they are 'natural' qualities, naturally present, and therefore persistent, in bodies: but also, on the other hand, non-natural, violently impressed, and therefore transitory. So in order to give us a clearer idea of this Galileo offers 'a better example'[104], that of a bell. When an impact sets it ringing the bell thereby acquires *a sonorous quality*, and it rings, i.e., it emits a sound, by virtue of this quality which is impressed in it. This is the explanation of the fact that the bell, under the action of an instantaneous impact, can emit a sound with a certain duration. The sonorous quality impressed or introduced in the bell by the impact is not natural to it; no more natural than the motive quality introduced into the stone when it is thrown. But once introduced or impressed, it is there; it belongs to the bell or to the stone and not to the clapper or to the hand. So once there it has an independent existence and no longer needs to be in continuous contact with its source. The motion of the moving body is the effect of the force (the motive quality) which propels it. There is no need at all for an external motor to sustain it.

We can see that the analogy goes a long way, even a very long way. It goes very much further, in fact, than some historians of science would be willing to credit. The motive virtue or quality is no more natural to the stone (a body is naturally at rest) than the emission of sound is to the bell. The motive virtue, just like the sonorous quality, is something 'impressed' on the object. Moreover, it is something which by its very nature is active.[105] The sonorous quality is the cause of the sound as the motive quality is the cause of the motion. And both of them are used up in producing their effects, sound and motion. Thus the bell does not ring indefinitely but in the end becomes silent. Similarly the thrown stone does not travel indefinitely but once the motive force is exhausted, it stops and returns to rest.[106]

Galileo is quite insistent on this point: the idea of the motive quality or force, impressed in various ways by the motor on the moving body, makes it possible to give a complete explanation of the phenomenon of the projectile. We have no need to be burdened with Aristotle's inept invention, the reaction of the medium.

But does not the idea of the motive force impressed on the moving body imply that the motion would continue indefinitely? In other words, does it not allow the principle of inertia to be formulated? We know that many famous historians have held this view. It was not, however, that of Galileo.[107] In opposition to some of those of an older generation (Cardan, Piccolomini, Scaliger), who stated that in certain conditions (in fact when motion takes place on a horizontal surface) the *impetus* is immortal,[108] Galileo resolutely insisted on its essentially perishable character. Eternal motion is impossible and absurd, precisely because motion is produced by the motive force, where this is used up in producing the motion.[109] Therefore there is always a slowing down as the motion is produced, and it is impossible to find two points at which the speed of the moving body is the same. Galileo, although he had read Benedetti and knew that it was possible for a motion to slow down indefinitely,[110] believed that this was sufficient proof of the necessity for motion to cease. This error is doubtless explained by his implicit substitution of space for time; he concludes that the distance covered, rather than the duration, is finite. However, this is not of great significance. The lesson we learn from Galileo remains just as crucially important for the history of science: *impetus* physics is incompatible with the principle of inertia.

Certainly more or less everyone agreed that violent motion progressively slows down, that the *impetus* is gradually used up. At least, everyone agreed that this was the normal case. But this did not at all prevent these very same people from holding with the utmost firmness to the view, mentioned above, that all motion, and in particular that of projectiles, starts with a phase of acceleration. Even Renaissance artillerymen firmly believed that a cannon-ball, as it is fired from the cannon, starts by increasing its speed and arrives at its maximum effectiveness at some distance from the muzzle.[111]

We will not stop here to consider the explanations of varying degrees of ingeniousness which were thought up for this imaginary phenomenon. But they do afford us additional proof of the intuitive character of the idea of *impetus*. For it does seem that as soon as the idea of force was *conceptualised*, more or less clearly, as soon as motion was *understood* as the effect of a force (natural or impressed), then it became impossible to accept that there could be a spontaneous acceleration of motion. Quite the contrary, one was forced to accept, as did Galileo, that motion—or at least violent motion, motion produced in a body by an 'impressed force'—could by itself only slow down. It is a curious fact that of all the partisans of *impetus* physics Galileo (together it seems with Hipparchus and Gaetano of Thien[112]) was the only one who fully understood this, the only one who dared to reject this phenomenon as impossible while his predecessors and contemporaries limited themselves to trying to explain it.

But he was thereby led to deny another phenomenon, one which was indubitable this time, namely the acceleration of the motion of a falling body. A body's fall is brought about by a constant force, its weight; therefore it cannot have anything but a constant speed.

Galileo says this quite clearly: the speed or slowness of the motion of fall depends on one cause only, the greater or lesser weight of the falling body.[113] Speed is not something which determines the motion, as it were, externally, a sort of addition to it as Aristotle thought since he assigned one cause to the motion and another to the speed. Speed is not a function of the resistance of the medium: it is something inherent in the motion itself. 'Speed cannot be separated from motion. For whoever asserts motion necessarily asserts speed; and slowness is nothing but lesser speed.'[114] Thus to a greater weight there corresponds a greater speed, to a lesser weight a lesser speed; and the inverse of this holds for lightness.[115] Therefore, the speed of fall of a body (a) is strictly proportional to its weight and (b) is a constant value for each given body.

Galileo has spelled out quite clearly here the inevitable theoretical consequences of *impetus* dynamics. In our view this is enough to show that in itself this dynamics was a dead-end.[116] And this also provides us with an explanation of a problem which greatly troubled Duhem, the problem of why it was that Nicole Oresme did not apply to the motion of falling bodies the theoretical (mathematical) ideas which he developed in the analysis of the 'latitude of forms'. The answer seems quite simple: Oresme understood himself better than do his historians.

We have said that Galileo denied the acceleration of the motion of fall. But this is not entirely true. He was obliged just like everybody else to admit that a stone goes ever more rapidly as it falls. However, this acceleration, he says, only takes place at the beginning of the fall, until the falling body has acquired its proper speed which we know to be strictly proportional to its weight. From this moment on, then, it remains constant and, Galileo adds, if we were able to perform an experiment, that is if there were a sufficiently high tower available, we would clearly see, by throwing heavy bodies from the top of the tower, accelerated motion change into uniform motion.[117]

But why is there an acceleration at the beginning? And, on the other hand, what is this proper speed? The answer to this second question is very simple, as we have already seen: this speed is a function of the weight. However, it is not the absolute weights of bodies that is involved here but their specific weights. A piece of lead will fall faster than a piece of wood. But two pieces of lead will fall with the same speed.[118]

Moreover, Galileo, on this point once again following the example of Benedetti, introduces a new element into his dynamics here, an element which, once it is fully understood, will bring this dynamics tumbling down: for in practice it is not a matter of absolute specific weights of bodies but of their relative (specific) weights.[119]

We will come back shortly to this important addition to the classical theory. For the moment let us return to the problem of the acceleration.

According to *impetus* theory as developed by Galileo bodies should fall with speeds which are constant and proportional to their relative weights.[120] They *should* do this but, in fact, they fall with accelerating speeds; and these speeds are not proportional to their weights, not even to their relative weights. On the contrary, it is *light* bodies which, at the beginning of the fall, move the fastest. It is only later that heavy bodies catch them up and overtake them. One can easily be convinced of this, according to Galileo, by *experiment*.[121]

This divergence between theory and practice is explained by the fact that the theory is constructed, as it were, in the abstract. It holds for the pure case, the case of bodies acted on solely by heaviness or gravity [*pesanteur*], a case which we do not come across in reality. For in the real world heaviness never acts alone but always in combination with lightness. It is the latter's modifying action that we must now investigate.

For example, take the case of a heavy body thrown vertically into the air. If it rises this is because we have impressed on it a lightness *praeter naturam* which, precisely, carries it upwards.[122] But in addition to this lightness *praeter naturam* which we have impressed on it the moving body still has its natural heaviness and this pushes it downwards. Therefore the lightness *praeter naturam* must from the very start act counter to the natural action or resistance of the weight. In general the body will not rise unless the lightness impressed on it is greater than its weight; furthermore, it will only rise to the extent that the lightness is greater than the weight. In effect it is only this surplus, the difference between the lightness *praeter naturam* and the natural weight, which can effectively act to produce the upwards motion.

Now, as the lightness *praeter naturam* produces the upwards motion it is used up (as with all impressed forces) in and by this very action. At a given instant the 'surplus' will all have been consumed. At this point the body will stop rising and will begin to descend by virtue of its own natural gravity.[123]

However, and this is the important point, at this instant not all of the lightness *praeter naturam* will have been used up, but only the 'surplus' of it. The instant at which descent begins is, in fact, that at which the lightness *praeter naturam* and the natural heaviness are exactly balanced. The falling body is, therefore, not acted on only by heaviness but also by the previously impressed lightness, or more accurately, by that amount of it which still remains. Now, a non-negligible quantity (equal to the weight) does still remain, and although this is not enough to cause the body in question to rise it is enough to *slow down* its motion of descent. For the force which moves the body downwards is not *the whole* of its heaviness but only the surplus of the heaviness over the impressed lightness. As this surplus increases (as a result of the lessening of the impressed lightness which is used up in and by its action of slowing the body down) the speed of fall increases proportionally. This happens up until that moment when, the lightness having been completely used up, the body, now acted on by heaviness alone, moves from then on at uniform speed.[124] So we see that the acceleration of the speed of fall is, in reality, only a speed which is less and less slowed down.

But, it will be said, this solution only holds for bodies which have in fact had a 'lightness *praeter naturam'* impressed on them, i.e., it only holds for bodies which have been thrown upwards. Not so, Galileo replies; it holds for all bodies. Let us suppose that a body's motion is stopped at that very instant when, having been thrown upwards, it stops rising and begins its descent. Is it not obvious that it will retain and in some way store up all of the lightness *praeter naturam* which it still has at that instant: and that if, after some interval, we now release the body, its downward motion will in no way be affected by the fact that it had been stopped? We can, therefore, consider a body placed at the top of a high tower as the same as a body thrown up to an equal height.[125] Moreover, does not the body at the top of the tower feel that whatever is supporting it exerts an upward pressure (which prevents it·from falling down), a pressure exactly equal to its weight?[126] It is precisely this pressure which impresses on it the non-natural lightness which will slow down its motion of fall. We can consider all those bodies on earth which are separated from its centre to be in a situation analogous to that of bodies at the top of a tower.[127]

Now we have already seen above that bodies are not all equally capable of taking in and conserving *impetus*, motive quality, lightness *praeter naturam*. In particular light bodies take in less and conserve it less well. This, precisely, is the reason why they fall *faster* at the beginning of fall than do dense, heavy bodies which are impregnated with lightness and only give it up reluctantly.[128]

The truth is that this theory that we have just summarised, and of which Galileo was very proud, was in fact much less original than he thought; it had already been sketched in by Hipparchus.[129] It was also less elegant than he thought, since it gave rise to flagrant contradictions. But it demonstrates well for us the spirit of *impetus* dynamics, and this is its interest and value for our purposes. Therefore we have no need to go into the details of this dynamics as it was developed by Galileo, but we will turn now to another aspect of his thought, an aspect that we have already had occasion to mention in passing above; its Archimedean aspect.

We have already mentioned above that Galileo identifies lightness (natural and *supra naturam*) as the cause of upwards motion, and that on the other hand the speed of a body's fall is determined, according to him,[130] not by its absolute weight but by its specific and relative weight. These important and interdependent points (already stated by Benedetti) in the end enabled Galileo to go beyond both Aristotelian and *impetus* dynamics, and to substitute for them—or rather to attempt to substitute for them—a quantitative physics for which the model was provided for him by Archimedes. Lightness is that which makes a body rise:[131] at first glance this seems to be no more than the classical definition of lightness, the cause of the upwards motion of bodies. In fact it is just the opposite. Lightness and heaviness are no longer understood as causes which produce determinate effects; they are, on the contrary, defined on the basis of their effects. Lightness is that which makes a body rise; heaviness is that which makes a body go downwards. But a 'heavy' body placed on the pan of a balance rises when the other pan goes down. And a piece of wood which falls in the air rises when it is placed in water. Contrary to the view of Aristotle and in agreement with the doctrine of 'the ancients', 'heavy' and 'light' are not absolute qualities,[132] but are relative properties, or rather are simple relations. A body is light or heavy, i.e., goes up or down, depending on the circumstances and depending on the medium it is in. If it is heavier than the medium it goes down, if it is less heavy it rises (as in the case of wood in air and in water). The force (and consequently also the speed) with which it goes downwards or upwards is measured, precisely, by the difference between its own (specific) weight and the weight of an equal volume of the medium it is in.[133] This implies that every body has an *absolute weight*, determined by the quantity of matter that it contains in a unit volume. This clarifies the doctrine of the 'ancients' according to which all bodies have weight and that there are, strictly speaking, no light bodies. Aristotle has got it wrong once again.[134]

Galileo's reasoning here, which actually follows that of Benedetti, is clearly a transposition of Archimedean reasoning.[135] Now this extension of hydrostatics has extremely important consequences: in particular it implies the substitution of a quantitiative scale for an opposition of qualities.

The importance of this substitution, which Benedetti had already attempted, and which is implicit in the doctrine of 'the ancients' was understood perfectly well by Galileo. Indeed he places great emphasis on it. Lightness is not a quality (and nor is heaviness, as distinct from *weight*): it is an effect.[136] Therefore upwards motion is not natural motion.[137] Bodies which rise never do so by themselves, spontaneously. If they rise it is because they are thrust upwards by other, heavier bodies. The only natural motion which Galileo accepts from now on is that of bodies with weight (and this means all bodies, even the air, even fire) downwards, towards the centre of the world. This is also the only motion which still has a natural goal, for upwards motion has none.

This distinction between absolute and relative weight (the weight we are used to measuring on our balances is always relative), and the repeated assertion that the speed of fall of a body is a function of its relative weight in a given medium (and not of its absolute weight), inevitably lead us to the conclusion—already adopted by Benedetti for similar reasons—that in a vacuum, and only in a vacuum, bodies will weigh their absolute weight,[138] and fall with speeds which are, in effect, being functions of their absolute weights, their proper speeds.[139]

This conclusion, fundamentally opposed to the most basic tenets of Aristotelian physics,[140] once it is established, can then be related to the idea of motion as the effect of a motive force impressed on or contained in the moving body. We have already pointed out that motion, according to this conception, is no longer what it was for Aristotle, a process, a passage from one place to another or from one state to another. In itself it is still even far from being a 'state'; this is the reason why it is not automatically conserved. It is, as we have seen, the effect of a force. But since this force is entirely contained in the moving body, it does not follow from the body's motion itself that anything external to it is involved.[141] This conception is perfectly compatible with the idea of a moving body in isolation from the rest of the universe. It also allows for a body moving in a vacuum. Since the speed of the body's motion is a function of the force which is moving it, the absence of resistance does not at all imply an infinite speed. If, on the one hand, the body's motion is violent then it will be always *aliter et aliter* both in relation to itself (since its speed will be different at each instant) and also in relation to the centre of the world (since it will be constantly changing position); but if, on the other hand, the body's motion is natural then although it will certainly be *aliter et aliter* in relation to the centre of the world, it will in relation to itself, by contrast, remain *idem et idem*, because its speed (in a vacuum) will be constant.

What we are observing here is the liberation of motion. The Cosmos is breaking up; space is being geometrised. We are on the route which will lead to the principle of inertia. But we have not arrived there yet. Indeed, we are still very far from it. So far from it, in fact, that before we can arrive we will have to abandon on the journey the idea of motion as an effect, the distinction between 'natural' and 'violent' motions,[142] and also the idea (and even the word) of 'place'. So the journey is a long and difficult one; and we know that Galileo himself never entirely completed it.

But that is another story, one which we are not concerned with here.[143] In the period we are studying now Galileo had only just begun the journey. There was still, for him, a 'natural place', and one only—the centre of the world. There was still one, and again only one, natural motion—that towards this centre.[144] There was, even, still a residue of Cosmic order; heavy bodies are located at, or near to, the centre of the world, and lighter bodies are located around them in concentric circles. This curious idea shows us clearly just how difficult it was for Galileo to liberate himself from the frameworks of the traditional representations of the world. He retains the concentric order of the elements, but explains it by reference to geometrical considerations; the heaviest bodies being also the most dense are naturally located at the place where there is least space available for matter to be contained in, i.e., at the centre of the sphere of the Universe,[145] which is assumed to be real.

Notice, however, just how vague and imprecise this sphere of the Universe has already become. In his critique of the Aristotelian idea of natural motion, whereas Galileo accepts the *natural* character of downwards (*deorsum*) motion, he argues against the natural character of upwards (*sursum*) motion not only because this motion is always violent (since all bodies are heavy), but also because it can have no natural end-point. One cannot go downwards for ever, but one can, on the other hand, always go higher.[146]

This odd passage shows us how—no doubt influenced by Copernicus[147]—Galileo's thought was undergoing a gradual transformation. The centre of the Universe is still there. But the sphere of the Cosmos has been enlarged, has become indefinite, has so to speak lost its circumference. Once it has become infinite[148] then, space being homogeneous, no trace will remain of the ancient Cosmos, there will no longer be any 'place' or direction with special status. But what an effort of thought is implied in this move! It is a move that Galileo did not make. Only Bruno, who was neither an astronomer nor a physicist, was able to take the decisive step.[149]

Let us go back a little. Where do these curious mechanical physics (all the motions of bodies, Galileo says often, can be reduced to the principle of the balance[150]) and hydraulics, that we have discovered in Benedetti and that we have just encountered again in Galileo, come from? We have already said often enough, they come straight from Archimedes, whose name Galileo never writes without surrounding it with praise and to whose authority he appeals.[151] Without any doubt he is right to do so. Besides, Galileo was not the only one who admired Archimedes above all others. From the time when Tartaglia (who was actually not able to get much out of him) published a Latin translation of Archimedes' works, his influence never ceased to grow. To such an extent that Cardan, who in all seriousness enjoyed ranking the great men in order of importance, placed Archimedes above Aristotle, all by himself in the highest category.[152] It is true that this immediately called forth a protest from Scaliger: to put this artisan above Euclid, above Aristotle, above Duns Scotus and Occam, what folly! But Cardan's opinion was very significant. It indicates the growing importance of Archimedes. As for his influence, it is quite manifest that the two best writers on mechanics of this period, Guido Ubaldo del Monte and Giovanni Battista Benedetti, owed what was most lucid in their thought to Archimedes. As for Galileo, one can say that it was, in a manner of speaking, in learning from Archimedes that he matured.

In fact Galileo's first contribution to science was his *Bilancetta*,[153] a treatise on the hydrostatic balance. And he obtained his first Chair of mathematics at the University of Pisa as a result of a work on the centre of gravity of solids, a work entirely Archimedean in inspiration and technique. It was in consciously and resolutely learning from Archimedes, in adhering to the tradition of thought which he represented (for 'the ancients'[154] and against Aristotle), that Galileo succeeded in going beyond the physics of impressed force, in achieving the level of a mathematical physics which was precisely an Archimedean dynamics.

It is worth repeating what we have already often pointed out that *impetus* physics, the physics of impressed force, of *elan*, was a reaction of common sense, of ordinary everyday experience, against the theoretical, cosmological physics of Aristotle. The ideas which it introduced were no more than abstract extensions of common sense. Therefore they could not be forged into a unity with the mathematical ideas which were being developed at the very same time, in spite of the mathematical genius of Nicole Oresme and in spite of the geometrisation of ultracosmic space accomplished by the Parisian school.

The ideas which Galileo, from his time in Pisa onwards, began to use in his analysis of motion, following but going beyond the example of Benedetti, were of quite a different order. For example, when he studies the motion of a body on an inclined plane (which he reduces, moreover, to the study of the lever); when he demonstrates that, on a horizontal plane, any force however small is sufficient to set a sphere however large in motion;[155] or when, in his critique of Aristotelian dynamics, and in support of his own theory of the fall of bodies in a vacuum, he demonstrates that the increase in the body's speed, a result of the decreased resistance, never goes beyond certain finite magnitudes (the increase being asymptotic), and that therefore the complete elimination of resistance in a vacuum does not result in the speed's becoming infinite;[156] when, in general, he studies motion in a vacuum, etc., he immediately and quite consciously does so from a position outside of reality. An absolutely smooth plane, a perfectly spherical sphere, both absolutely hard: these are things that are not found in physical reality.[157] These are not concepts drawn from experience; they are postulates that one makes about it. Therefore one should not be surprised to observe that the reality of 'experience' cannot be in complete agreement with one's deductions.[158] It is the latter, however, which are correct. It is deduction, it is 'imaginary' concepts which enable us to understand and to explain nature, to put questions to it and to interpret its replies. Galileo argues for the superiority of Platonist mathematicism over abstract empiricism.

However, he does not yet invoke the authority of the divine Plato[159] in support of this 'mathematical licence' of the new physics (for example in support of the use of the hypothesis that the lines of force gravity are parallel), but the example of the 'superhuman' Archimedes.[160]

Could there be any clearer indication of the historical line of descent? Could we be given any sharper vision of the direction of the scientific revolution in progress? Having rejected Aristotle's physics, having unsuccessfully tried to build a physics of common sense, it is an Archimedean physics that Galileo attempts to found from now on.[161]

An Archimedean physics means a deductive and 'abstract' mathematical physics, and it was just such a physics that Galileo was to develop at Padua. A physics of mathematical hypotheses; a physics in which the laws of motion, the law of fall of bodies, are deduced 'abstractly', without involving the idea of force, without recourse to experiments with real bodies. The 'experiments' which Galileo, and others after him, appealed to, even those which he did actually perform, were not and could never be any more than thought experiments.[162] These are the only kind that could be performed on the objects of his physics. Because the objects of Galilean physics, the bodies of his dynamics, are not 'real' bodies. After all, one cannot introduce 'real' bodies—if 'real' is taken to mean 'of common sense'—into the unreal space of geometry. Aristotle had understood this perfectly well. But he had not understood that one can postulate abstract bodies, as had been recognised by Plato, and as had been done by the Platonist Archimedes.[163] But Archimedes had not himself ever managed to set these abstract bodies in motion. This was the task of the Archimedean Galileo.

Galilean dynamics only holds for these abstract bodies located in geometrical space, in short for these Archimedean bodies. It is also only to them that the principle of inertia applies. And it is only when the Cosmos is replaced by the reality of the void of Euclidean space, when the bodies of Aristotle and common sense, defined by essence and quality, are replaced by these abstract 'bodies' of Archimedes, that space will no longer have a physical role to play, and that bodies will no longer be changed by their motions. They will then be unaffected by their state (of rest or motion) and motion, having become a state, will be able, just like rest of which it will have attained the ontological status, to be indefinitely conserved, with there being no need for us to find a cause in explanation of this phenomenon.

NOTES

AT THE DAWN OF CLASSICAL

SCIENCE

Introduction

1 See also the fine short work of Federigo Enriques, *Signification de l'histoire de la pensée scientifique*, Paris, Hermann, 1934.

2 We owe the idea and the term 'mutation' to G. Bachelard. See his *Le nouvel esprit scientifique*, Paris, 1934 and *La formation de l'esprit scientifique*, Paris, 1938.

3 Given the scientific revolution of the last ten years it seems preferable to reserve for this revolution the word 'modern' and to label prequantum physics as 'classical'.

4 This quite widely used concept should not be confused with that of Bergson for whom all physics, that of Aristotle as much as that of Newton, is in the last analysis the work of *homo faber*.

5 See V. Laberthonnière, *Etudes sur Descartes*, vol. II, Paris, 1935, pp. 288-289, p. 297, p. 304: 'Physique de l'exploitation des choses'.

6 'Bacon, the founder of modern science' is a joke, and a bad one at that, that one can still find in the text books. In fact Bacon understood nothing about science. He was credulous and completely uncritical. His manner of thinking was closer to alchemy and magic (he believed in 'sympathies'), in short to that of a primitive or to a thinker of the Renaissance than to that of a Galileo or even a Scholastic.

7 Cartesian and Galilean science was certainly a boon to the engineer and has been put to use by technology with well known success. But it was not created *by* technologists nor *for* technology.

8 'Descartes the artisan': this conception of Cartesianism was developed by M. Leroy in his *Descartes social*, Paris, 1931, and taken to an absurd extreme by F. Borkenau in his *Der Uebergang vom feudalen zum bürgerlichen Weltbild*, Paris, 1933. Borkenau explains the development of Cartesian philosophy and science by the appearance of a new form of production, namely manufacture; cf. the critique of Borkenau's work, far more instructive than this work itself, by H. Grossmann, 'Die gesellschaftlichen Grundlagen der mechanistischen Philosophie und die Manufactur', *Zeitschrift für Sozialforschung*, Paris, 1935.

   As for Galileo, it is to the tradition of Renaissance artisans, builders and engineers that he is related by L. Olschki, *Galilei und seine Zeit* (Geschichte der neusprachlichen wissenschaftlichen Literatur, vol. III), Halle, 1927. Now although it is true that Renaissance engineers and artists contributed a great deal to breaking the hold of Aristotelianism, and even sometimes—as in the cases of Leonardo da Vinci and Benedetti—endeavoured to develop a new, anti-Aristotelian dynamics, this dynamics, as Duhem has shown, was in its overall conception that of the Parisian nominalists. And whenever Benedetti—by far the most notable of these precursors of Galileo—went beyond the level of 'Parisian' dynamics this was not thanks to his work as an engineer and artilleryman, but thanks to his study of Archimedes.

9 Some have even contrasted the experimentalist Galileo to the theoretician Descartes: this is quite incorrect as will be shown below. Cf. the present author's 'Galilée et Descartes' IXth Congrès International de la philosophie, *Travaux*, vol. II, p. 41 ff, Paris, 1937.

10   For example, nobody has ever observed inertial motion and this for the simple reason that it is only possible under unrealisable conditions. Emile Meyerson has pointed out (*Identité et Réalité*, Paris, 1926, 3rd edition, p. 156) just how little agreement there is between experiment and the principles of classical physics.

11   It corresponds to a renewed primacy of Being over Becoming.

12   See most notably E. Duehring, *Kritische Geschichte der allgemeinen Principien der Mechanik*, Berlin, 1875, p. 24 ff.

13   See Kurd Lasswitz, *Geschichte der Atomistik*, Hamburg and Leipzig, 1890, vol. II, p. 23 ff.

14   V. J. Wahl, *Le rôle de l'idée de l'instant dans la philosophie de Descartes,* Paris, 1920.

15   Galileo's famous deduction of the law of falling bodies (see *Opere*, Ed. Nationale, vol. II, p. 261 ff and *Discorsi* in *Opere*, vol. VIII, p. 222, = *Discourses*, p. 174 ff) consists, in fact, in a purely kinematic study of the simplest form of accelerated motion and does not involve the idea of force, nor those of mass or attraction: see 'The law of falling bodies', chap. 1 below.

16   See E. Cassirer, *Das Erkenntnisproblem in der Philosophie und Wissenschaft der neueren Zeit*, Berlin, 1911, vol. I, p. 394 ff. Also Lasswitz, *op. cit.*, E. Mach, *Die Mechanik in ihrer Entwicklung,* Leipzig, 1921, p. 117 ff and E. Wohlwill, 'Die Entdeckung des Beharrungsgesetzes', *Zeitschrift für Völkerpsychologie und Sprachwissenschaft*, vol. XIV and XV.

17   E. Meyerson (*Identité et Réalité,* p. 124 ff) has correctly drawn attention to this point.

18   It is most interesting to compare Galileo and Kepler in this respect: Kepler is still a cosmologist, whereas Galileo no longer is. (See *Dissertatio cum Nuntio sidereo* in Galileo, *Opere*, vol. III, p. 97 ff = *The Starry Messenger*, p. 21 ff); cf. the present author's 'Rapport' in *L'Annuaire de l'Ecole pratique des Hautes Etudes*, 1934.

19   P. Duhem, *Le Système du Monde*, Paris, 1913, vol. I, pp. 194-5: 'In fact this dynamics seemed to be so very well in agreement with common observation that it could not but be accepted immediately by those who first speculated about forces and motions . . . Before physicists could bring themselves to reject Aristotle's dynamics and to construct modern dynamics, they had to understand that the phenomena which they witness everyday are not at all simple, elementary phenomena, to which the fundamental laws of dynamics should be directly applicable; that the motion of a boat which is pulled along, or that of a horse-drawn cart along a road, must be seen as motions of extreme complexity; in a word, in order to formulate the principle of the science of motion it is necessary, by abstraction, to consider a moving body acted on by a single force moving in a vacuum. Now Aristotle went so far as to conclude from his dynamics that such a motion is inconceivable.'

20   Caverni, *Storia del metodo sperimentale in Italia*, 5 vols., Florence, 1891-1896; see especially vols. III and IV.

21   P. Duhem *Le Mouvement absolu et le mouvement relatif*, Paris, 1905; 'De l'accélération produite par une force constante', *Congrès international d'histoire des sciences*, 3rd session, Geneva, 1906; *Etudes sur Léonard de Vinci. Ceux qu'il a lus et ceux qui l'ont lu,* vol. III; *Les Précurseurs parisiens de Galilée*, Paris, 1913.

22   One could sum up the scientific work of the 16th century by saying that it was the taking up again and gradual assimilation of the work of Archimedes. As far as the history of scientific thought is concerned the popular conception of the 'Renaissance' can be shown to be profoundly true.

23   See P. Duhem, *Etudes sur Léonard de Vinci*, 3 vols., Paris, 1909-13; F. J. Dijksterhuis, *Val en Worp*, Groningen, 1924; and Ernst Borchert, *Die Lehre von der Bewegung bei Nicolaus Oresme* ('Beitrage zur Geschichte der Philosophie und Theologie des Mittelalters', vol. XXX, I/3), Münster, 1934.

24   It has been said (cf. E. Mach, *op. cit.*, p. 118 ff, and E. Wohlwill, *Galilee und sein Kampf für die Koperikanische Lehre,* Hamburg and Leipzig, 1909, vol. I, p.115), that in these youthful works of Galileo, and in particular in *De Motu* which was written at Pisa, Galileo did no more than repeat without acknowledgement what had been taught by G. B. Benedetti (cf. his *Diversarum speculationum mathematicarum et physicarum liber,*

Taurini, 1585). This judgement, as we shall see below, is not entirely fair: while he does follow Benedetti (which, moreover, is explained by the fact that Benedetti's thought, like that of the young Galileo, is a strange mixture of Parisian 'empiricism' and Archimedean mathematicism) he does diverge from him on occasion, and when he does he is always right to do so. At such points, we would say, he shows himself to be more profoundly both 'empiricist' and Archimedean than Benedetti. It is this that makes a study of Galileo so instructive.

Aristotle

25 Published in vol. I of the admirable edition of the works of Galileo, *Opere*, Edizione Nazionale.

26 Francisci Bonamici Florentini, e primo loco philosophiam ordinariam in Almo Gymnasio Pisano profitensis, *De motu, libri X, quibis generalia naturalis philosophiae principia summo studio collecta continentur* . . . Florentiae, apud Bartholomeum Sermartelium, MDCXI. Bonamico's work is usually mentioned by Galileo's biographers. It seems, however, that none of them, not even Favaro or Wohlwill, have had the courage to open this enormous volume (1,011 pages *in folio*).

27 On this point the failure of Duhem, who is the only one to have seriously wished to revive it, is very significant.

28 Aristotelian physics is essentially non-mathematical and it cannot be mathematised (for example, by presenting it as based on the principle that speed is proportional to force and inversely proportional to resistance; this proportionality is only a *consequence* of Aristotelian principles) without being false to its spirit.

29 It was pointed out long ago that there has never been a god of gravity.

30 See E. Mach, *Mechanik*, chap. 8, p. 124 ff.

31 One could even say that Aristotle's greatness consisted precisely in his wish to *explain* 'natural' phenomena.

32 It is interesting to note that terms such as Cosmos, Universe—in the sense of totality—etc., which had completely lost all meaning during the classical period of physics, seem to have received a new meaning since Einstein.

33 It is only in 'its place' that a being is completed, perfected, is fully itself; this is the reason why it has a tendency to go there.

34 The idea of 'natural place' implies the finiteness of motion and thus the finiteness of the Universe. Or one could put this by saying that the concept of a finite, bounded Universe is expressed in the idea of natural place.

35 Natural motion in an upwards direction *proves* that the Universe is finite: for Galileo's criticism of this see above p.34, and note 137.

36 The geometrical order within the spherical Universe corresponds to the qualitative order (heavy-light), and it follows from this that violent or natural motion is movement of separation or approach of a given body to its proper place; and it also follows that the two kinds of motion are incompatible. Cf. Galileo, *Juvenilia, Opere*, vol. I, p. 61 ff.

37 Since motion can only be produced by other motion, any particular motion implies an infinite series of anterior causes.

38 Circular motion is the only uniform motion which can have indefinite duration in a finite Universe; it is also that which—if we attribute it to the sphere as a whole—changes nothing. Consequently it is the nearest thing there is to a natural state. Therefore Aristotle's opponents made every effort either to prove the natural character of circular motion in general, for all bodies and not only for celestial bodies (and this led in the end to the dynamics of Copernicus), or like Galileo, in this respect misunderstanding Aristotle, to prove that circular motion 'around the centre' is neither violent nor natural because 'in circular motion bodies neither come nearer nor get further from the centre.' Cf. below, note 155.

39 Thus motion takes place in a body and has an effect on it. Therefore it is understandable that a body can have only one natural motion and that if it is affected by two different motions—natural and violent—each of these motions is a hindrance to the other.

40 Aristotelian physics is usually said to be dominated by biological categories. While this is correct of course (its concept of motion can be understood as expressing the intermediate

situation of life between the immutability of Spirit and the immobility of death) this interpretation seems nevertheless to miss the point that the distinction between state and process (being and becoming) is a completely general one and is not limited to living beings.

41 In the extremely complicated mediaeval discussions about the nature of motion this is generally considered as a special type, *forma fluens*. Cf. the works of Duhem, Dijksterhuis and Borchert, *op. cit* and S. Moser, *Grundbegriffe der Naturphilosophie bei Wilhelm von Occham* (Philosophie und Grenzwissenschaften, vol. IV, fasc. 2-3), Innsbruck, 1932.

42 Local motion is therefore always both relative and absolute: relative because it necessarily implies a reference to a limit and cannot be conceived 'in itself', in relation to nothing, like Newton's absolute motion; absolute because the 'places' between which motion takes place form an absolute system, which has an essentially immobile boundary.

43 Aristotle is absolutely right. No process (becoming) continues by inertia. Motion only continues because it is not a process.

44 There is no force of attraction in Aristotelian physics.

45 From a strictly mechanical point of view there are, in effect, no others. Cf. E. Meyerson, *Identité et Réalité*, 3rd edition, p. 84.

46 Aristotle's theory is such a good one that it was imitated and used right up to the seventeenth century, in particular by Descartes and Huyghens.

47 Cf. *Timaeus*, 79 b.

48 Aristotle's theory consists in explaining the continuation of the motion by a vortex-like process in the surrounding medium; this acts on the moving body both drawing it along and pushing it. The theoretical 'trick' is the invention of a medium which is particularly suited to motion (an elastic medium, we would say nowadays), the air, Cf. Aristotle, *Physics*, IV, 8, 215a, VIII, 10, 267a.

49 See the history of this much discussed question given by Duhem, *Etudes sur Léonard de Vinci*, Paris, 1909-1913.

50 It should be remembered that the impossibility of a vacuum is also a Cartesian thesis. On this point as on many others Descartes is in agreement with Aristotle and against Galileo.

51 An infinitely fast motion, an instantaneous relocation of a body from one point to another, is in fact absurd.

52 In the homogeneous space of geometry all 'places' are similar and nothing new can result from relocation.

53 It is well known that Aristotle was very much against any confusion between different *genera*; the geometer's thought must not be confused with that of an arithematician nor that of a physicist with that of a geometer. This is a perfectly reasonable demand; as long as there are different kinds of thought one should not mix them up. But the division of thought into these kinds can be demolished.

Mediaeval discussions: Francisco Bonamico

54 On John Philoponus see E. Wohlwill, 'Ein Vorgänger Galileis im VI Jahrhundert', *Physicalische Zeitschrift*, vol. VII, 1906.

55 Cf. the works cited in notes 23 and 41 above. The study of the problem of motion is infinitely instructive—the study of failure always is—and it alone enables us to appreciate and understand the meaning and the importance of the Galilean revolution.

56 Bonamico's work, *De Motu*, Pisa, 1591, is both very instructive (on the one hand it demonstrates just how puzzled mediaeval thought was by the phenomena of falling bodies and projectiles, and on the other hand it reveals just how familiar knowledge of *impetus* physics was in university circles) and also practically unknown to historians. His book is, moreover, very rare: even the British Museum does not have a copy. Therefore we give extensive extracts from it, taken from the copy in the Bibliothèque Nationale.

57 Bonamico, *De Motu*, lib. V, c.xxxv, p. 503 « *De motibus praeter naturam et de projectis contra Platonem.* Quoniam vero oppositorum una est eademque methodus et scientia: motui vero: secundum naturam opponitur motus praeter naturam; postquam de motu

naturali satis dictum est: postulat nunc instituta ratio de motu, ut aliqua dicamus de eo qui est praeter naturam, qui item nascitur ex violentia: hic vero duplex est, vel simpliciter, vel quodammodo: vi autem moveri illa dicuntur quandocunque id quod movetur non confert vim, hoc est non habet illo propensionem, quo movetur, quia. s. non perficiatur ex eo motu, locum illum adipiscens in quo conservetur: hic autem est qui convenit suae formae; sed ab eo forma potius corrumpitur. Ideo quod unumquodque suae neci resistit, quantum potest; tantum abest ut eo properet, ut nisi virtus moventis resistentiam mobilis superet nunquam moveatur; et nisi praevaleat facultas violans, in pristinum locum semper retrocedat ; neque ullo modo conatum moventis adiuvat, sicut adjuvaret saxum, si magno impetu deiiceretur : nam virtus eiusmodi facultati accedens longe velociorem motum faceret. Itaque principium talis motus omnino externum alienumque est, solumque socium sui laboris habet medium, quod impetum a movente excipiens mobili impertit. Verum quod praeter naturam absolute movetur ; omnino et simpliciter nullam vim confert : immo renititur; sed ita vincitur a movente, ut simpliciter eandem illam lineam metiatur quam permearet, si moveretur secundum naturam : ideoque movetur ocyus ab initio, quam ad extremum. Quod vero aliqua ex parte praeter naturam movetur, non omnino resistit ; licet eo non propendeat, quo movetur, necque eandem lineam peragrat violatum ac si secundum naturam moveretur ; sed ad latera quodam pacto deflectitur. Quam ob rem etiam medium illi motui magis inservit, ob id velocius et ad maius spatium idem lapis in latera proiicitur, quam sursum directo et ad perpendiculum. Attamen neutrum illo simpliciter vergit quo agitur ; necque ibi manet secundum naturam; sed posteaquam vis movens contabuerit ad suum motum locumque naturalem sese recipit, describens lineam secundum quae est ad perpendiculum inter centrum mundi et extremum, et movetur aliquanto celerius in progressu. Principia vero quae violant varia esse queunt et contraria, quae materiam affligunt, ut apparet in fulmine, quod cum sit ignis, ab acqua circumstante expellitur et propter vim agitati corporis, ut fit, ubi venti extollunt aliqua pondera et raptu mobilis cujusdam, ut forte evenit in hyppeccaumate, impetu item aquae, aut aeris in gyrum acti, ut accidit in vorticibus et generatim pulsu, tractu, vertigine et vectione quae plurimum fiunt ab animatis.

Sed cum supra de caussa violenti motus universe satis dictum sit, agamus nunc de ipso speciatim et in praesentia vestigemus caussam alterius illius motus quem solent nobis significare nota projectorum. Quae longe abstrusior est et antiquitus etiam varias ostendit opiniones. Nam Plato quemadmodum eius verba sonant, asserebat caussam talis motus antiperistasim : quanquam quo pacto caussa haec accipienda sit, nec multum declarat Aristoteles, neque satis e Platone colligitur. Etenim vox est ambigua. Siquidem sit proprie contrariorum ambitus ; quando unum contrariorum ambit, et alterum velut in centrum adducit quemadmodum calor centrum versus aestate cogit frigus, unde multa poma oriuntur, quibus frigus insigniter dominetur ; et contra frigus hyeme centrum versus calorem propellit, unde ventres hyeme calidiores : secundo etiam communius accipiatur in latione sola, cum ambiens efficit lationem in eo quoe ambitur, ex eo ducens originem, ut Plato volebat ; quia movens omne, dum moveret, una quoque moveretur ; nec ullam vim, nisi qua corpus esset, mobili communicaret, aut in aliud a se transferret ; quapropter eodem motu quo mobile ipsum ageretur, ut, si animus res esset corporea, idemque corpus agitaret, ipse quoque primum pari ratione ferretur.

Ita igitur in projectione partes circumstantes in locum posteriorum succedunt, ut, A. si moveat B. subit in ejus locum et si B. propellat C. locum eius occupat et sic cetera deinceps. Hoc autem dubitatur, an sit per extensionem eius corporis quod ambitur ; an potius sit per successionem quae fit propter vacuum : nanque huiusmodi sensum ex eius verbis colligebat Simplicius, et haec item sententia ab Aristotele sub hac ratione confutata deprehenditur, quoniam ex eo quod a tergo rei mobilis coiret medium (hoc. n. liquidum esse oportet et facile coire posse) ne detur vacuum : facta autem illa coitione mobile procederet ulterius. Sed quocunque accipiatur a tergo medium convenire, sive impleat solum id spatii quod a mobili relictum fuerat, sive etiam id quod congreditur, ipsum promoveat, multa sunt quae nos ab ejus opinione avertant. Ac quantum de secunda est, quam de verbis Platonis Simplicius ipse profitetur, satis haec illus fallaciam significant. Primum quia ratio reddi non potest, cur primo cessante, reliqua moverentur :

ubi nam fiat motus per solum contactum, veluti fieret in hac hypothesi, uno moto deinceps omnia moverentur, eoque manente quiescerent ; quod omnia in alterius locum successione quadam subingrederentur. Quod si id non eveniret, omnia quoque manere opus est : talis nam motus est antiperistaseos, si credere dignum est Aristoteles quod unum quidem primum movetur et movens in eius locum subit ; ita ut una movens et mobile concitentur ; neque velocitate maiore partes in progressu q. ab initio moverentur : oppositum tamen apparet. Quod si de experientia dubites, vide item id evenire, si segnius in progressu concitetur quod in parte quadam motus illius negari non potest nanque idem tenor a natura servabitur, dum vacuum propulsare contendit, hoc studet, ut arceatur inane, id semper eodem instanti praestat quo motus efficitur ; nec potest effici motus, nisi movens succedat. Itaque idem est successionis instans et motus, atqui vacui pulsio perpetuo sui similis est ; et motus igitur. Praeterea natura solam intenderet coitionem, utputa, ut exploderet vacuum : ubi igitur aere in saxi locum subingresso, adepta illam fuisset ; non esset certe, quod amplius laboraret ; si ergo post primam saxi motionem coivit aer, cur motus procedit ulterius ? Quantum vero pertinet ad primum illum modum antiperistaseos qui affert extrusionem : habet et hic contra se multas experientias. In primis. n. ecquid erit caussa, quod vetet lapidem ad celum usque concitari ? nam, si aer in eius locum succedet, et lapidem idcirco propellit, quanto continue sit ea successio, continue quoque lapidis propulsio fiet, quousque suppetat aër, aut corpus aëri quod propter coëundi facultatem valeat idem atque aër. Tum item facilius palea, quam saxum proiici posset, tum quod palea levior est, et sursum magis propendet quam saxum tum etiam, quid maior est aeris impellentis ad paleam proportio, quam ad saxum : ex maiore autem proportione velocior motus procedeat necesse est. Rursus, si filum saxo appendatur, ob eandem caussam a fronte saxi ponderet : cum videamus igitur ipsum a tergo porrigi in longitudinem, et quasi trahi a saxo, potius quam ab aëre propelli; dicamus oportet extrusionem non esse caussam tali motus. Sic undique Platonis opinionem lubricam esse comperimus. »

[58] Bonamico, *De Motu,* lib. V. c. XXXVI, p. 504: « *Aristotelis sententia de proiectorum motu recensetur, et ea quae contra illam afferi solent exponuntur :* Repudiata Platonis opinione, decrevit Aristoteles a movente vim imprimi aëri sive medio, propter eius naturam quae anceps est, nec gravis tantum, aut levis : ob eamque caussam impetum quoquo versus excipere potest. Quia tamen impetus ille simpliciter eo versus non est, licet, ut alias a nobis dictum est, eius naturae minus hoc adversetur, quam si simpliciter sursum, aut deorsum moveatur : quia non tantum levis est, sed etiam gravis, tantisper item resistit, atque ubi seiunctus est aliquantum a primo motore, vim ab eo sibi impressam paullatim amittit, demum deferiscitur, et contabescit et ita proiectum ab alio non violatum, pristinas conditiones recuperat et secundum illas ad eundem locum festinat, unde coactum discesserat, quasi ferrum, quod ubi segregatum ab igni fuerit, ad propriam frigiditatem revertit. Verumtamen Philoponus, et alii Latini in Aristotelem acerrime invecti sunt, usque adeo, ut praeceptorem deserverint. Primum quia neque item eius positio difficultatem illam evitat quam Platoni paullo ante obiecimus ; nunquam. s. eius motum cessaturum, quoniam ab aere vehitur saxum, aër autem, hic ubi impetum excepit, non habet unde quiescat : quoniam impetus ille sit ei naturalis non secus atque descensus saxo secundum naturam sit : quare non modo saxum per aërem totum agitabitur, sed etiam tempore infinito, si infinitus fuerit aer. Nam dicere ipsum aërem fieri per se mobilem, ut moveri simul et manere possit, quod animatorum proprium est, longe aberret a versimili. Neque sufficit id quod adscribebat Averroes, medium a sua naturali forma moveri, eum tamen motum ab extrinseco sumere occasionem. Nanque esto hoc. At unde quies in medio? iam. n. adfuit occasio movendi; mediumq. secundum naturam movetur. Deinde si ab impetu iam indito et impresso a primo movente sit iste motus ; quo mobile propinquius erit moventi, eo quoque maior impetus erit saxi projecti, et motus ipse velocior. At hoc falsum est, quia proiectorum motus augetur per aliquantum spatii in progressu, quod item experientia testatur cum funda, aut balista, aut etiam quodvis tormentum ex distantia quadam vehementius feriat, quam cominus. Adde etiam quia saxum contra ventum moveri non posset. Etenim maiore impetu moveretur aër contra saxum, cum maior sit impetus venti quam proiicientis ipsius. Accedit eodem q. per aequalem distantiam moveretur lapis a tangente et a remoto,

quoniam aequalis impetus aeri posset imprimi ab utroque. Tum postremo eadem velocitate proiiceretur hasta oblonga ac brevis : quoniam aequalem impetum impertiri possis utranque proiiciendo. Quamobrem Philoponus, post ipsum vero Albertus, D. Thomas et alii complures opinati sunt, vim sane imprimi a primo movente non aeri quidem, sed mobili, utputa saxo ; et prout maior, aut minor vis illi imprimeretur, ita per maius spatium atque velocius agitari. Huiusmodi autem vim interdum expeditius ac promptius excipi. Nonnunquam aegrius et lentius ; propter illa quae motui solent auxiliari, utputa, figuram, magnitudinem, materiae multitudinem et caetera, quae supra caussas lationis socias appellavimus, sic longius fertur hasta, quam corpus quadratum, et chorda tenta, quia melius excipit impetum, retinetque diutius, quam remissa, diutius quoque tremit, atque ictum facit maiorem. Si quaeratur etiam, cur aër in iactu non agitur in immensum : respondent : quia communicatur ille motus a lapide partibus proximis, et ab hisce subinde reliquis contiguis, ut etiam vel eodem Aristoteles, teste et auctore (8° phys.) non sit unus ille motus, at lentius, at vero cum motus ille non sit neque lapidi, neque aëri naturalis, sed utrique eveniat ab externo praeterea circumferentiam versus dilatetur, quemadmodum fieri conspicimus ubi, lapis in acquam proiciatur, facit. n. rotationes in principio minores, sed velociores ; et ob maiorem proportionem quam habet tum movens ad mobile : et quia citius peragi solet spatium quo brevius est, in processu maiores quidem, sed tardiores : et aucto spatio et proportione moventis ad mobile imminuta : sic facit lapis in aërem proiectus: ideo motus segnior evadit ; ut demum fatiscat ; et interposita quiete ; quia motus aut contrarii sunt, aut contrariis respondent, semoto impediente moveatur secundum naturam. Reddi etiam caussa potest, cur pila lusoria facilius repercutiatur, quam lapis : in motu. n. ante reflexionem valde comprimitur : postquam reflexa, est dilatatur; ita quaerens innatam dimensionem (consequitur autem ipsam, non secus atq. suum locum elementum genitum assequatur, cum ablatum fuert impedimentum) ex repulsione maiorem impulsum adipiscitur. Quo fit, ut cum positio haec illa praestet quod bona quaestionis explicatio debet efficere : consentit.n. cum ratione, non oppognat sensum : satisfacit omnibus problematis quae de re proposita quaeri possunt : et inhaerentium caussas reddit : alacriter etiam a Latinis contra Arist. ipsum defendatur. Et quoniam ita potest in methodo naturali experientia, ut ceteris neglectis machinis ingenii et rationis, illi standum sit, statuamus ad opinionis huiusce confirmationem levissimam tabulam, ex qua torno, aut circino incidente orbis eximatur : ita ut sine mutuo attritu orbis ille intra illud cavum circumagi possit, et tabula alicubi defixa, vectis cum manubrio illi orbi infigatur, quod manubrium singulae utrinqfurcillae, seu cervi sustineant. Tunc manifesto apparebit circumactum orbem intra illud spatium tabulae orbiculatum moveri á moto motore, nullo aëre impellente. Neque tunc, quia motus ille in orbem est, locus erit aëri impellenti. Nam quamvis aër inter orbem et tabulam existat, adeo est exiguus, ut nullas vires ad eum motum habiturus sit ; eoque maxime, quod ipsius orbis politissima laevitas ad aëre circunstante, neutiquam agitationis instigationem accipere valebit. Quo.n. laevius quid est, eo magis agglutinationem respuit.

Quanquam quid aliud erat, quod a nobis in hac caussa reddenda posset afferri, quam auctoritas ipsa Arist. qui aut hanc caussam omnino recipit, aut si aliam probavit, evidentissima repugnantia concluditur? Habet.n. Q. Mech. tantum ferri id quod fertur. i. proiicitur et pellitur, quantum aëris moverit ad profundum, ideoque caussam reddebat, cur neque magna nimis, neque valde parva proiici possent. Monstrant haec omnia igitur impetum aëri in motu projectorum a movente primo non committi, contra q. ab ipso Arist. contra Platonem decretum fuerit. Ita magnum opus erit; si summus ille praeceptor a calumniis hisce purgetur, id quod nos pro veritate ipsa mox aggrediemur, oppugnatores enim acerrimi sunt. »

[59] Once again it is to Duhem's *Etudes* that we owe our knowledge of these discussions.

[60] We have pointed out already that this was also the solution adopted by Descartes.

[61] Bonamico, *De Motu*, lib. IV, cap. XXXVII, pp. 410 sq.: « Aggredimur questionem qua de cremento naturalis motus *in fine disseritur* . . . facile reddi potest caussa quaestionis illius ; cur ea quae moventur secundum naturam ocyus in fine moveantur, quam in principio motus. De qua sane quaestione multa dicta fuerunt tum Arist. ipsius temporibus, tum etiam usque ad haec nostra, caussaeque complures allatae, cum per se,

vel natura, vel locus, tum per accidens, ut impedimenti sublatio, calor rarefaciens, adventitia quaedam gravitas, atque haec vel seorsum vel coniunctim, eademque admodum verisimiles, ut nisi Argi oculos adhibeamus, facile decipi possimus. Idcirco praestat, ut singulas caussas curiosius requiramus. . .

Nam antiquitas (etenim nos Graecorum sententias primum recitabimus). Timeus, Strato Lampsacenus et Epicurus existimaverunt, omnia quidem esse gravia, nihil per se leve : duos autem esse terminos motus, alterum supremum, atque alterum oppositum illi infimum, sed unum nempe deorsum et infimum esse locum in quem omnia properent secundum naturam ; alterum vero ad quem vi ferantur : etenim cum omnia gravia sint, deorsum suapte natura feruntur, quod si quis ex his inferius est, aut superius, noc non aliunde proficisci quam, quod corpora graviora minus gravia premunt, et ideo subeunt illa, non quidem quia leve aliquid sit ; propterea suopte nixu sursum feratur, sed utraque corpora sunt in genere gravium ; alterum vero ex illis leve apparet, quoniam hoc gravissimum est, illud minus grave, et quoniam hoc gravissimum est, ideo premens illud quod est minus grave, subit ipsi, quod autem minus grave est, sic supereminet : quasi vero motus hic fit per extrusionem, quare, quo gravius est, magis extrudit, magisque opprimens id quod est minus grave, eo etiam velocius fertur. Ob id velocitas huius motus non quidem ab interna caussa derivabitur, verum ab externa, et erit violenta, non autem naturalis.

Ceterum in hos invectus est Aris. ab his quae monstrat sensus in aliquo genere motuum, atque conclusit nonnullum esse quoque motum naturalem in omni corpore et sursum etiam, tum quod ubi movetur aliquid vi, citius fertur, si minus sit, quam si fuerit maius, tum praeterea quia quicquid vi movetur in sui motus nitio velocius est ; evanescente vero illo moventis impetu, etiam deficit eius motus, ac naturalis illi succedit, qui quidem in principio segnior est, vegetior vero fit in progressu, ac postremum prope finem velocissime fertur : nam id quod aliquo fertur vi, movetur inde secundum naturam. At nos in elementorum motu, verbi gratia quando terra descendit, cernimus quo maius est illius moles, etiam ferri velocius. Praeterea conspicimus ipsam initio segnius agitari, quam in progressu et tum velocissime concitari cum fuerit prope finem motus, atque ubidemum pervenerit ad medium, ab ipso non moveri, nisi cogatur, idem quoque iudicandum de nonnullis quae sursum ferunt. Ergo non oppressione, aut extrusione, aut ulla denique vi moveri dicemus haec corpora, sed natura.

Veruntamen dicet quispiam. Esto motus hic naturalis, idemque in fine velocissimus, idque ab Aristotele contra philosophos illos optime sit conclusum. At non ob id huius eventi caussam tenemus, haec ergo superest inquirenda in qua etiam multum est laboratum, atque adeo ut septem opiniones circunferantur, et caussa quedam ab Aristotele allata, tanquam parum idonea repudiata fuerit.

Nanque Hipparchus ita referente Simplicio, in opusculo quodam, quo sigillatim disquirit hoc ipsum problema, censuit motum naturalem esse velociorem in fine, quia mobile prohibeatur aliena vi ab initio motus : ex quo efficiatur, ut vim suam nativam exercere non possit, ideoque pigerrime citetur : ceterum evanescente paullatim aliena illa, et extrinseca vi reficitur naturale robur, et quasi liberum impedimento efficacius operatur. Ita fieri ut gradum accelerent in progressu, non secus atque ubi conferbuerit aqua et amoveatur ab igne : namque ab initio paullatim tepescit, et vix ullum progressum facere videtur fatiscente vero calore, pristinam facultatem recuperat, celerius refrigeratur et eo usque demum procedit, ut etiam longe frigidior evadat, quam ipsa foret ante calefactionem. A qua item sententia non abhorrere censeas Arist. ipsum qui tali hypothesi nixus caussas grandinis indagavit et experientia piscatorum ipsas approbavit. Nota res est.

Contra Hipparchum haec dixit Alexander. Cum. n. duae sunt caussae propter quas elementa feruntur in propria loca ; prima quidem, quando generantur ; nanq. eo tempore quantum contrahunt de forma tantundem etiam assequuntur de ipso ubi : altera vero quando iam genita extra locum proprium ab aliquo detineantur, quemadmodum ignis apud nos, et amoveatur impedimentum. Esto igitur quod cum gignuntur, quia tunc perfecta non sunt, non possunt exercere facultatem illam suam nativam ; at postquam a genitis arceatur impediens, quid illa vetat, quominus secundum summum suae naturae concitentur ?

Fortasse poterat hoc adversus Hipparchum, quia non urget id positionem nostram : eo, quo adest semper impedimentum, quousque fuerint in loco proprio, atque ubi remotum fuerit universum, iam non moveantur sed in proprio loco quiescunt. Idcirco existimarunt alii nescio quod, multos autem in eam venisse sententiam.

Simplicius ipse testatur : eorum velocitatem ex illo amplificari, quod resistentia medii minor esset in fine motus, quam ab initio : quandoquidem minor medii portio relinqueretur a mobili superanda motu ad finem tendente, eaque minus resisteret. Talis. n. est conditio virtutum, quae in materia consistunt, quod ceteris paribus in maiore corpore sunt robustiores : medium vero motui resistere, immo vero caussam essé, cur tempus in loco mutando consumatur, ante docuimus quam ob rem ubi medium rarius est major solet esse celeritas, atque adeo ut in vacuo non futurus sit motus. Attamen caussa talis non est quam reddidit Arist. inquiens augeri velocitatem in fine motus ex additione gravitatis, non autem ex eo, quod minor portio medii supersit. Sed quoniam revocatur hic locus in controversiam, ne forte petitionem principii committamus, etiam sic urgeamus illos. Quia majori corpori ceteris paribus, utputa figura, et insigni parvitate molis, excepta, plus aeris obsistit quam minori. Nanque omnia haec motus evariare possunt, seu naturales sint, sive animales, sive etiam violenti. . .

Plus igitur aer obsistit majori corpori, quam minori, et tamen corpus maius citius delabitur quam minus. Non ergo medii resistentia potuit esse caussa cur motus ab initio pigrior sit. Deinde quoniam caussa eadem intercedit, medii nimirum imminutio ubi motus violentus sit, sicut etiam ubi naturalis, quare item effectus idem contingere plane deberet. Cum igitur hoc ipsa experientia non confirmet ; sed oppositum potius doceat, credibile item non est eam esse causam cur intendat motus naturalis in fine. » [62] Bonamico, De Motu, lib. IV, cap. XXXVIII, p. 412 sq.: « *Latinorum sententie de cremento naturalis motus in fine ex ordine recitantur.* Apud Latinos interpretes legimus opinatos fuisse nonnullos aerem a motu calefieri ; calefactum vero fieri rariorem : ob id cedere facilius iis quae per ipsum moventur, inde consequi unde quo longius aliquid moveatur, quia magis calefiat medium, et quoque rarefiat magis atque magis, subinde afficiatur ad rarefactionem. Quare per ipsum promptius, expeditius et denique velocius obiri possit motus. Ceterum etiam multo velocius in processu sagitta movebitur : praesertim si ex motu concalefacta fuerit, quam, si plumbea sit ; ita excalefieri testatur Arist. ut eliquescat : nihilosecius eo segnius assidue movetur.

Praeterquam quod his mihi videntur ordinem naturae prorsus pervertere. Nam prius est motus quam calefactio medii ; ipsi tamen priorem faciunt rarefactionem quam motum, et idcirco ponunt effectum qui suae caussae natura praecedat, quo certe nihil ineptius.

Tribuunt complures huiuscemodi eventi caussam viribus ipsius loci quas tamen interpretes non eodem modo omnes accipiunt, sed duobus modis ipsos de viribus loci differe comperimus. Aliqui, quemadmodum supra nos constituimus quia locus habeat vim conservandi mobile : omnia vero appetitu naturali suam ipsorum conservationem quaerant ; ex hoc effici ut plantae et animalia magis hoc quam illo coelo fruantur ; is autem esse debet huius modi, ut partim similis sit, ut ab eo locati materia conservetur partim contrarius ut emendetur exuperantia.

Sic unumquodque elementum cum illo cui contiguum est, in altera qualitate convenit, in altera vero differt, quod sane ab Averroë videtur, exceptum qui locum appeti dicebat a mobili, tanquam finem motus et quod in ipso sit eius quies. Alii dicunt in loco vim inesse trahendi mobile, quemadmodum est in magnete vis attrahendi ferrum. At ut aliqua contra posteriores dicamus. Nonne quo maius est corpus, eo quoque magis viribus attrahentis resistit ? Utique. Ergo maiora descenderent tardius quam minora. Neque item ex quacunque distantia moveretur gleba terrae, sicuti nec ex quacumque distantia ferrum moveri potest a magnete, cuiusque enim facultatis naturalis robur finitum est. Quare nec ullum esset robur Aristotelicarum rationum quibus acceptum est, e centro alterius mundi, quantumvis distaret ad centrum nostri ferri posse terram. Neque. n. moveretur huc nisi trahendi facultas, quae inest in medio nostri, posset eo pervenire. In caeteris vero, nisi per certum spatium procedere non apparet ; in quibuscum eveniat id nisi ratio varietatis efferri possit, idem omnino iudicium faciendum sit. Et quamvis antea docuerimus quantum sit illi rationi tribuendum ; tamen valeat apud eos, qui vim loco undecunque trahendi concedunt.

Quod si propensionem adieceris ; iam tecum ipse confliges.

Contra Averroëm invehuntur nonnulli, quanquam argumento fallaci, dicentes, quo magis caret res, eo quoque magis appetere. Sed tum caret magis, ubi longius absit quam ubi prope. Ubi igitur aberit longius ipsa res a suo loco, suaque forma tanto quoque citius eo properabit, atque perveniet. Sed certe non vident isti, appetitum, qui caussa motus est, esse maiorem in ea materia, quae propinquior est, quam in illa, quae longius a fine abest. Nam sicuti planta non appetit visum, neque talpa desiderat lumen, homo autem si fuerit caecus, appetit maxime, quia prope est, ut videat ; sic materia, nisi bonum experiatur quod ipsi per affectiones preavias offert efficiens, illud non appetit. Tum magis appetit, quo magis ipsi obiicitur, tum vero obiicitur ; magis, quo magis affecta, et provecta est in potentias propinquiores. Nec secus accidit, ut mea fert opinio, ac in amatoribus qui puellam expectantes, quo vicinior est hora, magis anguntur et hora una pro longissimo tempore habetur. Nec ab huiusmodi sensu abhorret iudicium Arist. quod item in iis qui usu comparantur, profectum in forma docet habilius reddere subiectum ad motum ; tanto magis in natura ; quanto etiam subiectum habet in seipso propensionem. Semper. n. bene mobilior, inquit, ad virtutem fit etiam quodcunque incrementum sumpserit a principio.

Nec video quemadmodum auctores huius rationis evitare possint, quin ab initio cum major adsit potestas ; velocius etiam concitentur, sed imprudentes in eo lapsi sunt, quod parem gradum privationis et potentiae fecerint, tametsi una existunt. Et illud plane verum ab initio plus privationis inesse, sed minus potestatis ; in progressu amplificari potentiam, quia privatio minuatur et ut alibi ostendetur commutant latitudinem potestatis cum gradu : maius est. n. ab initio motus spatium potestatis, ut in summe calido ad frigidum ut octo, in processu maior gradus : nam facilius summe frigidum fiet quod frigidum est, ut quinque, quàm summe calidum, amplificatur ergo potestas atque propensio non propter latitudinem, sed propter gradum. Ideo tantum huic tribuatur argumento quantum quisque patitur. Quam ob rem veniamus ad alia.

Quam vero nonnulli putant, efficacitatem universam esse tribuendam gradui formae, non autem multitudini materiae (quanquam nos ita non credimus) quia par gradus appetitus est in maiore, et minore gleba ; necesse item fuerit, utrasque pari gradu concitari, parem vero gradum appetitus in utraque ponere licet, ut si fingantur utraeque in eodem esse gradu perfectionis, aut potestatis. Sed illud apud nos plurimum valet. Quoniam imperfecta est haec opinio, quamvis caussam ab eius auctoribus allatam veram esse concedamus. Neque enim administratur ille motus ab ea caussa solum, sed aliae multae concurrunt praeter finem : efficiens. s. et alia principia per accidens, ut removens impedimentum et ipsa mobilis rei natura quae cuncta motus in actu caussa sunt.

Divus Thomas et post ipsum Albertus Saxon. arbitrati sunt, geminam esse gravitatem, ac levitatem in elementis : alteram sane quam inquiunt esse per se et naturalem atque alteram quam adventitiam reputant, illam inquiunt, sequi vim generantis et in proprio loco servari, hanc in processu motus acquiri ex eoque fieri, ut maiore impetu moveantur in processu corpora naturalia. Rem vero sic esse persuadent experientiis illis, quae supra a nobis allatae sunt, cum doceremus etiam in absentia moventis adhuc in mobili conservari vim quandam a qua mobile concitetur, ac si primum movens adesset. Igitur intermisso primi moventis impulsu fit adhuc motus, non ob aliud, nisi quod etiam superest in ea vis quaedam, propter quam eodem motu cietur quo pridem movebatur. Verum quoque aliena est illa vis et adsciticia, remittitur assidue, sed in iis, quae secundum naturam moventur, amplificatur : idcirco velocius agitantur. Ita quando nos cursum maiore quodam nixu arripuimus, etiam in eius fine vix continere nos possumus.

Quod si quis interroget auctores huius opinionis : undenam proficiscatur, et quid impetus iste sit. Ad hoc respondent ipsum esse qualitatem quandam, atque illam quidem potestatem quippe potestatem ad motum, ad illud vero dicunt ; eam a forma comparari per motum. Attamen in exponenda quaestionis huiusce caussa videntur ipsam iterum cum effectu commutare : quaeritur. n. caussa velocitatis in motu; eam vero dicunt ipsi facultatem esse, atque habilitatem, si rursus eos interroges, undenam habilitas ista proficiscatur ; aiunt a motu, hic autem, aut accipitur, quatenus velox, aut simpliciter, quod si simpliciter accipiatur : ergo motus ipsemet erit sibi caussa suae velocitatis, quod si quia velox. Erit igitur caussa, quam tamen ipsi quaestioni pro effectu supponunt. Inter iuniores Lud. Buccaf. statuit mobile agitare et quasi impellere medium ea ratione quia primam medii partem commoveret, atque propelleret. Haec vero postea contiguis suum motum communicaret. Ab his autem ita commotis mobile ipsum ferri. Quoniam vero mobili prevenerit, reddere motum eius faciliorem. Sed cum in fine motus impetus maior a mobili comparatus sit, aër etiam magis affectus ad excipiendum motum : hinc fieri ut velocior ille motus in fine reddatur.

Addunt alii praeter haec aëris illius impulsum qui iugiter mobili succedens ipsum magis expellit, ideoque effici, ut eius motus sit velocior, corrogant hic more consueto loca multa ex Arist. cum ex 8 Phys. tum etiam ex 4 de Coelo, quibus de hoc impulsu mentio facta est, ut opinionem suam confirment. Quoniam vero contra faciunt verba contextus Aristotelici quibus significatur ex additione gravitatis fieri motum velociorem in fine ; respondent hanc non esse veram mentem Aristotelis, sed eum ita pro hominum vulgique opinione fuisse locutum, neque ullo modo recipiunt auctoritatem Aristotelis in eo loco. Caeterum de loci illius veritate mox : interea monstremus eam esse falsam quam ipsi profitentur. Primum. n. in idem absurdum videntur incidere, atque D. Thomas et Albertus, qui impetum illus adventitium caussam esse velocitatis asseverant, nam cum effectu caussam commutant : siquidem velint impulsum aëris huiusce rei caussam esse, qui quidem fit a mobili. At quearere licet, undenam mobile vim habeat impellendi aërem et magis impellendi, quo longius fertur. Et cum maior impulsus sit ex maiore velocitate, caussa igitur eius eventi non erit impulsus, ut aiunt, sed velocitas. Et quomodocunque erit gravitas quam ipsi repudiant, nam quod velocius agitur, est gravius quod item medium magis opprimatur est ex gravitate, quae item magis operabitur in eo subiecto quod est grave aut leve simpliciter, quam in eo quod est tale quodammodo. Verum sit haec adscititia quaedam velocitas, seu gravitas. cur in processu non minuitur ? Accedit eodem quod pari pacto pellunt partes medii quo pelluntur, et minus in progressu quod magis distant a virtute movente : naturale. n. movens in progressu debilitatur, nisi afficiat ad formam, quod sane huic adscititiae virtuti non conceditur. »

63   *Ibid.*: « Obiicies hic ventos qui vires acquirunt eundo, et velociores vehementioresque fiunt. An eius eventi caussa non habet locum in elemento ; siquidem eius motionis quam vulgus ventum vocat, duae sunt partes, prima quae vere ventus est, exhalatio videlicet, quae propter diversa principia motus agitur in latus et quodammodo praeter naturam. Altera est aër contiguus et movetur quidem aër ea velocitate qua cietur exhalatio et in principio vehementius ; eius signum quod apud nos die prima boreae sunt vehementiores: at vero propter continuitatem aëris in progressu multae partes eius concitantur ; itaque maior est motus, neutiquam tamen velocior nisi forte in angustum contrahantur, comque contineri nequeant magno impetu erumpant, aut quod cum in angusto parva materiae copia consistat, ab eadem vi vehementius agatur. Non igitur aër commotus agit velocius exhalationem, sed ab ea semper agitur. Ergo etiam et in motu elementi non magis agent elementum, quam ab ipso agatur. Quam ob rem impetus in mobili praecedat oportet. Praeterea nonne reiicit Aristoteles illorum dicta qui putant impulsum facere motum velociorem, quod in fine langeret, non autem augeretur et quia facilius impelleretur mobile minus, quam maius ? Videtur etiam gravitas esse caussa velocitatis, quoniam id quod gravius est, fertur velocius. Quod sicubi impulsum illum in aere collocavit Arist. ille est quo natura utitur in motu proiectorum : at nos de motu naturali nunc agimus. Mitto quod dum student defendere motum illum in elemento per se inesse, caussam faciunt quae moveat per accidens : volunt enim mobile a medio ferri : atqui haec est vectio ; ea vero est motus per accidens. Ita fit ut cum ab Aristotele discedere cupiunt, turpissime quoque labantur. »

64   For example, he attributes to St. Thomas, the most orthodox of Aristotelians (cg. *Comment, in quattuor libros de Coelo,* lib. III, lect. 7), the doctrines of Albert of Saxony.

65   Mediaeval discussions on the nature of motion and acceleration were of far greater richness and complexity than Bonamico's treatment would suggest. Cf. the works cited in notes 23 and 41 above.

66   The passage about the little disc rotating in a cavity is borrowed from a text by J. C. Scaliger, *Exotericarum exercitationum liber XV, De subtilitate ad Hieronimum Cardanum,* Lutetiae MDLVII, *exercitatio* XXVIII, *De motu projectorum.* Cf. Duhem, *Etudes sur Léonard de Vinci,* vol. III, p. 200.

Impetus Physics: Benedetti

67 Giovanni Battista Benedetti is better known than his contemporaries and predecessors; cf. K. Lasswitz, *Geschichte des Atomismus,* vol. II, p. 14 ff.; G. Vailati, 'Le speculazione di Giovanni Benedetti sul moto de gravi' in *Rendiconti dell'Academia Reale delle scienze di Torino,* 1897-1898, reprinted in his *Scritti,* Leipzig-Florence, 1911; E. Wohlwill, 'Die Entdeckung des Beharrungsgesetzes' in *Zeitschrift für Völkerpsychologie,* etc., vol. XV, p. 394 ff.; G. *Galilei und sein Kampf für die Kopernikanische Lehre,* vol. I, p. 111 ff.; P. Duhem, 'De l'accélération produite par une force constante', *Congrès international d'histoire des sciences,* 3rd session, Geneva, 1906, p. 885 ff.; *Etudes sur Léonard de Vinci,* vol. III, p. 214 ff.; G. Bordiga, 'G. B. Benedetti', *Atti de R. Istituo Veneto,* 1925-1926.
    Although he is better known, however, he is still not known well enough. Therefore it has seemed right to devote several pages to him.

68 G. B. Benedetti, *Diversarum speculationum mathematicarum et physicarum liber,* Taurini, 1585, p. 184 [The English translation given in the text is a modified version of that which can be found in eds. Stillman Drake and I. E. Drabkin, *Mechanics in Sixteenth-Century Italy: Selections from Tartaglia, Benedetti, Guido Ubaldo, and Galileo,* Madison, Wisconsin, 1969]: « Aristoteles in fine. 8. physicorum sentit corpus per vim motum et separatum a primo movente, moveri aut motum esse per aliquod tempus ab aere, aut ab aqua, quae ipsum sequuntur. Quod fieri non potest, quia imo aer, qui in locum desertum a corpore subintrat ad fugandum vacuum, non solum hoc corpus non impellit, sed potius id cohibet à motu, quia aer per vim a corpore ducitur retro, et divisus a parte anteriori a dicto corpore, resistit similiter et quantum dictus aer in dicta parte condensatur, tantum in posteriori rarefit, unde per vim sese rarefaciens non permittit, ut dictum corpus cum ea velocitate fugiat, cum qua aufugeret, quia omne agens in agendo patitur. Quam ob rem cum aer a dicto corpore rapiatur, corpus quoque ipsum ab aere rapitur. Huiusmodi autem rarefactio aeris naturalis non est, sed violenta ; et hanc ob causam resistit, et ad se trahit, sed non sufferente natura, ut inter unum et aliud ex dictis corporibus reperiatur vacuum ; idcirco sunt haec semper contigua et mobile corpus aerem deserere cum nequeat, eius velocitas impeditur. Huiusmodi igitur corporis separatim a primo movente velocitas oritur quadam naturali impressione ex impetuositate recepta à dicto mobili, quae impressio et impetuositas, in motibus rectis naturalibus continuo crescit, cum perpetuo in se causam moventem, id est propensionem eundi ad locum ei à natura assignatum habeat. »

69 G. B. Benedetti, *Ibid,* p. 286 : « *Epistola,* Illustr. Joanni Capra Novarensi Sabaudiae Ducis . . ., *De revolutione rotae putealis et aliis problematibus.* Omne corpus grave, aut sui natura, aut vi motum, in se recipit impressionem et impetum motus, ita ut separatum a virtute movente per aliquod temporis spatium ex seipso moveatur ; nam si secundum naturam motu cieatur, suam velocitatem semper augebit, cum in eo impetus et impressio semper augeantur, quia coniunctam habet perpetuo virtutem moventem. Unde manu movendo rotam ab eaque ; eam removendo, rota statim non quiescet, sed per aliquod temporis spatium circunvertetur »; cf. the text cited in the previous note, and p. 184, cap. XXIV: « *Disputationes de quibusdam placitis Aristotelis. Idem vir gravissimus an bene senserit de motibus corporum violentis et naturalibus* : Huiusmodi igitur corporis separatim a primo movente velocitas oritur a quadam naturali impressione, ex impetuositate recepta a dicta mobili. »

70 *Ibid,* p. 160.

71 G. B. Benedetti, *ibid. De Mechanicis,* cap. XVII, p. 160 : « Vera ratio cur multo longius corpus aliquod grave impellatur funda, quam manu, inde oritur, quod circumvolvendo fundam, maior impressio impetus motus fit in corpore gravi, quam fieret manu, quod corpus liberatum deinde cum fuerit a funda, natura duce, iter suum a puncto, a quo prosiliit, per lineam contiguam giro, quem postremo faciebat, suscipit. Dubitandumque non est, quin dicta funda maior impetus motus dicto corpori imprimi possit, cum ex multis circumactibus, maior semper impetus dicto corpori accedat. Manus autem eiusdem corporis motus, dum illud ipsum circumvolvitur (pace Aristotelis dixerim) centrum non est, neque funis est semidiameter. Immo manus quam maxime fieri potest

in orbem cietur ; qui quidem motus in orbem, ut circumagatur etiam ipsum corpus, cogit, quod quidem corpus, naturali quadam inclinatione, exiguo quodam impetu jam incepto vellet recta iter peragere, ut in subscripta figura patet, in qua *e* significat manum, *a* corpus, *ab* lineam rectam tangentem girum *aaaa* quando corpus liberum remanet. Verum quidem est, impressum illum impetum, continuo paulatim decrescere unde statim inclinatio gravitatis eiusdem corporis subingreditur, quae sese miscens cum impressione facta per vim, non permittit ut linea *ab* longo tempore recta permaneat, sed cito fiat

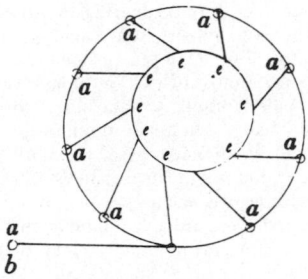

curva, cum dictum corpus a duabus virtutibus moveatur quarum una est, violentia impressa, et alia natura, contra opinionem Tartaleae qui negat corpus aliquod motibus violento et naturali simul et semel moveri posse. Neque est silentio praetereundus hac in re quidam notatu dignus effectus qui eiusmodi est quod quanto magis crescit impetus in corpore *a* causatus ab augmento velocitatis giri ipsius *e* tanto magis oportet, ut sentiat se trahi manus a dicto corpore *a.* mediante fune, quia quanto maior impetus motus ipsi *a* est impressus, tanto magis dictum corpus *a,* ad rectum iter peragendum inclinatur, unde ut recta incedat tanto majore quoque vi trahit. »

72 Renaissance artillerymen and writers on ballistics, on the basis of this firmly fixed idea that no two motions could coexist in a moving body without struggling against each other, believed that a cannon-ball (like any other projectile) starts by moving *in a straight line*, and then when its motion, or motive force, decreases in strength, its falls *vertically* to earth, the two rectilinear parts of the trajectory being connected by the segment of a circle. Tartaglia, who took an interest in ballistics, and even published elevation tables for cannon fire, put forward the traditional theory in his *Nuove Scienza* (1537), even though he also asserted, on the other hand, that the trajectory is *sempre curva.* In fact it was Galileo and not, as has often been said, Tartaglia, nor even Benedetti, who was the first to clearly recognise that the trajectory of a cannon-ball is curved right from the beginning. Cf. below, pp. 155 and 195.

73 G. B. Benedetti, *ibid*, p. 285: « De motu molae et trochi, de ampullis aquae, de claritate aeris et Lunae noctu fulgentis, de aeternitate temporis et in finito spacio extra coelum, coelique figura. *Illustr. Ioanni Paulo Capra Novarensi Sabaudiae Ducis Hospicij Magistro.* . . Quaeris a me litteris tuis, an motus circularis alicuius molae molendinarie, si super aliquod punctum quasi mathematicum, quiesceret, posset esse perpetuus, cum aliquando esset mota, supponendo etiam eandem esse perfecte rotundam et levigatam. Respondeo huiusmodi motum nullo modo futurum perpetuum, nec etiam multum duraturum, quia praeterquam quod ab aere qui ei circumcirca aliquam resistentiam facit stringitur, est etiam resistentia partium illius corporis moti, quae cum motae sunt, natura, impetum habent efficiendi iter directum, unde cum simul iunctae sint, et earum una continuata cum alia, dum circulariter moventur patiuntur violentiam, et in huiusmodi motu per vim unitae manent, quia quanto magis moventur, tanto magis in iis crescit naturalis inclinatio recta eundi, unde tanto magis contra suam et naturam volvuntur, ita ut secundum naturam quiescant, quia cum eis proprium fit, quando sunt motae, eundi recta, quanto violentius volvuntur, tanto magis una resistit alteri, et quasi retro revocat eam, quam antea reperitur habere.

Ab eiusmodi inclinatione rectitudinis motus partium alicuius corporis rotundi fit, ut per aliquod temporis spacium, trochus cum magna violentia seipsum circumagens, omnino rectus quiescat super illam cuspidem ferri quam habet, non inclinans se versus mundi centrum, magis ad unam partem, quam ad aliam, cum quaelibet suarum partium in huiusmodi motu non inclinet omnino versus mundi centrum, sed multo magis per transversum ad angulos rectos cum linea directionis, aut verticali, aut orizontis axe, ita ut necessario huiusmodi corpus rectum stare debeat. Et quod dico ipsas partes non omnino inclinare versus mundi centrum, id ea ratione dico, quia non absolute sunt unquam privatae huiusmodi inclinatione, quae efficit ut ipsum corpus eo puncto nitatur. Verum tamen est, quod quanto magis est velox, tanto minus premit ipsum punctum, imo ipsum corpus tanto magis leve remanet. It quod aperte patet sumendo exemplum pilae alicuius arcus, aut alicuius alterius instrumenti, seu machinae missilis, quae pila quanto est velocior, in motu violento, tanto maiorem propensionem habet rectius eundi, unde versus mundi centrum tanto minus inclinat, et hanc ob causam levior redditur. Sed si clarius hanc veritatem videre cupis, cogita illud corpus, trochum scilicet, dum velocissime circumducitur secari, seu dividi in multas partes, unde videbis illas omnes, non illico versus mundi centrum descendere sed recta orizontaliter ut ita dicam, moveri. Id quod a nemine adhuc (quod sciam) in trocho est observatum. Ab huiusmodi motu trochi, aut huius generis corporis, clare perspicitur, quam errent peripatetici circa motum violentum alicuius corporis, qui existimant aerem qui subintrat ad occupandum locum a corpore relictum, ipsum corpus impellere, cum ab hoc, magis effectus contrarius nascatur.

Illud, nihil, Aristotelis extra caelum nullomodo nobis inservit pro eiusdem Coeli spherica rotunditate, cum cuiusque alterius ex in finitis figuris Coelum ipsum esse possit secundum suam superficiem convexam. Nam Coelum ea ratione sphericum non est, quod magis sit capax, quia ei innumerabiles alias figuras adeo magnas poterat concedere causa divina : sed sphaericum est effectum, ne partem aliquam haberet sui termini superfluam, quia nullum corpus a breviori termino quam a spherico terminari potest. »

[74] G. ᛒ. Benedetti, *ibid.*, p. 168 sq.: « *Disputationes de quibusdam placitis Aristotelis : Tanta est certe Aristotelis amplitudo atque authoritas, ut difficillimum ac periculosum sit quidpiam scribere contra quam ipse docuerit, et mihi praesertim, cui semper visa est viri illius sapientia admirabilis. Veruntamen studio veritatis impulsus, cuius ipse amore in seipsum si viveret excitaretur, in medium quaedam proferre non dubitavi, in quibus me inconcussa mathematicae philosophiae basis, cui semper insisto ab eo dissentire coegit.* cap. II : *Quaedam supponenda ut constet cur circa veloci tatem motuum naturalium localium ab Aristotelis placitis recedamus.*

Cum susceperimus provinciam probandi quod Aristoteles circa motus locales naturales deceptus fuerit, sunt quaedam primo verissima et objecta intellectus per se cognita praesupponenda, ac primum quaelibet duo corpora, gravia aut levia, area aequali similique figura sed ex materia diversa constantia, eodemque modo situm habentia, eandem proportionem velocitatis inter suos motus locales naturales, ut inter suam et pondera aut levitates uno in eodemque medio, servata. Quod quidem natura sua notissimum est si considerabimus non aliunde maiorem tarditatem, aut velocitatem gigni, quam a. 4 causis (dummodo medium uniforme sit et quietum) idest a maiori aut minori pondere aut levitate ; a diversa figura ; a situ eiusdem figurae diversae respectu lineae directionis, quae recta inter mundi centrum et circunferentiam extenditur ; et ab inaequali magnitudine. Unde patebit, quod figuram non variando, nec in qualitate nec in quantitate, neque eiusdem figurae situm, motum fore proportionatum virtuti moventi, quae erit pondus aut levitas. Quod autem de qualitate, de quantitate et situ eiusdem figurae dico, respectu resistentiae ipsius medii dico : Quia dissimilitudo aut inequalitas figurarum, aut situs diversus non parum alterat dictorum corporum motus, cum figura parva facilius dividat continuitatem medii, quam magna ; ut etiam celerius idem facit acuta, quam obtusa ; et illa quae cum angulo, qui antecedat movebitur velocius quam illa quae secus. Quotiescunque igitur duo corpora unam eandemque resistentiam ipsorum superficiebus, aut habebunt aut recipient, eodem motus inter seipsos eorum plane modo proportionati consurgent quo erunt ipsorum virtutes moventes ; et e converso, quotiescunque duo corpora unam eandemque gravitatem aut levitatem et diversas resistentias habebunt, eorum motus inter seipsos eandem proportionem sortientur, quam habebunt eorum resistentiae converso modo ; quae quidem resistentiae inter seipsas eandem proportionem quam ipsarum superficies habebunt, aut in qualitate sola figurae, aut in quantitate sola, aut in situ, aut in aliquibus ex dictis rebus, eo tamen modo qui superius positus fuit, ut scilicet corpus illud quod alteri comparatum, aequalis erat ponderis, aut levitatis sed minoris resistentiae, existet velocius altero, in eadem proportione cuius superficies resistentiam suscipit minorem ea quae alterius est corporis, ratione facilioris divisionis continuitatis aeris, aut aquae. Ut exempli gratia, si proportio superficiei corporis maioris superficiei minoris sesquitertia esset, proportio velocitas dicti corporis maioris, velocitati corporis minoris, esset subsesquitertia, unde velocitas minoris corporis maior esset velocitate corporis maioris quemadmodum quaternarius numerus ternario maior existit. »

[75] *Ibid.*, p. 169 : « Aliud quoque supponendum est, velocitatem scilicet motus naturalis alicuius corporis gravis, in diversis mediis, proportionatam esse ponderi ejusdem corporis in iisdem mediis ; ut exempli gratia, si pondus totale alicuius corporis gravis significatum erit ab. *a. i.* quo corpore posito in aliquo medio minus denso, quam ipsum sit (quia in medio se densiore si poneretur, non grave esset, sed leve, quemadmodum Archimedes ostendit), illud medium subtrahat partem *ei* unde pars *ae* eiusdem ponderis libera maneat ; et, posito deinde eodem corpore in aliquo alio medio densiore, minus tamen denso quam ipsum sit corpus, hoc medium subtrahat partem, *u. i* dicti ponderis, unde pars *a. u* eiusdem ponderis remanebit. Dico proportionem velocitatis eiusdem corporis per medium minus densum, ad velocitatem eiusdem per medium magis densum futuram ut *a.e.* ad *a.u,* ut est etiam rationi consonum magis quam si dicamus huiusmodi velocitates esse ut *ui* ad *ei* cum velocitates a virtutibus moventibus solum (cum figura una, eademque in qualitate, quantitate situque erit) proportionentur. Quae nunc diximus, plane similia sunt iis, quae supra scripsimus, quia idem est dicere proportionem velocitatum duorum corporum heterogeneorum, sed similium figura, et magnitudine aequalium, in uno solo medio, æqualem esse proportioni ponderum ipsorum, ut si dicamus proportionem velocitatum unius solum corporis per diversa media eandem esse cum ea quae est ponderum dicti corporis in iisdem mediis. »

[76] Cf. above, p. 15.

[77] This is why the resulting speed is given by the division of the weight by the resistance.

[78] G. B. Benedetti, *ibid.*, cap. XXVI, p. 185. « Manifeste indicat (Aristoteles) se causam nec gravitatis, nec levitatis corporum naturalium nosce, quae est densitas aut raritas corporis gravis, aut levis, maior densitate aut raritate medii permeabilis, in quo reperitur. »

[79] G. B. Benedetti, *ibid.,* p. 172. « *Disputationes, de quibusdam placitis Aristotelis,* c. VI: *Quod proportiones ponderum eiusdem corporis in diversis mediis proportiones eorum mediorum densitatum non servant. Unde necessario inaequales proportiones velocitatum producuntur,* cap. VII. *Corpora gravia aut levia eiusdem figurae et materiae sed inaequalis magnitudinis, in suis motibus naturalibus velocitatis, in eodem medio proportionem longe diversam servatura esse quam Aristoteli visum fuerit* », i.e., the proportion will be arithmetical and not geometrical.

[80] *Ibid.,* cap. VIII; cf. cap XVIII.

[81] *Ibid.,* cap. XXV, p. 184: « *Motus rectus corporum naturalium sursum aut deorsum non est naturalis primo et per se.* »

[82] *Ibid.* cap XXIII, p. 183: « *Motum rectum esse continuum vel dissentiente Aristotele* ». It is sufficient to consider the rectilinear motion produced by the rotation of a circle about its centre; if C is a fixed point on the circumference of the circle and CD a line of constant length intersecting AB, then the point D will move backwards and forwards along AB without rest.

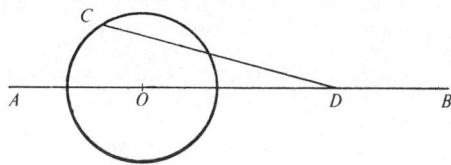

[I have changed Koyré's presentation of this demonstration in order to make it clearer. Benedetti's point is as follows (see eds. Drake and Drabkin, *Mechanics in Sixteenth-Century Italy*, p. 215): 'Aristotle says that it is impossible for a body to move in a straight line, first in one direction and then in the opposite direction, i.e., going and returning on the same line, without [an interval of] rest at the extremities. I hold, on the contrary, that it is possible.' As Professor Drabkin points out, 'the distinction between an instantaneous velocity of zero and a non-instantaneous interval of rest is the key to the problem—a distinction that was not clearly made before the seventeenth century. But Benedetti seems to have understood Aristotle in the sense I have indicated'. *Trans.*]

83 *Ibid.*, cap. XXIX, p. 186: « *Dari continuum infinitum motum super rectam atque finitam lineam* ». It is sufficient to represent the motion of the point of intersection I on the line RX as the line AO turns about the fixed point A. The point O moves towards T but the point I can never reach R.

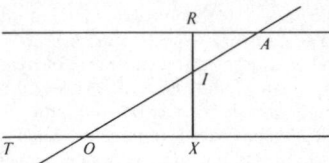

84 *Ibid.*, cap. XXXVI, p. 195: « *Minus sufficienter explosam fuisse ab Aristotele opinionem credentium plures mundos existere* ».

85 Cf. above, p. 7.

86 G. B. Benedetti, *ibid.*, cap. XIX, p. 179: « *Quam sit inanis ab Aristotele suscepta demonstratio quod vacuum detur*. . . Ut igitur idem facilius ostendamus, comprehendamus imaginatione infinita media corporea, quorum unum altero rarius sit, in qua placuerit nobis ex proportionibus, incipiendo ab uno, imaginemur etiam corpus *Q*. densius primo medio, cuius corporis totalis gravitas sit *a. b.* et positum in ipso medio . . . »: it follows that the speed of bodies in a vacuum would be finite and not infinite.

87 *Ibid.*, cap. IX, p. 174: « *An recte Aristoteles disseruerit de proportionibus motuum in vacuo.* Cum vero Aristoteles circa finem cap. 8 lib. 4 physicorum subiungit quod eadem proportione dicta corpora moverentur in vacuo, ut in pleno, id pace eius dictum sit plane erroneum est. Quia in pleno dictis corporibus subtrahitur proportio resistentiarum extrinsecarum a proportione ponderum, ut velocitatum proportio remaneat, quae nulla esset, si dictarum resistentiarum proportio, ponderum proportioni aequalis esset, et hanc ob causam diversam velocitatum proportionem in vacuo haberent ab ea, quae est in pleno. »

88 'Of the same material'—bodies of different material fall with different speeds. Cf. above, p. 31.

89 G. B. Benedetti, *ibid.*, *Disputationes*, cap. X, p. 174: « *Quod in vacuo corpora eiusdem materiae aequali velocitate moverentur.* Quod supradicta corpora in vacuo naturaliter pari velocitate moverentur hac ratione assero.

Sint enim duo corpora *o* et *g* omogenea et *g* sit dimidia pars ipsius *o*. Sint alia quoque duo corpora *a* et *c* omogenea primis, quorum quolibet aequale sit ipsi *g* et imaginatione comprehendamus ambo posita in extremitatibus alicuius lineae, cuius medium sit. *i.* clarum erit tantum pondus habiturum, punctum *i.* quantum centrum ipsius *o.* quod *i* virtute corporis *a* et *e* in vacuo eadem velocitate moveretur, qua centrum ipsius. *o* : cum autem disiuncta essent dicta corpora *a.* et *e* a dicta linea, non ideo aliquo modo suam velocitatem mutarent, quorum quodlibet esset quoque tam velox quam est *g* : igitur *g* tam velox esset quam *o.* » Cf. *ibid.*, cap. XVIII, p. 179.

90 Cf. *ibid.*, cap. XXXVII, p. 196: « *An recte loquutus sit Philosophus de extensione luminis per vacuum.* » Benedetti, of course, argues that the propagation of light is not at all prevented by a vacuum.

91 Cf. *ibid.*, cap. XXXIX, p. 197: « *Examinatur quam valida sit ratio Aristotelis de inalterabilitate Coeli.* Similiter de terra dici posset quando ipsa ita eminus prospiceretur. »

92 Cf. below, note 147.
93 G. B. Benedetti, *ibid.*, *Disputationes*, cap. XX, p. 181: « Hoc modo nullum est corpus, quod in mundo aut extra mundam (dicat Aristoteles quidquid voluerit) locum suum non habeat. » Moreover, space is not at all an all-enclosing surface; it is an *interval*. Bruno said the same thing. Cf. below, Part 3, note 41.
94 *Ibid*, cap. XXI, p. 181: « *Utrum bene Aristoteles senserit de infinito* : Nullum inconveniens sequeretur, quod extra coelum reperiri possit corpus aliquod infinitum, quam vis id ipse, nulla evidenti ratione inductus perneget. Sensit quoque, absque eo, quod aliquam rationem proponat . . . infinitas partes alicuius continui esse solum in potentia, non item in actu, hoc non est illico concedendum, quia si omne totum continuum et re ipsa existens, in actu est, omnis quoque eius pars erit in actu, quia stultum esset credere, ea quæ actu sunt, ex iis quae potentia existunt, componi. Neque etiam dicendum est continuationem earundem partium efficere, ut potentia sint ipsae partes, et omni actu privatae, Sit, exempli gratia, linea recta *a. u* continua quae deinde divadatur in puncto *e* per aequalia, dubium non est, quin ante divisionem, medietas *a. e.* tam in actu (licet coniuncta cum alia *e. u)* reperiretur, quam totum *a. u.* licet a sensu distincta non esset. Idem affirmo de medietate *a. e* id est de quarta parte totius *a. u* et pariter de octava, de millesima, et de quavis, ita ut essentia actualis infiniti hoc modo tota concedi possit, cum ita sit in natura. . . »
95 *Ibid.*, « multitudo non minus infinita quam finita intelligi potest. »

Galileo

96 Published in vol. 1 of the National Edition of the *Opere di Galileo Galilei*, ed. Antonio Favaro. [Written about 1590 the texts collected together by Favaro under the title *De Motu* consist of several drafts of an essay, a dialogue and a series of notes all on the subject of motion. English translations have been published in the University of Wisconsin Publications in Medieval Science series as follows: the essay 'On Motion' is in Galileo Galilei, *On Motion and On Mechanics*, translated by I. E. Drabkin and Stillman Drake, Madison, Wisconsin, 1960; the 'Dialogue on Motion' and 'Memoranda on Motion' are in eds. Drake and Drabkin, *Mechanics in Sixteenth-Century Italy*, Madison, Wisconsin, 1969. *Trans.*]
97 One could say that *De Motu* was conceived as a critique of Aristotelian dynamics from the point of view of the dynamics of impressed force. The critique is often violent and is not always fair. In effect Galileo detaches Aristotle's dynamics from his metaphysics; motion for Galileo is solely *local* motion [whereas for Aristotle it is a broader category of processes of change understood in terms of 'movement' from potentiality to actuality. If this meaning of the concept is ignored it is easy to misunderstand or ridicule Aristotle's arguments. *Trans.*] Therefore he does not always understand Aristotle's thought. But one could say that his way of understanding (or misunderstanding) is itself a sign and an effect of a new mental attitude.
98 Cf. *De Motu*, pp. 265, 276, 285, 302 and *passim*.
99 *De Motu*, p. 307: « *A quo moveantur projecta*? : Aristoteles, sicut fere in omnibus quae de motu locali scripsit, in hac etiam quaestione vero contrarium scripsit. . . Non poterat Aristoteles tueri, motorem debere esse coniunctum mobili, nisi diceret proiecta ab aere moveri. »
100 *De Motu*, p. 307 ff. The *sphaera marmorea* is Tartaglia's favourite example.
101 *Ibid.*, p. 307. This objection, we know, is not a fair one; air is a medium which is especially liable to motion. Cf. *ibid.*, p. 314: « Concludamus igitur tandem, proiecta nullo modo moveri a medio, sed a virtute motiva impressa a proiciente. » Cf. above, pp. 13, 19, 21.
102 *De Motu*, p. 310: « *Virtus motiva*, nempe levitas conservatur in lapide, non tangente qui movit; calor conservatur in ferro ab igne remoto: *virtus* impressa successive remittitur in proiecto, a proiciente absente; calor remittitur in ferro, igne absente: lapis tandem reducitur ad quietem; ferrum, similiter, ad naturalem frigiditatem redit : motus ab eadem vi magis imprimitur in mobili magis resistenti quam in eo quod minus resistit, ut in lapide magis quam in pumice levi. »

103 Taking *impetus* to be like a quality, and in particular like heat, was classical and goes
     back to Themistius. Cf. Wohlwill, 'Die Entdeckung des Beharrungsgesetzes', *loc. cit.*, p.
     379; and above, pp. 10 and 12.
104 *De Motu*, p. 310.
105 *De Motu*, p. 314: « Privatur lapis quiete : introducitur in campanam *qualitas sonora*
     contraria eius naturali silentio ; introducitur in lapidem *qualitas motiva* contraria illius
     quieti. »
106 *De Motu*, p. 314 : « Nunc . . . prosequamur ostendere, hanc virtutem successive
     diminui. » Benedetti also held that the *impetus impressus* gradually weakens. But, like
     his predecessors, he failed to draw all the conclusions: thus he believed, as did everyone
     else, in the initial acceleration of violent motion.
107 *De Motu*, p. 314 ff: « cap . . . *in quo virtutem motivam successive in mobili debilitari
     ostenditur.* » The main reason given by Galileo is precisely the impossibility of inertial
     motion: « Quare, eadem argumentatione repetita, demonstrabitur, motum violentum
     nunquam remitti, sed eadem velocitate semper et in infinitum ferri, eadem semper
     manente virtute motiva ; quod certe absurdissimum est : non ergo verum est, in motu
     violento posse duo puncta assignari, in quibus eadem maneat virtus impellens. Quod
     demonstrandum fuit. »
108 P. Duhem, 'De l'accélération . . .' *op. cit.*, (note 21), p. 887. Also see below, p. 75.
109 *De Motu*, p. 314: « Nec posse dari in eo motu duo puncta temporis, in quibus eadem sit
     virtus motiva. »
110 *De Motu*, p. 328: « caput . . . *in quo contra Aristotelem probatur, si motus naturalis in
     infinitum extendi posset, eum non in infinitum fieri velociorem.* . . Velocitas augetur vel
     minuitur asymptotice. » Cf. above, p. 26.
111 The extent to which this absurd belief was firmly entrenched can be seen from the fact
     that Descartes, in a letter to Mersenne (letter of January, 1630, Descartes, *Oeuvres*, ed.
     C. Adam and P. Tannery, vol. 1, p. 110) wrote: 'I would also like to know whether you
     have done any experiments at all on whether a stone thrown with a sling, or a ball fired
     from a musket, or an arrow fired from a cross-bow, move faster and have more force at
     the middle of their motion than they have at the beginning, and whether they have a
     greater effect. Because this is what is popularly believed, and yet I have reasons for not
     agreeing with it: and I find that things which are pushed and which do not move by
     themselves, must have more force at the beginning than they have immediately
     afterwards.' In 1632 (see Adam and Tannery, vol. I, p. 259) and again in 1640, Descartes
     explained to his friend what there is that is true in this belief (letter to Mersenne, March
     11, 1640, Adam and Tannery, vol. II, p. 37 ff.): « *In motu projectorum* I do not at all
     believe that the projectile ever goes less fast at the beginning than at the end, counting
     from the first instant that it is no longer impelled by the hand or by the machine: but I do
     believe that a musket, when it is only a foot or half a foot away from a wall, would not
     have such a great effect as it would if it were at a distance of fifteen or twenty paces,
     because the ball, as it leaves the musket, cannot so easily expel the air which is between
     itself and the wall, and must therefore move less fast than if the wall were not so close.
     However, it is for experiment to determine whether this difference is appreciable, and I
     have strong doubts about any which I have not performed myself.' Beeckman, on the
     other hand, firmly denied the possibility of an acceleration of the projectile. (See
     Beeckman's letter to Mersenne of 30th April 1630, *Correspondance du Père Mersenne*,
     Paris, 1936, vol. II, p. 437: « Funditores vero ac pueri omnes qui existimant remotoria
     fortius ferire quam eadem propinquiora, certo certius falluntur. » However, he too
     admitted that there is some truth in this belief and that it needs to be explained what it is.
     (« Non dixeram plenitudinem nimiam aeris impedire effectum tormentarii globi, sed
     pulvarem pyrium extra bombardam jam existentem forsitan adhuc rarefieri, ideoque
     fieri posse ut globus tormentarius extra bombardam nova vi (simili tandem) propulsus,
     velocitate aliquandiu cresceret. »
112 P. Duhem, *Etudes sur Léonard de Vinci*, vol. III, p. 111.
113 *De Motu*, p. 260: « Caput . . . *Unde causetur celeritas et tardistas motus naturalis* . . . ex
     eadem causa pendere tarditatem et celeritatem, nempe ex maiori vel minori gravitate. »
114 *De Motu*, p. 261: « Attendendum est celeritatem non distingui a motu: qui enim ponit

motum, ponit necessario celeritatem ; et tarditas nihil aliud est quam minor celeritas. » A quantitative scale is thus substituted for an opposition of qualities. Cf. *ibid.*, p. 289 ff.

[115] *De Motu*, p. 251: « Lationem omnem naturalem, sive dorsum sive sursum illa sit, a propria mobilis gravitate vel levitate fieri. »

[116] An impasse in itself. But there is no doubt that it was the dynamics of *impetus* which was, at first anyway, the vehicle for, or provided the imaginative habit for, Archimedean thought. It was conceptually confused and contained a mixture of disparate elements, the opposition between which Galileo was to come to see clearly.

[117] *De Motu*, p. 334 ff.

[118] *De Motu*, p. 263: « Dicimus ergo mobilia eiusdem speciei . . . quamvis mole differant, tamen eadem celeritate moveri, nec citius descendere maior lapis quam minor. » As for the opposite view, according to which a large piece of iron would fall more swiftly than a small piece, 'it is clearer than daylight how ridiculous this view is.' For otherwise (and this is Benedetti's argument), if there are two bodies of which one moved faster than the other in natural motion, then a combination of the two bodies would move more slowly than that part which, by itself, moved the faster (*ibid*, p. 265). Cf. *ibid.*, p. 275: « Ex his quae in hoc et superiori capite tradita sunt, colligitur universaliter, mobilia diversae speciei eandem in suorum motuum celeritatibus servare proportionem, quam habent inter se gravitates ipsorum mobilium, dum fuerint aequales mole ; et hoc quidem non simpliciter, sed in eo medio ponderata in quo fieri debet motus. »

[119] *De Motu*, p. 254: « Ex hoc autem patet, quomodo in motu non sit solum habenda ratio de mobilis gravitate vel levitate, sed de gravitate etiam et de levitate medii per quod fit motus: nisi enim aqua levior esset lapide, tunc lapis in aqua non descenderet. »—*Ibid.*, p. 262: « Diversa mobilia in eodem medio mota aliam servare proportionem ac quae illis ab Aristotele est tributa. » In particular, the ratio is arithmetical and not geometrical. Galileo, following Benedetti, applies theorems from hydrostatics to the problem of falling bodies. Cf. *ibid.*, p. 272: « Excessus quibus gravitas sua mediorum gravitates excedit. »

[120] *De Motu*, p. 272: « Erunt enim inter se talium mobilium velocitates, ut excessus quibus gravitates mobilium gravitatem medii excedunt. »

[121] *De Motu*, p. 334: « Experientia tamen contrarium docet : verum enim est, lignum in principio sui motus ocius ferri plumbo : attamen paulo post adeo acceleratur motus plumbi, ut lignum post se relinquat, et, si ex alta turri demittantur, per magnum spatium praecedat : et de hoc saepe periculum feci. » It is clear from this that one cannot accept Galileo's claims about his 'experiments' uncritically.

[122] *De Motu*, p. 311: « Cum enim leve illud dicamus quod sursum fertur, lapis autem sursum fertur, ergo lapis levis est dum sursum fertur. Sed dices, leve illud esse quod sursum naturaliter fertur, non autem, quod vi. Ego autem dicam, leve id naturaliter esse quod sursum naturaliter fertur ; leve autem id praeternaturaliter aut per accidens aut vi esse, quod sursum praeter naturam, per accidens et vi fertur. Talis autem est lapis a virtute impulsus. »

[123] *De Motu*, p. 314: « Sic proiectum levi impellante liberatum suam veram et intrinsecam gravitatem descendendo prae se fert. » Following Benedetti, but in an original way, Galileo shows that contrary to the general view the body does not come to rest at the instant when the direction of motion is reversed. Cf. *ibid.*, p. 323: « Caput. . . *In quo contra Aristotelem et communem sententiam ostenditur in puncto reflexionis non dari quietem* ». Ibid., p. 323: « si enim semel quiescerent, semper deinde quiescerent. ».

[124] *De Motu*, pp. 315 f.: « Cap . . . *in quo causa accelerationis motus naturalis in fine longe alia ab ea quam Aristotelici assignant, in medio affertur.* » Cf. p. 329: « Naturalis resumatur gravitas, atque idcirco remota causa, acceleratio desinat. » It is worth remembering that Descartes also said that the acceleration of fall only occurs at the beginning, and that the falling body finally moves with an almost uniform speed. In fact, without gravity the acceleration cannot be explained.

[125] See *De Motu*, p. 336 ff.

[126] *De Motu*, p. 296.

[127] This is (although Galileo did not see this) incompatible with a *constant speed* of fall.

[128] *De Motu*, p. 313: « Mobile, quo levius erit, eo quidem facilius movetur dum motori est

coniunctum. Sed, a movente relictum, brevi tempore impetum receptum retinet ; facilius moveri, sed minus *impetum* receptum retinere. » Cf. p. 333 sq. « caput . . . *in quo causa assignatur, cur minus gravia in principio sui motus naturalis velocius moveantur quam gravia.* »

129 Cf. above, p. 16. It is in fact difficult to believe that Galileo could have considered himself original.

130 Cf. the passages quoted in notes 119 and 120 above.

131 See *De Motu*, p. 289: « Cum gravia definiantur ea esse quae deorsum feruntur, levia vero quae sursum. »

132 See *De Motu*, p. 289: « Caput . . . *in quo contra Aristotelem concluditur, non esse ponendum simpliciter leve et simpliciter grave: quae etiam si darentur, non erunt terra et ignis ut ipse credidit* » Cf. above, pp. 15 f, 24, 25.

133 See *De Motu*, p. 289: « Grave et leve non nisi in comparatione ad minus gravia vel levia considerarunt qui ante Aristotelem ; et hoc quidem, meo iudicio, iure optimo : Aristoteles autem 4°, Caeli, opinionem antiquorum confutare nititur, suamque huic contrariam confirmare. No autem, antiquorum in hoc opinone secuturi. » Cf. the passage quoted in note 122 above.

134 *De Motu*, p. 289: « Quod si . . . per se, simpliciter et absolute . . . quaeratur utrum elementa gravia sint, respondemus, nedum aquam aut terram aut aerem, verum etiam et ignem, et si quid igne sit levius, gravitatem habere et demum omnia quae cum substantia quantitatem et materiam habeant coniunctam. » *Ibid*, p. 355: « Gravitate corpus nullum expers esse, contra Aristotelis opinionem. » This thesis is, in the last analysis, Democritean, and can be found already in Nicole Oresme and Copernicus. Galileo appeals here to 'the ancients' (p. 289) and to Plato (p. 292). Cf. p. 293: « gravissimum non possit definiri aut mente concipi nisi quatenus minus gravibus substat . . . neç corpus levissimum esse id quod omni careat gravitate, hoc enim est vacuum, non corpus aliquod. »

135 *De Motu*, p. 275: « Eadem vi, qua sphaera plumbea resistit ne sursum trahatur deorsum etiam fertur : ergo sphaera plumbea fertur deorsum tanta vi quanta est gravitas qua excedit gravitatem sphaerae aqueae. Hoc autem licet in lancis ponderibus intueri. » Cf. p. 342.

136 *De Motu*, p. 270: « Motus sursum fit a gravitate, non quidem mobilis, sed medii ; . . . celeritas motuum sursum, esse, sicut excessus gravitatis unius medii super gravitatem mobilis se habet ad excessum gravitatis alterius medii super gravitatem eiusdem mobilis. » *Ibid*, p. 259: « in mobilibus etiam naturalibus, sicut et in ponderibus lancis, potest motuum omnium, tam sursum quam deorsum, causa reduci ad solam gravitatem. Quando enim quid fertur sursum, tunc attollitur a gravitate medii. » Cf. *ibid.*, p. 361 ff. On the reduction of lightness to a difference of weights, and of upwards motion to a motion of 'extrusion', (a conception adopted by Nicole Oresme and, in a different way, by Copernicus) cf. above, p. 15.

137 *De Motu*, p. 352 sq: « *Motus sursum nullum naturalem esse* : Conditio ex parte motus . . . est ut non possit in infinitum esse et ad indeterminatum, sed ut finitus et terminatus . . . ad aliquem terminum, in quo naturaliter quiescere possit . . . ut non ab extrinseca sed intrinseca moveatur causa . . . motum sursum, ratione qua elongatio quaedam est a centro, non posse esse naturalem. » *Ibid.*, p. 359: « At simpliciter sursum, quo nihil magis sursum et quod etiam ut deorsum esse non possit, non solum actu non datur, verum neque ipsa cogitatione concipi potest. » *Ibid.*, p. 361: « Motum sursum ex parte mobilis naturalem esse non posse »; p. 363: « Corpora sursum per extrusionem moventur. » p. 359: « talem motum posse dici violentum. »

138 Galileo gives over a whole chapter to refuting Aristotle on the impossibility of a vacuum. Cf. *De Motu*, p. 276: « Quod si in vacuo ponderari possent, tunc certe, ubi nulla medii gravitas ponderum gravitatem minueret, eorum exactas perciperemus gravitates. Sed quia Peripatetici, cum principe suo, dixerunt, in vacuo nullos fieri posse motus et ideo omnia aeque ponderare, forte non absonum erit hanc opinionem examinare et eius fundamenta et demonstrationes perpendere: haec enim quaestio est una eorum quae de motu sunt. »

139 *De Motu*, p. 294: « Caput . . . *in quo contra Aristotelem et Themistium demonstratur, in*

vacuo solum differentias gravitatum et motuum exacte discerni posse.» Themistius, like Aristotle, asserts the equality of bodies' speeds in a vacuum: « Quanto autem haec falsa sint mox innotescet, cum, quomodo in solo vacuo possint vera gravitatum et motuum discrimina dari, et in pleno nulla haec inveniri posse, declaraverimus. »

[140] *Ibid,*, p. 282: « Dicere ex. gr. in vacuo non magis huc quam illuc, aut sursum quam deorsum, movebitur mobile, quia non magis versus sursum quam deorsum cedit vacuum sed undique æqualiter, puerile est : nam hoc idem dicam de aere ; cum enim lapis est in aere, quomodo magis cedit, deorsum quam sursum, aut sinistrorsum quam dextrorsum, si aeris ubique eadem est raritas ? . . . cum dicunt : in vacuo non est neque sursum neque deorsum, quis hoc somniavit ? Nonne, si vacuus esset aër, vacuum prope terram esset centro proprinquius vacuo quod esset prope ignem . . . Et, primo, Aristoteles peccat in hoc, quod non ostendit quomodo absurdum sit, in vacuo diversa mobilia eadem celeritate moveri, sed magis peccat. . .quare nec celeritates erunt aequales. » Cf. above, pp. 24, 37.

[141] Absolute motion is, therefore, possible in *impetus* dynamics.

[142] Galileo criticises this distinction: *De Motu,* p. 304: « caput . . . *in quo de motu circulari quaeritur, an sit naturalis an violentus.* Motus . . . naturalis est dum mobilia, incedendo, ad loca propria accedunt ; violentus vero est dum mobilia, quae moventur, a proprio loco recedunt. Haec cum ita se habeant, manifestum est, sphaeram super mundi centrum circumvolutam neque naturali neque violento motu moveri. » *Ibid.*, p. 305: « si sphaera esset in centro mundi, nec naturaliter nec violenter circumageretur, quaeritur, utrum, accepto motus principio ab externo motore, perpetuo moveretur, necne. Si enim non praeter naturam movetur, videtur quod perpetuo moveri deberet ; sed si non secundum naturam, videtur quod tandem quiescere debeat. »

[143] See Part 3 below, 'Galileo and the Principle of Inertia', chapter 1, section 2.

[144] This is an instructive example of the persistence of a 'natural' idea: that of the fall of heavy bodies. It is interesting that Copernicus managed to rid himself of it, whereas Galileo never managed to do this completely.

[145] *De Motu,* p. 252: « cap. *Gravia in inferiori loco, levia vero in sublimi a natura constituta esse, et cur.* Cum enim ut antiquioribus philosophis placuit, una omnium corporum sit materia, illa quidem graviora sint quae in angustiori spatio plures illius materiae particulas includerent, ut iidem philosophi, immerito fortasse ab Aristotele 4 Caeli confutati asserebant ; rationi profecto consentaneum fuit, ut quae in angustiori loco plus materiae concluderent, angustiora etiam loca, qualia sunt quae centro magis accedunt, occuparent. » Cf. *ibid.,* p. 345.

[146] Cf. the passages quoted above, p. 34.

[147] The influence of Copernicus on the development of Galileo's thought has been emphasised by P. Tannery, *Galilée et les principes de la dynamique,* (Mémoires scientifiques, vol. VI), Paris, 1926, p. 400 ff. Cf. below, Part 3, p. 154 ff. Galileo was in a manner of speaking Copernican *ab initio.* This is understandable in the light of the fact that Benedetti was a confirmed Copernican. See E. Wohlwill, *Galilei und sein Kampf . . .* vol. I, p. 19 ff.

[148] S. Hessen quite rightly sees this as the meaning of the Galilean revolution. See S. Hessen, 'Die Entwicklung der Physik Galileis und ihr Verhältnis zum physikalischen System von Aristoteles', *Logos,* vol. XVIII, p. 339 ff. However, Hessen does not seem to us to appreciate the significance of the fact that Galileo himself *did not conceive the Universe as infinite.*

[149] See P. Duhem, *Etudes sur Léonard de Vinci,* vol. III, p. 257 ff, and below, 'Galileo and the Principle of Inertia', chapter 1, section 2. It is worth pointing out in passing that this is a rare case of philosophy being in advance of science.

[150] *De Motu,* p. 259: « . . . naturalium mobilium notus ad ponderum in lance motum congrue reducatur. » Cf. above, p. 34.

[151] *De Motu,* p. 300: quoted in note 160 below.

[152] Cf. P. Duhem, *Etudes sur Léonard de Vinci,* vol. III, p. 199.

[153] See *Opere,* vol. I, p. 210 ff.

[154] *De Motu,* p. 359: « Haec Aristoteles contra antiquos et nos pro antiquis. » According to E. Goldbeck, *Galileis Atomistik,* Bibliotheca Mathematica, N. F., vol. III, chapter 1, 'the

ancients' were the Greek atomists. This is true, but they were also the 'ancients' of the Scholastic tradition; cf. above, p. 15. They include Plato and Archimedes.

155   *De Motu*, p. 296 and p. 298. Therefore motion on a horizontal plane is neither natural nor *contra naturam*. Cf. p. 299: « Amplius : mobile, nullam extrinsecam habens resistentiam, in plano sub horizonte quantulumcunque inclinato naturaliter descendet, nulla adhibita vi extrinseca . . . et idem mobile in plano quantulumcunque super horizontem erecto non nisi violenter ascendit : ergo restat, quod in ipso horizonte nec naturaliter nec violenter moveatur. Quod si non violenter movetur, ergo a vi omnium minima moveri poterit. Quod etiam aliter demonstrare possumus : nempe, quodcunque mobile nullam extrinsecam resistentiam patiens, a vi quae minor sit quacunque vi proposita, in plano quod nec sursum nec deorsum tendat, moveri posse. » Cf. note 142 above.

156   *De Motu*, p. 276 ff: « Caput . . . *ubi, contra Aristotelem, demonstratur, si vacuum esset, motum in instanti non contingere, sed in tempore*. Posuit enim ejusdem mobilis motus in diversis mediis eam, in celeritate, inter se proportionem servare, quam habent mediorum subtilitates : quod quidem falsum esse, supra abunde demonstratum est. . . Et quod eodem loco scribit Aristoteles, quod impossibile est numerum ad numerum eam habere proportionem quam numerus ad nihil, verum quidem est de proportione geometrica, et non solum in numeris sed in omni quantitate. . . Attamen hoc non est necessarium in proportionibus arithmeticis : potest enim in his numerus ad numerum eam habere proportionem quam numerus ad nihil. Quare . . ., si celeritas ad celeritatem non geometrice sed arithmetice dictam proportionem servaret, iam nullum absurdum sequeretur. At certe quidem celeritas ad celeritatem [se habet] sicut excessus gravitatis mobilis super huius medii gravitatem. . . Quapropter in vacuo quoque eadem ratione movebitur mobile, qua in pleno. » Of course, motion will be fastest in a vacuum. For, since « excessum super nihil est maius quam in medio », fall will be *velocissima*. Cf. above, pp. 25, 26.

157   *De Motu*, p. 296: « Caput *in quo agitur de proportionibus motuum eiusdem mobilis super diversa plana inclinata* . . . manifestum est, grave deorsum ferri tanta vi, quanta esset necessaria ad illud sursum trahendum ; hoc est fertur deorsum tanta vi, quanta resistit, ne ascendat », p. 298: « Haec demonstratio intelligenda est nulla existente accidentali resistentia . . .: supponendum est, planum esse quoddammodo incorporeum . . . mobile esse expolitissimum, figura perfecta sphaerica. Quare omnia si ita disposita fuerint, quodcunque mobile super planum horizonti aequidistans a minima vi movebitur, imo et a vi minori quam quaevis alia vis. Et hoc, quia videtur satis creditu difficile . . . demonstrabitur hac demonstratione. »

158   *De Motu*, p. 300: « Et haec quae demonstravimus, ut etiam supra diximus, intelligenda sunt de mobilibus ab omni extrinseca resistentia immunibus : quae quidem cum forte impossibile sit in materia invenire, ne miretur aliquis, de his periculum faciens, si experientia frustretur, et magna sphaera, etiam si in plano horizontali, minima vi non possit moveri. Accedit enim, praeter causas iam dictas, etiam haec : scilicet, planum non vere posse esse horizonti aequidistans. Superficies enim terrae sphaerica est, cui non potest aequidistare planum : quare plano in uno tantum puncto sphaeram contingente, si a tali puncto recedamus, necesse est ascendere. . . »

159   This came later, once Galileo had understood that his mathematicism was a kind of Platonism. See below, Part 3, p. 203 ff. and p. 207 ff.

160   *De Motu*, p. 300: « Hic autem non me praeterit, posse aliquem obiicere, me ad has demonstratione tanquam verum id supponere quod falsum est : nempe, suspensa pondera ex lance, cum lance angulos rectas continere ; cum tamen pondera ad centrum tendentia concurrerent. His responderem, me sub suprahumani Archimedis (quem nunquam absque admiratione nomino) alis memet protegere. » Cf. below, Part 3, p. 242.

161   Cf. E. Meyerson, *Identité et Réalité*, 3rd edition, Paris, 1926, p. 145 ff.

162   Cf. Galileo, *Le Mechaniche*, in *Opere*, vol. II, p. 159: « Quello che in tutte le scienze demostrative è necessario diosservarsi, doviamo noi . . . in questo trattato seguitare : che è di proporre le diffinizioni dei termini proprii di questa faculta, e le prime supposizioni, delle quali, come da fecondissimi semi, pullulano e scaturiscano conseqentemente le cause e le vere demostrazioni delle proprietà di tutti gl' instrumenti mecanici. . .

Adimandiano adunque gravità quella propensione di muoversi naturalmente al basso, la quale nei corpi solidi, si ritrova cagionata della magiore o minore copia di materia dalla quale vengono constituiti. . . Momento è propensione di andare al basso, cagionato non tanto dalla gravità del mobile, quanto dalla dispozisione che abbino tra di loro diversi corpi gravi ; mediante il qual momento si vedra molte volte un corpo men grave contrapesare un altro di maggior gravità : come nella stadera si vede un picciolo contrapeso alzare un altro peso grandissimo. . . E dunque il momento quell'impeto di andare all'basso, composto di gravità, posizione e di altro, dal che possa essere tal propensione cagionate. » (English translation in eds. Drake and Drabkin, *Galileo, On Motion and On Mechanics*, p. 151.)

[163] For the whole doxographic tradition Archimedes was a 'Platonist philosopher'.

PART II
THE LAW OF FALLING BODIES:
DESCARTES AND GALILEO

INTRODUCTION

The law of falling bodies—the first of the laws of classical physics—was formulated by Galileo in 1604.[1] Fifteen years later, in 1619, it was formulated again by Beeckman.[2] Beeckman, it is true, did not accomplish this by himself. A good physicist but a mediocre mathematician,[3] he needed to enlist the help of Descartes. It was to him that Beeckman sent the problem of integration which he had been unable to resolve himself. It would be wrong, however, to minimise Beeckman's role, to see it as having been purely incidental, and to attribute to Descartes all the glory for the discovery. In fact Beeckman's role was much more important. He not only posed the problem, he also suggested to Descartes the basis for its solution. And it was he who, misinterpreting Descartes' reply, gave the correct formula for the law (the very same as that found by Galileo fifteen years earlier) putting it forward, however, as Descartes' own work.

In fact, in his reply Descartes had got it wrong. The formula which he sent to Beeckman was incorrect. What is interesting, however, is that his error was the very same—or more accurately was the complement of—that made by Galileo fifteen years earlier. For he also had made a mistake.[4]

This kind of 'coincidence' is not uncommon in the history of scientific thought. The very same ideas crop up in different places and in quite different minds. Everyone is familiar with disputes about who was first . . .; and everyone agrees that these amazing cases of simultaneous discovery are of the greatest interest for the history of scientific thought.

But none of these 'coincidences', not even the most famous, that of Newton and Leibniz in the invention of the infinitesimal calculus, or that of Carnot and Clausius in the discovery of the law of entropy, seems to us to be as fascinating as that in which the work of Galileo doubly coincided with that of Beeckman and Descartes. It is the only case in which one finds a coincidence not only of truth but also of error.

The law of falling bodies is a law of great importance: it is the fundamental law of modern dynamics.[5] At the same time it is an extremely simple law: it is entirely contained in a definition: *bodies fall with a uniformly accelerated motion.*[6]

Now, in the discovery of this law, a law so simple that nowadays children can grasp it without difficulty, Descartes and Galileo made a serious mistake. How are we to explain their error? Historians of Galileo, and equally those of Descartes, by and large do not dwell upon this mishap. This is understandable. Every historian, and especially every biographer, is something of a hagiographer. Therefore they often gloss over the errors and failures of their heroes. They mention them only to explain them away. What is served, after all, by dwelling on error? Is not the important thing the successful outcome, the discovery, and not the difficult, winding paths that had to be followed, and on which there was always the possibility of going astray. Of course, the historian-hagiographer is right. What matters from the point of view of posterity is the victory, the discovery, the invention. However, for the historian of scientific thought, at least for the historian-philosopher, failure and error, especially the error of a Galileo or a Descartes, can sometimes be just as valuable as their successes. They can, perhaps, be even more so. They are, in fact, very instructive. They sometimes enable us to grasp and to understand the hidden processes of their thinking.

No doubt the objection could be made that one should not look for a rational explanation of error. Error is a consequence of the weakness and limitations of the human mind, a function of its psychological, and even biological, conditioning. Everyone is capable of falling into error. Anyone can make mistakes. Nobody is an exception to this. It is enough to explain error by a lack of attention, by distraction, or by 'inadvertence'.[7] We cannot accept this objection, or at least not entirely. No doubt any mistaken reasoning is inadvertent. And when Galileo and Descartes made mistakes they were guilty of this. But that this duplicated inadvertence (this duplication being in itself already an extremely curious fact) should lead them to exactly the same error, it is this, it seems to us, that cannot have been the result of pure chance. Certainly this is not in itself an impossibility. Nevertheless it is far from plausible. There must be some reason for similarity in error.

Therefore the question we have raised remains unanswered. In deducing an extremely simple law Descartes and Galileo made a mistake. Could it not be that we have stumbled upon here an indication that this simplicity of the law is only apparent? Or, to put it another way, does not this indicate that the law of falling bodies is only simple if it is located within a particular system of axioms, and if one starts with a particular set of concepts? In other words it presupposes and implies a certain number of specific concepts—concepts of space, of causality, of motion—which, themselves, are not 'simple' at all: or, one might say, which are *too* simple—like all basic ideas—and for this very reason, extremely difficult to isolate and identify.[8]

1
GALILEO

The phenomenon of fall has always been a subject of meditation and astonishment for physics. It is, therefore, not surprising that Galileo, who already as a young man at Pisa had devoted his mental effort to the solution of the double problem of fall—i.e., of fall as downwards motion and of its acceleration—should continue to be concerned with this at Padua. He understood very well that a fundamental theorem, the fundamental theorem even, of the new science was involved.

Now this is what he wrote to Paolo Sarpi in that letter of 16th October 1604 which we mentioned above:[9] 'Reflecting on the problems of motion for which, for the demonstration of the accidents which I have observed, I lacked an utterly indubitable principle that I could take as an axiom, I have arrived at a proposition which is most natural and evident, and with it being assumed I can demonstrate the rest, namely, that *spaces traversed in natural motion are in the squared* [*doppia*, i.e., double] *proportion of the times*, and consequently *the spaces traversed in equal times are as the odd numbers ab unitate*, etc. And the principle is this, that the natural moving body increases its speed in the proportion that it is distant from the beginning of its motion, as for example, in assuming that a heavy body is falling from the point A through the line ABCD, I suppose that the degree of speed which it has in C to the degree of speed it has in B is as the distance CA is to the distance BA, and so consequently in D it will have a degree of speed more than in C according as the distance DA is more than CA.'

This is a fascinating text—we will compare it later with those of Descartes—and one which shows very clearly the characteristic feature of Galileo's logic. What he is looking for is not some kind of descriptive formula, a formula which will enable him to calculate observable and measurable magnitudes of the phenomenon of fall, its 'accidents' (speed, distance travelled by the moving body, etc.). Quite the contrary: Galileo *already has* such a formula (we will leave aside the question of how he came to have this).[10] He already knows that the *spaces* traversed in equal *times* are to each other as the series of odd numbers. He knows equally that the *space* traversed by the moving body is proportional to the *square of the times*. However, he is looking for something more, and this is not the logical or mathematical relation which connects these two propositions—it is clear that he is familiar with this relation—but a fundamental and evident 'principle' which will make it possible to deduce, or as Galileo says to 'demonstrate', the 'accidents' of the motion of fall. One could say, applying to Galileo the saying of a modern physicist, that he had no confidence in observations which had not been verified theoretically. Galilean epistemology is not positivist. It is Archimedean.[11]

In other words, Galileo already has the *law* of falling bodies. But his view is that that is not enough. For he only has this law as a fact. He does not understand *why* it is so. Bodies fall; this is a fact. Moreover when they fall their motion accelerates. The distances they cover in falling are to each other as the odd numbers. But why is all this so? Galileo believes that it is necessary to know this.

But we must be careful here. What it is necessary to explain or understand, on Galileo's view, is not the fact of the fall itself; it is not a matter of discovering the *cause* of the bodies' fall.[12] What he is looking for is the *essence* of the motion of fall. The motion of falling bodies is, in fact, a quite particular motion: it is a kind of motion which is highly specific and which is repeated, the same, wherever bodies fall. The problem is to discover the nature of this kind of motion, its essence, or one might say its *definition* (which comes to the same thing). It is this which would constitute that evident and indubitable principle, that fundamental axiom from which it would then be possible to deduce everything else.

As for why bodies fall, Galileo cannot answer this.[13] Nobody before Newton could explain it.[14] It has often been considered a major claim to fame on Galileo's part that he gave up any attempt at a causal explanation and aimed instead at discovering the essence, or as it has been called, the 'law'. However, in giving it up (and Galileo in fact only did so because he had to) he severed, or at least weakened, the link between his thought and reality, and thereby in a strange way made his task very much more difficult. Error, on the other hand, was able to creep in that much more easily.

We will return to these problems later. Whatever one might say about them the fact remains that in finding the essence of the motion of fall Galileo committed an error. For the 'principle' which he accepts as being sufficiently evident and natural—*the speed of the moving body* (in free fall) *is proportional to the distance covered*—does not at all lead to the law of fall as he had himself just formulated it. It leads to a quite different law, one which it would not have been possible for him to work out.[15]

It has been shown by Wohlwill[16] and Duhem[17] that the principle which Galileo wanted to postulate as the basis of his dynamics—*the speed of the moving body is proportional to the distance covered* (instead of the correct principle which is that *the speed of the moving body is proportional to the time elapsed,* which was already known by Leonardo da Vinci)—was not discovered by Galileo. One could try to explain why Galileo thought that this principle was self-evident by the influence, whether conscious or not, of a tradition. Galileo thought that he understood, whereas in reality he was merely remembering. This is, in fact, what Duhem's explanation amounts to. But this is really only to take the problem a step back. How is it possible that a principle which, however plausible, is not to us at all *self-evident,* could have been accepted as such by minds which, while not the equal of

Galileo's, were nevertheless by no means negligible? What was it about this 'principle' that was so attractive? We can get some inkling as to why this was so by taking a brief look at the history of the problem.

The principle on which Galileo tried to base his 'demonstration' was formulated, as clearly as one could wish, by G. B. Benedetti, who, it is generally accepted, was Galileo's immediate precursor. In his *Book of Various Mathematical and Physical Ideas* Benedetti wrote: 'Aristotle should not have said that the nearer a body approaches its terminal goal the swifter it is, but rather that the further distant it is from its starting point the swifter it is.'[18] Benedetti here asserts *expressis verbis* his opposition to Aristotle's view, and yet one can ask just how real this opposition is. For does not a body which goes from A to B, for example a body which falls to earth from a high tower, or even any body which is heading for the centre of the earth, approach its goal precisely to the extent that it gets further from its starting point? In other words, does it not get further away from its starting point just to the extent that it approaches its goal? The two expressions seem completely equivalent. Moreover, Niccolò Tartaglia, who seems to have been the first, at least among modern writers, to have introduced the consideration of the starting point into the discussion, says, quite sensibly: 'Every uniformly heavy body in natural motion will go more swiftly *the more it shall depart from its beginning or the more it shall approach its end*'.[19]

We should add that Benedetti himself by no means neglects the point of arrival, the natural goal of the motion. In fact at just the point when he is reproaching and correcting Aristotle[20] in the passage just quoted above he writes: 'This impression or impetus continuously increases in rectilinear natural motions, since the body always has within itself the moving cause, i.e., the tendency to go *toward the place assigned to it by nature*'.[21] And a few lines further on, in explaining the acceleration of the motion of fall, Benedetti adds; 'For the impression is always greater, the more the body moves in natural motion. Thus the body continually receives new impetus since it contains within itself the cause of motion, which is the tendency to go towards its own proper place, outside of which it remains only by force'.[22]

In these circumstances, i.e., in an exposition of purely Aristotelian cosmological physics, how can Benedetti think that he is making an innovation? What is the meaning of his complaint against Aristotle? Why does he not see the equivalence between his proposition and that which he is rejecting?

This is an important problem. But in order to resolve it we must take precisely these facts as our starting point: the fact that Benedetti, while upholding an Aristotelian viewpoint, feels himself to be in opposition to Aristotle: and the fact that while replacing Aristotle's proposition (or at least the proposition which he takes to be Aristotle's) by a formally equivalent one of his own, he distinguishes between the two and even, in contrast to Tartaglia, sees them as being opposed to each other.

It would be possible to argue that this question that we are raising is not, in itself, of any importance. Benedetti's thought is rather obscure, rather confused even, and this is itself enough of an explanation for his inconsistencies and his non-sequiturs. Now it is certainly true that the thought of Giovanni Battista Benedetti is no model of clarity. Nevertheless it is vigorous and honest. One must not forget, moreover, that a body of thought—in general, but especially in periods of transition—can be obscure and confused without, for all that, being worthless. Quite the opposite. As has been strongly argued by Duhem and admirably demonstrated by Emile Meyerson, it is through the obscure and the confused that thought makes progress. It proceeds from obscurity towards clarity. It does not proceed, as Descartes proposed, from clarity to clarity.

Benedetti's thought is certainly confused. But this is because in it the Aristotelian tradition meets up with the Parisian tradition (*impetus* physics) and because there is grafted onto this double tradition another, more recent, tradition of increasing importance, that of Archimedean physics. Thus Benedetti, while being, as we have pointed out, a firm defender of Copernicus,[23] cannot give up the general framework of Aristotelian cosmological physics—what would he replace it with?—and yet nevertheless has reason to present himself as an opponent of Aristotle. For *impetus* physics, which sees motion as the effect of a force contained within the moving body, makes it possible to separate the body's motion from the idea of the goal towards which it is directed, makes it possible to isolate the moving body from the rest of the Universe.[24] Benedetti is therefore right to deny the equivalence of separation from the *terminus a quo* with approach to the *terminus ad quem*, because his conception of motion precisely makes it possible to delete, in thought if not in fact, the *terminus ad quem*. The moving body, set in motion by a force, necessarily starts from somewhere: from the place at which it was at rest. Therefore one cannot, in defining the motion, do without the idea of the *terminus a quo*. But this terminus is enough. The moving body, propelled by a force, sets off—in rectilinear motion—*in a particular direction*. It does not head for *a particular goal* (whether or not this goal in fact exists). This is clearly so in the case of violent motion: when one hits a ball the *impetus* which is thereby impressed on it directly determines the speed and the direction of its motion. One can at the same time aim at a particular target. But this is not in any way necessary.

This conception can be extended to the case of natural motion. The moving body (heavy or light) moves (or is set in motion) in a particular direction: downwards or upwards. It does not go towards a goal. Therefore we must, in contrast to Aristotle, talk of its separation from the starting point and not of its approach to its point of arrival.[25] This has an extremely important consequence: the motion of a moving body is completely determined by its past state and not at all by its future state.[26]

Benedetti's way of conceiving motion is different from that of Tartaglia. Or, to put it another way, the underlying concept of space in Benedetti's reasoning—the same concept as that underlying the reasoning of the young Galileo[27]—is different from that of Tartaglia. The equivalence which exists for the latter does not exist at all for Benedetti and this is precisely because in his space, a space which is geometrical and no longer physical, (rectilinear) motion can go on indefinitely. This is not possible for Tartaglia; and even less so for Aristotle.

For Benedetti motion is an effect of the force (*impetus*) contained in the moving body, and space is not physical but geometrical. Therefore, as we have seen, for him motion in a vacuum is perfectly possible. But his space is not, however, *completely* geometrical: it is not completely homogeneous. For him there are still privileged directions; down and up. His space is Archimedean, or more accurately Epicurean.

Of course we cannot give an account here of the history of the problem of fall or give details of all the explanations (variations of resistance, reaction of the medium, etc.) which mediaeval theorists thought up to account for the surprising phenomenon of acceleration.[28] We must, however, recall the explanation that derives from the idea of *impetus*, which was as far as the immediate precursors of Galileo had reached.

The theory of *impetus*, as we have seen, consists in taking motion to be the effect produced by a cause internal to the moving body. The conceptualisation of this cause—the *impetus*—was very vague; it was thought of as similar to a form, or a quality, or a force. It is this force, impressed on the moving body by the action of the external mover—the impact—which, persisting in the moved body, explains the continuation of its motion. It is enough to take the natural heaviness (or lightness) of bodies as similar to such an *impetus* to be able to explain natural and violent motion in an analogous manner, to understand that natural motion and violent motion (or more accurately the *impetus* of these motions) can be added together in one and the same moving body. It is sufficient to take a moving body as being subjected, during its motion, to successive impulsions or impacts which impress new *impetus* on it, to be able to give a tolerably good explanation of the accelerated motion of fall.

This theory, developed by the Parisian nominalists, was very popular among sixteenth century thinkers. It was accepted by Piccolomini,[29] Cardan and Scaliger[30]—as it had been earlier by Leonardo da Vinci. Benedetti's exposition of it is as clear as one could wish.

The *impetuses* accumulate; they are impressed on the moving body before the influence of the first *impetus*—or of successive *impetuses*—has disappeared. This is an important point. *Impetus*, being an efficient cause producing motion as its effect, diminishes as it produces it. It follows that all *impetus* becomes exhausted, i.e., becomes weaker by the very fact of the motion of the body which it is propelling. Therefore the motion slows down and every body set in motion has a tendency to return to rest. For there to be an acceleration the new *impetus*, the new impact or push or pull, must come into play while the earlier *impetus* still remains, i.e., while the body is still in motion.

When applied to the problem of fall the most developed versions of the theory of *impetus* gave rise to one or other of the following ways of looking at the problem. (1) In the first version it is said that in the first instant of fall the gravity of the heavy body gives the body a specific motion (a certain speed); after this, at the second instant, it is as if the body in question were acted upon by both its natural (constant) gravity and in addition by a certain accidental gravity which is a function of the speed with which it is moving. The natural and the accidental gravities act together to give the body a particular new speed which is, of course, greater than the first; and this then happens again and again. It could be said, therefore, that the (total) gravity of the body continually increases as it falls and that this in turn explains the increase in speed.

(2) In the second version it is said that the natural gravity produces an *impetus* in the heavy body which takes it towards its goal, or in the natural direction of its motion, and that before this *impetus* is used up the gravity produces a second *impetus* which is added to the first, and so on, so that the heavy body 'always increases its speed because there is conjoined with it a perpetual motive virtue'.

These are, no doubt, very subtle ideas, but as the most dedicated Aristotelians[31] quite reasonably pointed out they are fundamentally illogical. In the first version, *impetus* (as cause of motion) is in fact identified with its product or effect. In the second version gravity is conceived not as a force or cause but as a source from which flow the *impetuses* which accumulate in the moving body.

In both versions the *impetuses* are produced at each instant, just as Leonardo da Vinci had already said more clearly than any of his successors: 'The heavy body in free fall acquires a degree of motion with each degree of time, and a degree of speed with each degree of motion'.[32]

How is it then that Leonardo himself (and after him Benedetti and then Michel Varro) asserted that the speed is proportional not to the time elapsed but to the distance covered? It is clear that they believed that these two propositions are equivalent, and this for a very simple reason: to each instant of time there corresponds in effect a point of space traversed. Now, as Duhem points out,[33] 'in order to derive from the law: *the speed of motion of a heavy body is proportional to the duration of fall*, this other law: *the distance covered by the heavy body is proportional to the square of the duration of fall*, it would have been necessary for Leonardo to have had the idea of instantaneous velocity, or in other words, the ideas of *fluxion* and *derivative*'. Similarly, in order to realise that in spite of the fact that there is a one-to-one correspondence between points of time (instants) and points of space (on the path of motion) there is a non-equivalence of their magnitudes it would, of course, have been necessary for Leonardo and his successors to have had the basic ideas of the integral calculus.

However, after Archimedes and after Nicole Oresme, perhaps it would not be too much to expect this of them. But we should not be too harsh; we should not blame Leonardo and Benedetti for sliding so easily, with the help of the equivocal idea of extended motion, from time to space, from the duration of the motion to its path. It is easier—and more natural—to *see*, i.e., to *imagine*, in space, than it is to *think* in time.

Duhem has explained very clearly for us just why neither Leonardo da Vinci nor Benedetti were able to formulate the correct law of fall, and why it was only Galileo who achieved this. However, he does not explain for us why, of two equivalent relations (or rather of two relations thought to be equivalent; speed proportional to time elapsed and speed proportional to distance covered) Leonardo, and also later Galileo and Descartes, chose the second with such conviction. The reason for this seems at once both simple and profound. It is entirely a matter of the role that geometry plays in modern science, of the relative intelligibility of spatial relations.[34]

The process which gave rise to classical physics consisted in an attempt to rationalise, in other words to geometrise, space, and to mathematise the laws of nature. In fact these come to the same thing because the geometrisation of space means precisely the application of geometrical laws to motion. For how could something have been mathematised— before Descartes—except by geometrising it?

Furthermore, it is, as we have just pointed out, more 'natural', 'easier', to imagine in space than to think in time. The ideas arrived at by Leonardo, Benedetti and Galileo do in fact seem to be 'natural' enough. For if one thinks, as does Benedetti, of heavy bodies falling through his Archimedean space, would one not be 'naturally' inclined to assert that the more distant they become from their starting points the faster they fall? i.e., that they fall from *higher up*, or that as they fall they get *lower*? Is it not natural to take their speed to be a function of the space traversed? For example, consider a body which falls from a height of one hundred feet. When it reaches the ground it has a certain speed. If, now, we let it fall from a height twice as great, its speed on reaching the ground will be greater. What could be more 'natural' than to take this speed as dependent on the only thing that is different in the two cases, on the height from which it has fallen, i.e., on the length of the path it has travelled? And what could be more natural than to postulate a relation between the variation in height and the increase in speed, to take the speed as a function of the height, even to postulate that there is a strict proportionality? i.e., to say: a body which falls from twice the height acquires twice the speed as it falls.[35] Now, compared with this idea, the idea of taking the speed with which a falling body passes through the space traversed to be dependent not on the distance but on the time it takes to cover that distance, on the time which is manifestly itself a function of the speed, this idea hardly seems 'natural'; it seems even to be extremely and unnecessarily complicated.[36]

What is it that forces thought to attribute to time, to the duration, an important role in fall? Clearly it is the fact that the idea of time is contained in that of motion; but also—perhaps above all—it is the *causal* analysis or explanation of fall. Impulses, *impetuses*, follow one another *in time*; they act, in the first instance, in time and only by implication in space. Let us for a minute forget causality, whatever it is that produces fall, motion and acceleration: if our thought is no longer inclined towards this problem it straight away turns quite 'naturally' towards space. Dynamics, unable to remain at the stage of kinematics, is transformed into geometry. This is the reason why Galileo, who had while at Pisa already recognised the impossibility of basing a mathematical dynamics on the notion of *impetus*, and who had, as we have seen above, replaced the search for a cause by the search for the essence, turned immediately to what we might call thorough-going geometrisation.

From his earliest work at Pisa the young Archimedean and Platonist[37] Galileo had a particular aim in mind: that of the mathematisation of physics. Nobody before him—not even Benedetti—had worked towards this goal so consciously, so patiently, so stubbornly. He tried in the first place to mathematise Aristotle's physics, but this attempt ended in failure. He tried again, this time taking the idea of *impetus* as his starting point, but once again he met with failure. Moreover, it is very easy to understand this—*post factum*. How could it have been possible to translate the idea of *impetus* into mathematics? It is a vague and confused idea, closely related to perceptual experience, the idea of a quality which is not in itself measurable. How would one go about calculating the progressive weakening of *elan*? This could only be done by replacing this obscure idea with those of motion and momentum [Fr. *force vive* = *vis viva*]—a radical transformation disguised, and aided by the persistence of an out-dated terminology.[38] How could one postulate the accumulation of successive *impetuses* in the moving body? Once again this was only possible by a radical transformation of the primitive concept of *impetus*, by replacing the idea of its production by an internal cause with that of the repeated action of external causes,[39] (attractions or impacts) each of which produces a lasting effect.

Of course Galileo was not to achieve these transformations completely: this had to wait for Descartes and Newton. But, as we have seen, from his first Pisan works the young Galileo uncovered the faults in the arguments of those such as Benedetti, Cardan and Tartaglia whose entire teachings were based on a paralogism or on an ambiguity. It is a contradiction to postulate that a constant *cause* could produce a variable *effect*. There is no way that the fall of a heavy body in an Archimedean space could be a motion which would increase its speed all by itself. To postulate this would be to postulate a creation *ex nihilo*. A constant *cause* can only produce an *effect* which is constant. Therefore the fall of heavy bodies takes place at constant speed. And if, as a matter of fact, a falling body's motion accelerates—up to the point at which it reaches its proper speed—this is because it is, at the start, slowed down.

This ingenious idea (in which we can no doubt recognise that of Hipparchus[40]) is unfortunately self-contradictory, or more accurately is incompatible with a geometrical conception of space, because it necessarily presupposes the idea that a heavy body tends towards a goal, the idea of the distance of a heavy body from its goal; it therefore rules out a *constant* speed of fall.[41]

Galileo, therefore, tries something different. He tries, this time directly inspired by Archimedes, to build a physics in terms of, or on the model of, hydrodynamics. Following the lead of the 'ancients' he gives up all qualitative distinction between 'heavy' and 'light' and drops the idea of natural motion in an 'upwards' direction. From now on all motion is to be explained in terms of the quantitatively determined reaction of the body with the surrounding medium.

At roughly the same time he tried a different approach, in attempting to assimilate the laws of motion to those of the lever. It could be said that Galileo tried to build a physics of rigid connections.[42]

We do not know why Galileo did not go further with his attempt to develop this hydrodynamical physics, nor why he did not pursue his attempt to found a physics of rigid connections. But we can suggest a hypothesis: both hydrodynamical physics and a physics of rigid connections require a physical space; they do not permit either the complete geometrisation of space nor even motion in a vacuum. Now, motion in a vacuum and the geometrisation of space are both essential aspects of Galilean physics. They are, for him, the decisive contributions made by *impetus* physics. Although he gave up *impetus* dynamics itself, Galileo was never to abandon its achievements.

We must emphasis the fundamental importance of the fact that Galileo gave up the idea of *impetus*, the internal cause of the body's motion. He does, of course, keep the word,[43] but with a completely different meaning: instead of being the *cause* of motion *impetus* becomes its effect. The idea of *impetus* as cause of motion purely and simply vanishes. This hybrid, confused and obscure idea has no equivalent in his thought. Or, which comes to the same thing, it is replaced by the ideas of *speed* and of *motion*. Already at Pisa, investigating the special, abstract cases of motion, i.e., the 'simple' cases (circular motion 'around a centre'; horizontal motion, i.e., motion which is at the limit between the accelerated motion of fall and decelerated upwards motion), Galileo had learned that, in such cases, contrary to the basic intentions of *impetus* theory, motion seemed to be able to last for ever.[44] It is true that *impetus* theorists, or at least some of them (for example, Piccolomini and Buridan) had stated that in certain cases (in particular in the case of circular motion) the *impetus* was everlasting (or immortal). The *impetus*, they said, has no resistance to overcome in such cases, so why should it become weaker? One could certainly claim to be able to find some glimmering of the truth in this argument, but it is not one which Galileo could accept as it stood. He saw clearly that *impetus*, if it is defined as the cause of motion, must be used up as it generates the motion. If it remains unchanged this is because it plays no role at all in the continuation of the motion. It is not the *impetus* which keeps motion going and makes it last: motion is conserved by itself. And since speed is the essential feature of motion, to say that motion conserves itself means equally that speed is also conserved. Both motion and speed (but the latter especially) undergo a kind of change of ontological status. From being effects produced by a cause, effects which can only exist or last as long as the action of the cause which produces them persists (e.g., pressure) they become relatively independent entities which are conserved by themselves in just the same way that rest is conserved in a body which is not moving.[45] This is the case for 'abstract' motion. As for 'concrete' or 'mechanical' motion, it was at Padua that Galileo developed the idea of this: it was progressively separated off and emancipated from the confused magma of the doctrine of *impetus*. It was at Padua, in the course on mechanics which he gave there, that Galileo formulated his idea of *momento*,[46] weight multiplied by speed. No doubt one could find earlier versions of this idea in the work of the Aristotelian author of *Mechanical Problems*[47] and even more among *impetus* theorists, in their concept of the accidental gravity which they thought was generated by the body's motion itself, and by its speed, or more accurately by its *impetus*. Duhem is quite right to emphasise this. However, Duhem did not point out the decisive transformation which this concept underwent in the work of Galileo.[48]

Galileo's idea of *momento* in fact implies just that elevation of the ontological dignity of concrete motion and speed which has just been mentioned above. There is no need for *impetus* as cause, no need for any intermediary whatsoever; motion is combined *directly* with weight. In short, motion, or speed, purely and simply take the place of *impetus*. This substitution quite clearly has the most profound consequences. For while *impetus* as producer of motion *could* not be conserved, and motion consequently *must necessarily* lose speed and eventually end in rest, it is on the contrary quite possible for motion and speed, promoted to independent entities, to be conserved *indefinitely*. A body, once in motion, no longer has any need to stop, nor even to slow down. This in itself provides the basis for the correct solution to the problem of fall.

When Galileo took up the problem of the fall of heavy bodies again in 1604 he had at his disposal, as we have seen, the formulae which relate the duration of fall to the distance covered: and also, as we have just seen, the most important principle of the conservation of motion and of speed. On the other hand he had given up any attempt at causal explanation and he restricted himself to looking for a principle, an axiom, which would allow the deduction of the descriptive laws of fall. Now, as we have also seen, it was being concerned with causes in the analysis of motion (of motion in general, and of that of fall in particular) that focused attention on the idea of time. Thérefore it is not surprising that giving up causal explanation should strengthen the tendency to geometrisation and thus to spatialisation. Instead of thinking about motion Galileo pictures it to himself. He sees the line, the distance covered with a variable speed. And it is this line, this trajectory, that he takes as the argument of the function, speed. The struggle for geometrisation, sustained and corroborated by the imagination, unfettered from causal thought, goes beyond the goal he had set himself. The goal of dynamics was to mathematise time; but Galileo eliminates it. His effort resulted in failure, a failure which Galileo does not notice straight away. For by reversing the argument, that which had led from the correct descriptive formulae to an incorrect principle, he is led back from this principle to the correct consequences from which he had started.

This is what he in fact says:[49] 'I postulate (and perhaps I will be able to prove this) that a heavy body falling naturally moves with a speed which continuously increases as the distance from the point at which it started grows larger. For example, if the body starts from point A and falls through the line AB then I postulate that the degree of speed at the point D will be as much greater than the degree of speed at the point C as the distance DA is greater than CA; thus the degree of speed at C will be to the degree of speed at D as CA is to DA, and similarly at each point on the line AB the body will have a degree of speed which is proportional to the distance of that point from the starting point A. This principle seems to me very natural, and it is in agreement with experiments on machines and instruments which work by impact, in which the effect of the impact is greater to the degree that the fall is from a greater heigh. If this principle is assumed everything else can be demonstrated.

'Make a line AK at any angle with AF, and draw parallels CG, DH, EI, FK through the points C, D, E, F. Since the lines FK, EI, DH, CG are to each other as the lines FA, EA, DA, CA, it follows that the speeds at the points F, E, D, C are as the lines FK, EI, DH, CG. Thus the degrees of speed increase at each point on the line AF in the same proportions as the increase in the parallels drawn from these same points. Moreoover, since the speed with which the moving body goes from A to D is composed of all the degrees of speed which it has acquired at each of the points on the line AD, and the speed with which it has covered the line AC is composed of all the degrees of speed acquired at each of the points of the line AC, it follows that the speed with which it has covered the line AD is to the speed with which it has covered the line AC in the same proportion as all the parallels drawn from all the points of the line AD up to AH are to all the parallels drawn from the line AC up to the line AG: and this proportion is that of the triangle ADH to the triangle ACG, i.e., that of the square of AD to the square of AC. Thus the speed with which the line AD has been covered is to the speed with which the line AC has been covered as the double proportion of that of DA to CA. And since the ratio between the speeds is the inverse of the ratio of the times (for to increase the speed is the same as

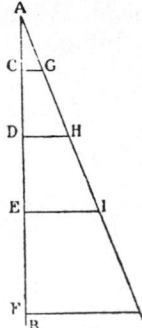

to decrease the time) it follows that the time of the motion through AD is to the time of the motion through AC as the sub-double proportion of that of the distance AD to the distance AC. The distances from the point of departure are thus as the squares of the times, and consequently the spaces covered in equal times are to each other as the odd numbers *ab unitate*: this is in agreement with what I have always said and with observed experiments. Thus all the truths are in agreement. And if these things are true I will demonstrate that the speed in violent motion decreases in the same proportions that it increases in natural motion along the same straight line.'

Galileo's argument is plausible. Nevertheless it is false because, as can be seen clearly, it contains a double error.[50] It is, of course, true that the ratios between speeds are the inverse of the ratios between times taken, but only on condition that the basis of comparison, i.e., the distance covered, remains constant, and not if different distances are covered as is the case here. Secondly, it is also true that the total speed of the moving body is the sum of the (instantaneous) speeds that it acquires at each point on its path, just as it is also the sum of the speeds that it acquires at each instant of its motion. But these 'sums' are not the same. When the increase is constant and uniform in relation to time it is not so in relation to distance, and conversely. In particular the 'sums' of speeds which increase as a linear function of distance covered cannot be represented by triangles. This representation is only valid for an increase which is uniform in relation *to time*. Once again Galileo's thorough-going geometrisation *transfers to space that which is valid for time*.

It is interesting to note that Galileo was to notice his error later[51] (i.e., the erroneous choice of principle-definition of the accelerated motion of fall) whereas, according to Duhem, Descartes never did. It is even more interesting to note that the argument by which Galileo tried to show the absurdity of the principle which had, at first, seemed so 'natural' to him, is itself thoroughly mistaken.[52]

But perhaps it was not this specious argument (which presupposes familiarity with the method of valid deduction) which guided Galileo's thought. It is more likely that his failure became apparent to him more directly, from the very fact that his 'axiomatic principle' could not play the role that he wished to give it. For it was, it goes without saying, impossible to deduce from it the descriptive formulae.[53] It was even impossible—for Galileo—to use it correctly. This was probably enough; repeated examination of the problem probably showed Galileo where his mistake was. There is no doubt that it came down to a matter of his having ignored the 'intimate relationship between time and motion'.[54] And perhaps also to his neglect of the causal aspect. This hypothesis seems quite reasonable[55] in the light of Galileo's later praise for the idea of attraction as formulated by Gilbert[56] and the admiration which he always expressed for this great English physicist.[57] The falling body's motion is accelerated because it is influenced at each successive instant by a constant instantaneous action (attraction) of the earth. Thus the formula (the essential definition) of accelerated motion must be based not on space but on time.

2

DESCARTES

Let us now turn to Descartes.

It was in 1618 that Isaak Beeckman, by accident, came to know 'Monsieur du Perron'. It was not long before he discovered the extraordinary gifts which nature had bestowed on the young Frenchman.[58] He therefore wrote to Descartes to ask for his help in resolving the formidable problem of the accelerated fall of heavy bodies.

The story of the collaboration between Beeckman and Descartes, a real comedy of errors, has already been told many times.[59] We believe, however, that it would not be unprofitable to go over it again.

Beeckman did not ask Descartes why bodies fall in general: this he already knew. No doubt he had learned it from Gilbert[60] or from Kepler. Bodies fall because they are attracted by the earth. Nor did he ask why they accelerate; this he also knew. Falling bodies accelerate because they are attracted by the earth again at each instant of their motion and because these additional attractions at each instant give them a new degree of motion while the previous motion still remains. In 1613 in fact Beeckman had formulated this important proposition: *that which is once set in motion remains in motion for ever.* Since 1613, then, he had been familiar with the law of conservation of motion.[61]

So Beeckman knew all of this (and it is a great deal; it is the complete structure of the physical problem[62]) *before* his contact with Descartes. But although he understood perfectly well (far better than Descartes) the *physical* aspect of the problem, he was not able to get a grip on it *mathematically.* He was not able to deduce the consequences from his principles. He could not discover the formula which would allow him to calculate the speed and the distance covered by the body.[63] It is this that he asked of Descartes.

So he asks him:[64] 'Given my principle, namely that *something once set in motion will keep moving for ever in a vacuum*, and assuming a vacuum between the earth and the falling stone, can one find out what distance a body will cover in one hour if one knows the distance it covers in two hours?'

The formulation of the question is rather strange. Beeckman does not ask what might seem to be the natural question, i.e., can one find out what distance will be covered by a falling body *in two hours* if one knows the distance covered in one hour? As we have seen, he puts the question the other way round.

Clearly Beeckman, who certainly no longer sees fall as a 'natural' motion, but as the effect of a terrestrial attraction on the body which, in itself, has no tendency to move one way rather than another, or indeed in general, to move at all (the body naturally remains at rest unless some *exterior* force sets it in motion, and then it remains in its new state of motion just as it had remained at rest) can still only conceive of fall as a motion having a normal and natural end—the earth—and not, as did Benedetti or the young Galileo, as a motion which can last indefinitely.[65] Thus he sees it as a motion which goes from A to B: from the top of a tower (or from some point or other above the earth) as far as the earth. It is this motion, the completed motion, which we can measure, i.e., we can measure the distance covered and the time taken. It is from this that we must start in order to reconstitute, by analysis, the previous phases of the motion.[66]

Descartes does not conceive of the motion of fall in quite the same way. Consequently his reply is incorrect, but Beeckman does not notice this.

According to Beeckman Descartes's reply, 'based on (Beeckman's) principles', to the question 'Why does a stone falling in a vacuum fall with ever greater speed?' is as follows:[67] 'When there is a vacuum between the body and the earth the body moves downwards, towards the centre of the earth, in the following way: in the first moment it covers as much space as possible as a result of the attraction[68] of the earth. In the second it keeps up this motion, to which a new motion of attraction is added, so that in this second moment it covers a double space. In the third moment the double space is maintained[69] and to it is added, by the earth's attraction, a third, so that in one moment it covers a space three times the first.'

These considerations, which as we shall see in a minute are Cartesian arguments translated into Beeckmanian terms, make possible a correct resolution of the problem and the calculation of the time of fall. So let us follow Beeckman's account:[70] 'If the moments are not divided up, the space covered by a falling body in one hour will be ADE. The space covered in two hours will be double the proportion of the times, i.e., will be ADE to ACB, which is double the proportion of AD to AC. Let the moment of space that the body covers in falling for one hour be of some magnitude, e.g., ADEF. In two hours it will cover three similar moments i.e., AFEGBHCD. But AFED contains ADE and AFE. And AFEGBHCD contains ACB with AFE and EGB, i.e., with the double of AFE.

'Thus, if the moment is AIRS the proportion of the spaces will be ADE with *klmn* to ACB with *klmnopqt,* i.e., once again, the double of *klmn*. But *klmn* is much smaller than AFE. Since, therefore, the proportion of space covered to space covered is composed of the proportion of one triangle with another triangle, with equal [magnitudes], added to these terms, and since these equal additions become ever smaller as the moments of space become smaller, it follows that these additions become null quantities when the moment has become a null quantity. Now, such is the moment of

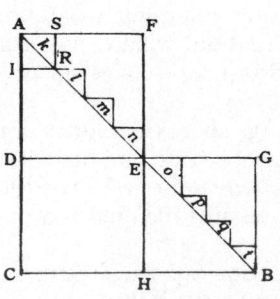

space through which the body falls. It remains, therefore, that the space through which the body falls in one hour is related to the space through which it falls in two hours as the triangle ADE to the triangle ACB . . .

'If, therefore, the experiment were performed and the body falling for two hours covered 1,000 feet, the triangle ABC would contain 1,000 feet.[71] The root of this is 100 for the line AC which corresponds to two hours. Bisected at D, AD corresponds to one hour. Since, therefore, it has the double proportion of AC to AD, ACB to ADE is 4 to 1, or 1,000 to 250.'

The solution is both elegant and correct: the distances covered are recognised to be proportional to the squares of the times. But it is not Descartes' solution. It is well known that Beeckman was mistaken in his interpretation of 'M. Perron's' reply.[72] Here is the second account, which we have from Descartes himself in the form of a short note in his *Cogitationes Privatae*:[73] 'It happened a few days ago that I came to know an extremely clever man who asked me about the following problem: a stone, he said, descends from A to B in one hour; it is perpetually attracted by the earth with the same force without losing any of the speed impressed upon it by the previous attraction. According to him, that which moves in a vacuum will move always. He asks, what time is taken to cover a given space?'

We should point out first of all that Descartes admits having received from Beeckman not only the problem but also the principles to be used in its solution,[74] principles which Descartes, unlike Beeckman, does not consider to be true. For Descartes they are only hypotheses, ones which, moreover, he does not completely understand. But this does not prevent him from solving the problem, nor even from giving two different solutions. Poor Beeckman was not asking for this. He wanted to know how stones do in fact fall. Descartes, not content with this, explains to him in addition how they could fall.[75]

Here, then, is his reply:[76] 'I have solved the problem. In the right-angled isosceles triangle, ABC represents the space (the motion); the inequality of the space from point A to the base BC [represents] the inequality of the motion.[77] Therefore, AD will be covered in the time represented by ADE; and DB in the time represented by DEBC; it being noted that the smaller space represents the slower motion. ADE is one third of DEBC; therefore AD will be covered three times as slowly as DB.

'But the problem could be posed in a different way: suppose the attractive force of the earth remains the same as at the first moment: and that a new one is produced, the first remaining. In this case the problem would be resolved by a pyramid.'[78]

This strange supplement shows clearly just how far from Descartes' mind is the problem of the physical mechanism of fall. He scarcely pauses over the fact that Beeckman had solved this problem, but goes on to invent another 'possible' case, in which *the attractive force* increases at each instant. Then, at the second instant, the body will be attracted by a double force, at the third by a triple force, and so on. In this case, of course, the body would fall very much faster.

How could such an increase in the 'attractive force' be possible? Descartes does not ask himself this. In fact he sees the problem not as a physicist but as a pure mathematician, a pure geometer. It is a matter of establishing a relation between two series of variable magnitudes. While one is at it one might as well try out an amusing hypothesis.

Descartes is a geometer, a pure mathematician. This, apparently, is the reason why he does not properly grasp Beeckman's 'principles' and gives an incorrect reply to his question. He sees the problem, and the phenomenon under investigation, in a quite different way from Beeckman.

He starts, as does Beeckman, from the completed fall. But, in contrast to Beeckman, he sees it, so to speak, statically. In other words the only aspect of the fall that he keeps in mind is its trajectory. Or, to put it yet another way, he instinctively eliminates time.

The line ADB, which for Beeckman represented the time elapsed,[79] for Descartes represents, as if naturally, the trajectory covered. So the problem is changed: a trajectory is covered with a 'uniformly variable' speed; therefore the problem is to determine the speed at each point on the path. The triangles ADE and ABC, which for Beeckman represented the space traversed (the path), for Descartes represent the body's motion, i.e., 'the sum of the speeds' acquired. Thus he concludes, quite reasonably, that the 'sum of the speeds' being triple, the space DB will be covered three times as fast. Time is reintroduced, but it is too late. Thorough-going geometrisation, spatialisation, the elimination of time—where it cannot be eliminated—the ignoring of the physical and causal aspect of the process, lead Descartes, as they had previously led Galileo and before him Benedetti and Michel Varro, to conceive of uniformly accelerated motion as motion in which the speed increases in proportion to the path covered, and not in proportion to the time elapsed.

Now, while we may be free to give whatever arbitrary definitions we like to our ideas, the lesson we have learned from Galileo is that in doing so we must try to grasp the essence of natural phenomena. In other words, we should not ignore causes nor forget time.

We have just said that Descartes did not fully grasp the 'principles' of Beeckman's physics. We could have gone further and said that he did not understand just what progress his friend had made.[80] It is true that Beeckman did not understand this very clearly himself. The relevant passage in the *Physico-mathematica*. which confirms our analysis of the sources of Descartes' error, seems to us to demonstrate his incomprehension quite clearly. Therefore we will give this text by Descartes in full.[81]

'In the proposed problem, in which it is imagined that at each instant[82] a new force is added to that with which the heavy body moves downwards, I say that this force increases in the same manner as do the transverse lines

DE, FG, HI, and the infinite other transverse lines that can be imagined between them. To demonstrate this I take as the first minimum or point of motion,[83] caused by the first attractive force of the earth that can be imagined, the square ALDE. For the second minimum of motion we have the double, namely DMGF: the force in the first minimum persists and a new, equal force is added to it. Thus in the third minimum of motion there will be three forces, namely those of the first, second and third time minima, and so on. This number is triangular, as I will perhaps explain more fully elsewhere, and it appears here to represent the figure of the triangle ABC. But, you will say, there are parts which protrude, ALE, EMG, GOI etc., which are outside the figure of the triangle. Therefore, the figure of the triangle cannot represent this progression. But I reply that these protuberant parts come from the fact that we have extended the minima which must be imagined as indivisible and as containing no parts. This is demonstrated as follows. I divide the minimum AD into two equal parts at Q; then ARSQ will be the [first] minimum of motion, and QTED the second minimum of motion, in which there will be two minima of force. Similarly we divide DF, FH, etc. Then we have the protuberant parts ARS, STE, etc. Clearly they are smaller than the protuberant part ALE.

Furthermore, if I take a smaller minimum such as Aα, then the protuberant parts will be yet smaller, such as Aαβγ, etc. If, finally, I take as this minimum the true minimum, i.e., the point, then these protuberant parts will be nothing, for they could not be the whole point clearly, but only a half of the minimum ALDE, and a half of a point is nothing.

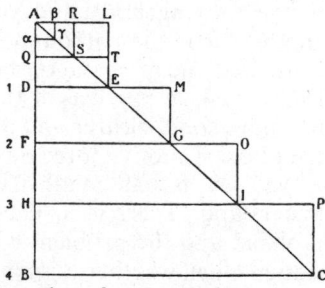

'From which it clearly follows that if we imagine, for example, a stone which is attracted by the earth, in a vacuum, from *a* to *b*, by a force which always remains equal to the first, persisting, force, then the first motion at *a* will be to the last at *b* as the point *a* is to the line *bc*. The part *gb*, which is half, will be covered by the stone three times as fast as the other half *ag*, because it will be drawn by the earth with three times the force. The space *fgbc* is three times the space *afg*, as is easily proved. And one can say this of the other parts proportionately.'

It is difficult to imagine a text which combines in the way this one does such supreme mathematical elegance[84] with the most hopeless physical confusion. Descartes has definitely not understood Beeckman's 'principles'. Thus he simply throws away Beeckman's intellectual prize, the principle of the conservation of *motion*. He substitutes

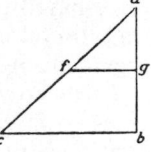

the conservation of force. He starts from the idea that *speed* is proportional to *force*[85] and from this concludes that a constant force produces a constant speed. Thus he is reverting to the traditional conception of *impetus* physics. He imagines that if the motion of the falling body accelerates this must be because it is attracted more strongly by the earth at the end of its motion than at the beginning, or to put it in his own terms, because the earth's attractive force produces a growing motive force in the stone. Thus he aggregates (the passage given here corresponding to the first of the hypotheses in the passage from the *Cogitationes Privatae* that was quoted above) the forces acting and not simply the speeds.[86]

One gets the impression that while Descartes accepts Beeckman's principle of the conservation of motion (as a hypothesis) he does not really trust it, that in trying to solve the problem of fall he prefers to do without the ideas developed by Beeckman, ideas which are clearly too novel for him, too unfamiliar, too difficult. In fact the idea of motion that is implicitly brought into play by Beeckman (it is the idea of motion of classical physics) is, so to speak, located at the border line between mathematics (geometry) and physics (time). It is very difficult to keep this idea of motion clearly in focus, and the fact that someone such as Descartes should find it so difficult to grasp, so difficult to keep to this precise point of contact between physics and the space of geometry, would itself be proof enough of this difficulty, even if there were no others. This is the reason why Descartes avoids this idea of motion—a paradoxical entity, for it is a state of the moving body and yet nevertheless is passed from one body to another, it is the embodiment of change and yet at the same time it is conserved and remains constant; it seems to him to be an illegitimate entity. Consciously as much as instinctively he replaces this idea with ideas which are more solid, clearer and more easily represented to himself as *images*[87], the ideas of motive force on the one hand and of trajectory on the other.

And yet his mathematical deduction is a brilliant success. This is easy to understand. There is, in fact, no *formal* difference between Beeckman's problem and the problem which Descartes substitutes for it. It does not matter much whether it is a problem of forces, distances or speeds; in each case what is involved is the same thing, namely the problem of calculating the rhythm of variation of a magnitude which increases uniformly in relation to *time*. So when Descartes considers the attractive force he inevitably thinks of a variation, or of something's being produced, in time. It is when he tries to express the results of his integration in terms of space that, carried away by the force of his imagistic representation and by his tendency to thorough-going geometrisation, he makes the mistake which, interestingly·enough, he could in principle have avoided in spite of his force-based physics.[88] He makes the mistake because he substitutes the trajectory for the motion, and makes the trajectory, and not time, the argument of his function.

This Cartesian translation or interpretation of Beeckman's ideas seems to us so fascinating and at the same time so revealing of the underlying tendencies of the human mind, of the difficulties that it had to overcome in order to arrive at that idea of motion which Descartes would announce, ten years later, to be so simple and clear that it neither needs nor can be given any definition, that we hope the reader will not begrudge it if we illustrate it with yet one more passage from the *Physico-mathematica*.[89]

'This problem can be solved in another, more difficult, way. Let us imagine the stone remaining at point *a*, the space between *a* and *b* being a vacuum. And that for the first time, for example today at nine o'clock, God creates at *b* a force which attracts the stone, and that at successive moments he creates ever new attractive forces, equal to that created at the first moment; and that combined with the previously created forces these pull the stone ever more powerfully, and even more powerfully given that in a vacuum a thing once set in motion moves for ever; and suppose that the stone, which was at *a*, arrives at *b* at ten o'clock. If we ask how long it takes to cover the first half of the path, i.e., *ag*, and how long the remainder, I reply that the stone descends through the line[90] *ag* in $\frac{7}{8}$ of an hour and through the line *gb* in $\frac{1}{8}$ of an hour. Thus we must make a pyramid on a triangular base and of height *ab*, and divide the whole pyramid in some way by horizontally equidistant transverse lines. The stone will pass through the lower parts of the line *ab* as much faster as these parts are contained in larger sections of the pyramid.'[91]

Descartes is right to consider that this way of setting up the problem is 'more difficult'. This time, in fact, he is adopting Beeckman's principle of the conservation of motion. But he supplements this principle with a constant increase of the attractive force (we can see why he relies on God's help here). The remarkable thing is that of all the possible cases that Descartes considers there is only one which he does not investigate; and this is, precisely, the case suggested to him by Beeckman.

How was it that Beeckman did not notice Descartes' error and did not claim for himself the fame of having found the correct solution? Of course we will never be able to give an exhaustive explanation for this. But this much is clear: Beeckman, looking for the solution to a problem in *physics*, posed in *physical* terms, and asking Descartes a specific mathematical question, naturally applies the answer he gets to his own problem. So that when Descartes speaks of space, Beeckman understands in terms of time.[92] Or more accurately, where Descartes moves surreptitiously from time to space, Beeckman avoids this move. Thus Beeckman makes in relation to Descartes (but in the opposite direction) the same mistake that Descartes makes in relation to Beeckman; the latter, therefore, manages in a way to re-establish the original situation. These are the main points of the explanation given by G. Milhaud,[93] and we must admit that we can see no other. We must accept the fact that Beeckman does not notice that Descartes' solution differs from that which he attributes to him. He does not notice that the physical principles of this solution are different from his own. So he ascribes to Descartes the solution which he writes down in his notes.

Does this not indicate that for Beeckman the problem was above all a mathematical one, and that he saw the merit of his young friend as having solved this mathematical problem, as having performed the integration?

One could go even further. If Beeckman does not see the difference between his solution (speed proportional to time elapsed) and that of

Descartes (speed proportional to distance covered) it is because for him it does not exist; it is because he believes these two solutions to be equivalent.[94]

No doubt the reader will find this extremely unlikely. However, it should not be forgotten that Beeckman, while he was certainly a good physicist, was a rather mediocre mathematician. And we shall see below that Descartes himself, even though he was a mathematician of genius, was never able either to recognise the mistake he had made, nor even, when he came across the correct formula in Galileo,[95] to recognise that it was different from the one that he had put forward earlier himself. From which we can see once again just how difficult it was to isolate and grasp those simple and clear ideas with which we are so familiar from classical physics and from Cartesian philosophy. Even for a Galileo. Even for a Descartes.

Ten years after his memorable encounter with Beeckman another opportunity arose for Descartes to take up the problem of falling bodies. This time it was his friend Mersenne who raised the question. Descartes' reply was different at every point from that which he had given to Beeckman,[96] with one exception: Descartes gave to his friend, just as he had ten years earlier, an incorrect formula; it was the very same formula as before, in which the speed of the moving body is a function not of the time elapsed but of the distance covered.

Descartes writes:[97] 'First I assume that motion once impressed in a body remains there always, unless it is removed from it by some other cause, i.e., that that which has once started to move in a vacuum moves for ever and with the same speed.[98] Assume, then, a body at A, pushed by its gravity towards C. I say that if, as soon as it has started to move, its gravity leaves it, it would nonetheless continue with the same motion until it arrived at C. But it would descend neither more nor less rapidly from A to B than from B to C. Now, since it is not like this, and since it keeps its gravity which pushes it downwards and which, at each instant adds a new force to the descent, it happens consequently that it covers the space BC much faster than AB, because in traversing it it retains all the *impetus* which moved it through the space AB and, in addition, it adds to this new [*impetus*], as a result of the gravity which pushes it over again at each new instant. As for the proportion in which this speed increases this is shown by the triangle ABCDE; for the first line denotes the strength of speed [*la force de vitesse*] impressed in the first moment [*moment*], the second the strength impressed in the second moment, the third the strength conferred in the third moment, and so on. Thus is formed the triangle ACD which represents the increase in the speed of the body in its descent from A to C, and ABE which represents the increase in speed in the first half of the space traversed by this body; and the trapezium BCDE which represents the increase in the speed in the second half of the space traversed by the body, namely BC. Since, as is clear, the trapezium BCDE is three times as large as the triangle ABE, it follows from this that the body will descend three times as fast from B to C as from A to B; i.e., if it descends from A to B in three moments it will descend from B to C in a single moment. That is to say, in four moments it will cover twice the distance covered in three and, consequently, in twelve moments twice as much as in nine, and in sixteen moments four times as much as in nine, and so on.'[99]

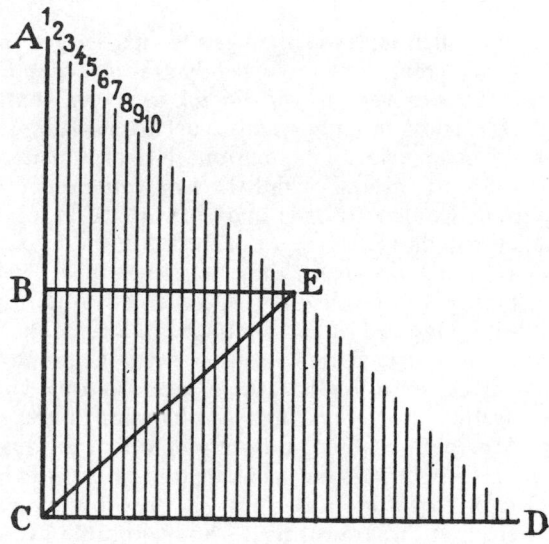

We have said above that the solution to the problem of fall sent by Descartes to Mersenne is quite different from those which he had worked out under Beeckman's influence. In fact the idea of attraction which had been so profitably employed by Beeckman had now disappeared completely. Descartes, in fact, reverts to the concept of *impetus,* and his interpretation of fall differs only slightly from those which had been given by Benedetti and Scaliger:[100] gravity, an essential quality of the body, which produces a new *impetus* at each instant, propels the body downwards; the acceleration comes from the fact that these *impetuses* are successively produced at each new instant—a conception of acceleration which amounts to a transposition into *impetus* terminology of the conception which had been developed in terms of attraction.[101] For each *impetus* produces a motion of *equal speed;* so it is only by the summation of new *impetuses* that the acceleration can be explained. It is true that Beeckman's principle—conservation of motion—is now asserted without qualification (though with no mention of Beeckman) but, and this is extremely odd, it is understood as the conservation of *impetus.*

The deduction of the formula for the motion of fall, for uniformly accelerated motion, is also different from the earlier deductions except, as we have already said, in that the final formula is the same. Descartes slides, just as he had earlier, from time to space, from physics to geometry.

In fact in as much as he thinks about the real, physical mechanism of acceleration, Descartes sees the *impetuses* springing up and being generated one after the other in successive moments of time. In contrast, as soon as he moves to a mathematical investigation of the *motion*, he immediately substitutes space for time, the distance covered for the time elapsed.

The diagram which serves as the basis for his deduction is certainly not very clear. It is different from the earlier diagrams except in one respect; the line AC, which goes *vertically from up to down,* represents the fall's trajectory. Descartes' thought yields, just as it had previously, to the temptation of geometrical imagination. His arguments seem to be as follows: at the first instant of fall the first *impetus* produces a motion which, by itself, would carry the body towards C at a given speed. This *impetus* acts throughout the length of the path; therefore it is represented by the line AC which stands for the whole trajectory. The second *impetus* produces a motion of an (absolute) speed equal to that produced by the first. But it is not involved right from the beginning of the motion; it seizes on the body, so to speak, at some distance from the point A; the third only becomes involved even further on,[102] and so on. Therefore, all the *impetuses* together are represented by the sum of the segments of the trajectory of the path covered, throughout which they each act.

Descartes has forgotten, so to speak, that the *impetuses* become involved *in succession*, or to put it another way, he sees this succession laid out in space along the motion's trajectory.[103] Not being able, even in 1629, to fully grasp the new idea of motion which is introduced by the principle of its conservation, he always divorces the causal conception of it from its mathematical analysis, the development of the fall through time from its geometrical representation.

Mersenne, and who can blame him, did not really understand Descartes' explanation. The latter, therefore, tried again. He wrote to Mersenne:[104] 'In your last letter you ask me why I say that the speed is impressed by the gravity as *one* in the first moment of fall, and as *two* in the second moment, etc. I reply, with all due respect, that this is not what I meant; rather, that the speed is impressed by the gravity as *one* in the first moment, and again as *one* in the second moment by the same gravity, etc. Now, *one* in the first moment, and *one* in the second, make *two,* and *one* in the third make *three*, and thus [the speed] increases in arithmetical proportion. Now, I consider that I have adequately proved that the gravity eternally accompanies the body it is in; and it cannot accompany the body without constantly propelling it downwards. Thus, if we assume, for example, that a piece of lead is falling downwards as a result of the force of the gravity, and that right at the first moment after the fall has started God takes away all gravity from the lead, so that from then on the piece of lead is no heavier than it would be if it were air or feathers, it would nevertheless continue to descend, especially in a vacuum, as it had started to descend; and one could give no reason why its speed should decrease rather than increase. Now if after a while God gives the lead back its gravity, but only for a single instant, after which He takes it away again, would not the force of the gravity propel the lead in this second moment just as much as it did in the first moment? And we could say the same for the other moments. From which it certainly follows that if you let a cannon-ball fall *in spatio plane vacuo* from a height of 50 feet, whatever it is made of it will always take exactly three times as long for the first 25 feet as for the last 25. But in the air things are quite different. . .'

This new explanation does not really add anything to what Descartes had said to Mersenne in his previous letter. Let us point out, once again, just how close is Descartes' way of conceiving things to that of the *impetus* theorists: gravity is a cause added to the moving body and propelling it downwards! This is pure Benedetti. [105] Note, moreover, that Descartes adds in the margin: 'It must be remembered that we have stated that a body, once set in motion, will move for ever in a vacuum, and I have just demonstrated this in my treatise'; finally, let us note that in this same letter, speaking of Beeckman, Descartes says: 'he states, *as I do*, [emphasis added] that once something has begun to move it will continue to move by its own force [*sua sponte*] unless it is stopped by some exterior force, and thus, in a vacuum, will move for ever. . . '

In the following years Descartes was to have many further opportunities to return to the problem of fall. However, he never again attempted to give its formula, he never again tried to establish its law. The reason for this is that around 1630 a very profound change occurred in Descartes' thought, a change so fundamental, so radical, that we can call it a revolution. Reflection on method, meditation on human thought and its relation to reality, concerns which found magnificent expression in the *Regulae ad directionem ingenii*, began to bear fruit. As a result Descartes' attempt to rebuild physics on new foundations, to rethink the order of the physical world, was from this point on based on 'the order of reason' and not on the order of material substances.

There is no need for us to stress the decisive importance of this intellectual revolution. [106] It is enough for us to point out that this inversion of order enabled Descartes to grasp, and to set out for us with unsurpassable clarity, the new concept of motion, the foundation of the new science, and to determine its structure and ontological nature. It enabled him to express with complete lucidity all that had remained implicit and had only been obscurely foreshadowed in the thought of those such as Beeckman and Galileo, and which we have had to disentangle and make explicit in the course of this study. Finally, it enabled him to formulate the principle of inertia. These are accomplishments which make Descartes the scientist the equal of Descartes the philosopher, in other words, of the highest rank.

What is remarkable, however, is that this very intellectual revolution was responsible for Descartes' losing touch with all the concrete achievements of the 'new science', of that mathematical physics which was developing right before his eyes, and to the creation of which he had himself so powerfully contributed.

It is well known that Descartes' physics, as it is set out for us in the *Principes* no longer contains mathematically expressible laws.[107] It is, in fact, no more mathematical than that of Aristotle. As for the problem of falling bodies, the *Principes* has nothing to say on the subject.

Was this a matter of chance, or was it inevitable? We believe this question to be of some importance.

The decision to proceed only from what is clear to what is clear, advancing methodically, and starting at the beginning, i.e., 'with the most clear and most distinct ideas', implies, as is well known, the complete mathematisation of nature—which in practice means its geometrisation.[108] It also implies the necessity for a development which is systematic and for a construction, or reconstruction, based on clear and distinct ideas, of all the ideas which are used or which are implicit in physics. Finally it implies giving up, once and for all, all 'obscure' ideas, ideas of which physics, even mathematical physics, had made rather abundant use.

Descartes' new convictions are expressed with perfect clarity in his letters to Mersenne. 'It is impossible to say anything good or solid concerning speed without having elucidated what gravity really is, together with the whole world system', he writes to him on 12th September 1638.[109] And in his famous critique of Galileo, in which Descartes grudgingly admits that Galileo is 'a philosopher quite out of the ordinary',[110] what Descartes holds against him above all is that he proceeded 'without method' and that he did not go to the limit in his analysis of the ideas he used;[111] thus he retained, and used just as they were, ideas (such as those of gravity and of the vacuum) which, so to speak, announced quite openly their origins in perception, instead of attempting to reconstruct them on the basis of the clear and distinct ideas of extension and motion, ideas of which the origin is entirely in the understanding.

In the autumn of 1631 Descartes wrote to Mersenne:[112] 'I do not retract any of what I have said concerning the speed of bodies falling in a vacuum: because if one assumes that there is a vacuum as everyone imagines then all the rest can be demonstrated; but I believe that one cannot assume a vacuum without error. I will try to explain *quid sit gravitas, levitas, durities*, etc., in the two chapters I have promised to send you at the end of this year; this is why I refrain from writing to you about these things now.' It is necessary to explain *quid sit gravitas, levitas, durities*, etc.; and the explanation must take as its starting point the idea of motion, *the most simple idea that we have*.[113]

A paradoxical statement: for had not the problem of motion been *the* problem of philosophy since Aristotle, if not before? Were not the shelves of philosophical libraries filled with heavy volumes called *De Motu*? Descartes is well aware that his statement is astonishing. He explains that what is involved is not at all the philosophers' motion. He is speaking of something entirely different. For 'the philosophers assume many motions which they think could take place without any body changing its place. . . For my part I know of none but that which is more easy to conceive than the lines of the geometers, which occurs when bodies pass from one place to another and successively occupy all the spaces between the two'.[114]

The philosophers were guilty of yet another misdeed. 'They attribute to the least of these motions an existence much more substantial and real than they attribute to rest, which they say is no more than privation. For my part I conceive that rest is just as much a quality, which must be attributed to matter as long as it remains in one place, as is motion, which is attributed to it when it is changing place.'[115]

From which it self-evidently follows that motion *is not a process*, but a *status*, and it is as such that it obeys in the new 'World' constructed in thought by Descartes the laws which applied to *states* in 'the old world'. Therefore the first of the *'rules' which God makes matter obey* is 'that each individual particle of matter remains always in the same *state*[116] as long as contact with others does not compel it to change it: that is to say . . . if it is halted at any place it will never leave this place as long as others do not expel it therefrom; and that once it begins to move it will continue to do so always and with the same force until others stop it or slow it down'.[117]

The philosophers were not unfamiliar with this conservation law. On the contrary, they accepted it for many things, including rest, but 'they considered motion an exception to it, and yet it is just this that I wish most expressly to be included under it. And', Descartes adds, 'do not conclude from this that I intend to contradict them: the motion which they speak of is so very different from that which I conceive that it can easily be the case that what is true of the one is not so of the other'.[118]

We know that Descartes was right. His motion-state, the motion of classical physics, has nothing in common with the motion-process of Aristotelian and Scholastic physics. This is the reason why it is of their essence that they obey completely different laws. While in Aristotle's well-ordered Cosmos motion-process *self-evidently* needs a cause to sustain it, in Descartes' world of extension motion-state *self-evidently* persists by itself, and continues indefinitely in a straight line into the infinity of the completely geometrised space which Cartesian philosophy has opened up before it.

As we have said there is no need to stress just how important and decisive was Descartes' work, and the fact that it accomplished with incomparable singleness of mind the destruction of the Cosmos, and laid down the framework for the new ontology. But we should now look at the other side of the coin.

Cartesian motion, this motion which is the easiest and clearest thing we can know, is not, Descartes tells us, the motion of the philosophers. But neither is it the motion of the physicists. Nor even that of physical bodies. It is a geometer's motion, the motion of geometrical entities—the motion of a point which traces out a straight line, the motion of a line which describes a circle. But motions such as these, in contrast to physical motions, have no speed and do not take place in time. Thorough-going geometrisation—the original sin of Cartesian thought—leads to the intemporal: space is retained but time is eliminated.[119] It dissolves the real entity into the geometrical. But reality has its revenge.

The law of falling bodies, as it had been formulated earlier by Descartes (leaving aside the fact that he made mistakes in formulating it) and Beeckman, and as it had in the meanwhile been formulated by Galileo, was without any doubt an 'abstract' law. It was a law which could not be realised as such in men's daily experience. It assumed the existence of the vacuum and strictly speaking only held in a vacuum since it abstracted from the resistance of the air. Furthermore it assumed, as was explicitly spelled out by Descartes, that the action of gravity does not vary. These assumptions could only be made in the absence of knowledge of the true nature of gravity. But Descartes now knew this: gravity is not at all a simple, basic quality of the body, and nor is it the consequence of the earth's attraction of the heavy body. It is the result of pressure, of the fact that the body is pushed towards the earth by a cloud of particles, by the subtle matter which swirls in a vortex around the earth.[120] So one can see how unreasonable it is to assume a vacuum. Not only is a vacuum impossible in itself; not only would admitting its possibility force us to accept the obscure and occult idea of action at a distance (attraction); but also, more concretely, the supposition of a vacuum would in no way make the fall of heavy bodies easier. Quite the contrary; it would make it impossible: 'It is certain' writes Descartes, 'that if the subtle matter which revolves around the earth did not in fact do so then no body would have weight. . .'.[121]

Now, in Descartes' earlier letters to Mersenne concerning falling bodies he not only 'assumed a vacuum, but also that the force moving the stone was constant, which is manifestly contrary to the laws of Nature: because all natural powers act to a greater or lesser degree depending on the extent to which the subject is disposed to receive their action; and it is certain that a stone's disposition to receive a new motion, or an increase of speed, is not the same when it is moving very fast and when it is moving very slowly'.[122] It follows that the acceleration is not uniform; therefore the very basis of the argument collapses.

One might find it surprising that Descartes appears here to misunderstand his own law of the relativity of motion, a law which he was to state *expressis verbis*.[123] One might also be surprised that he speaks of natural powers, because in Descartes's World, this World of reified Geometry, there is only one 'natural power': *Motion*. But for this power Descartes' proposition clearly holds true. For in Descartes' World there is only one manner of communication between substances: contact. And only one kind of action: impact. Now clearly the force of the impact of a body moving at a given speed on another body depends on the latter's own state of motion. Therefore the successive impacts on a falling body will be weaker and weaker and its speed, rather than increasing indefinitely, will approach a limit, namely the speed of the subtle matter itself. So this is how the acceleration of a heavy body in free fall is explained: 'The subtle matter pushes the descending body in the first moment and gives it a degree of speed; . . . this occurs in *rationem duplicatam* at the beginning of the body's descent. But this proportion is completely absent after it has fallen several *toises*, and then the speed no longer increases, or hardly increases, at all'.[124]

Now since the mechanism of fall is reduced to impact, it is clear that the nature, i.e., the physical constitution, of the body will have a crucial role in determining the fall. In the same way that bodies differ in the extent to which light can pass through them, similarly they resist to a greater or lesser extent the passage of particles of subtle matter; and this means that they are affected by the impacts of these particles to a greater or lesser degree. It follows that they fall at different speeds. Descartes in fact writes to Mersenne: 'As for the calculation made by Galileo of the speed of falling bodies which you ask about, it does not agree at all with my Philosophy, according to which two lead spheres, for example of one pound and of one hundred pounds, would not have the same ratio between them as two wooden spheres also of one pound and of one hundred pounds; nor even as two other lead spheres, one of two pounds and the other of two hundred pounds. These cases are not distinguished by him at all, and this makes me believe that he cannot have arrived at the truth'.[125]

No doubt. But what is this truth? How do bodies fall *in rerum natura*?

At first Descartes hoped to be able 'to determine now in what proportion increases the speed of a stone which falls not *in vacuo* but *in hoc vero aer*'.[126] But years go by, and Descartes sees that it is much more difficult than he had thought. He thinks, of course, that Galileo is wrong in believing that bodies all fall with the same speed; and that he is also wrong in believing in the mutual independence of motions. Things may well be like that in the abstract. But in reality . . . 'As for what he says about a cannon fired horizontally, I believe you would find some detectable difference if you were to perform the experiment accurately.'[127] And Descartes is right: the air resistance *supports* the body as it moves through it. But what is the correct solution? Descartes does not manage to give it and he writes sadly to Mersenne: 'I must ask you to excuse me if I do not reply to your question concerning the slowing down of a heavy body's motion by the air it is moving in; for this depends on so many things that I cannot give a clear account of it in a letter; all I can say is that neither Galileo nor anyone else could work out anything clear and demonstrative on this question unless they first knew what gravity is, and unless they have the true principles of physics'.[128] Certainly. But Descartes does have these 'true principles of physics', and he also knows what gravity is. So why then does he not give us the solution? Because it is too complicated. Because in a physics such as his own, a physics of the plenum and of the continuum, everything depends on everything else, everything acts *instantaneously* on everything else. One cannot isolate any phenomenon and as a result one cannot formulate simple laws in mathematical form.[129]

Phenomena cannot be isolated. Therefore one cannot have an 'abstract' physics like that of Galileo. Abstraction, which ignores the complexities of the real, concrete case is entirely legitimate in Galileo's world, an Archimedean world. It enables him to isolate out the simple case, the ideal case, and starting from this he can go on to explain the complex, concrete case. But for Descartes, only a physics of the concrete is possible. Galilean abstraction would not lead him to the simple case: it would lead him to a case which is *unthinkable*. In order to do something analogous to Galileo he would have had to investigate not the simple case but the normal case.[130] This, the investigation of the motion of a body in a perfect fluid, is infinitely beyond his mathematical means. Descartes puts this by saying that it is beyond the limits of human knowledge. Experimental investigation is equally impossible. For how could one measure the most important variable, the speed of the motion of the subtle matter?

So the extremely strange situation is that Descartes, who had not succeeded in deducing the correct law of fall because he had not grasped the new idea of motion that Beeckman suggested to him, and had not known how to bring together the physical (causal) investigation of the phenomenon of fall with his mathematical analysis, abandons the problem at that very moment when, having fully clarified the idea of motion, he succeeds in formulating the fundamental principle of modern science, the principle of inertia! The reason is that here, once again, he was unable to maintain a balance: in identifying matter with extension he substituted geometry for physics. Once again, thorough-going geometrisation and the elimination of time. And this is why the physics of clear ideas, a physics which was a return to Plato, ended in failure: failure analogous to Plato's own.[131]

3
GALILEO AGAIN

Let us now return to Galileo.

In a fragment which is included in the second volume of his *Opere*[132], a fragment which comes from the first draft of his 'new science' and which is, moreover, partly reproduced in the *Discorsi*, Galileo writes: 'The accidents of uniform motion have been discussed in the previous section. Now we must consider accelerated motion.

'First of all we should investigate and find an adequate explanation for the definition of that [accelerated motion] which is found in nature. For, although it is permissible to arbitrarily invent types of motion and to consider what properties they would have (as do those, for example, who have thought up various lines, conchoids and helices as described by various motions—even though these motions are not to be found in nature—and have very commendably studied their properties) nevertheless, the motions occurring in nature, and in particular that of falling bodies, are of a certain very specific type. Now we can investigate the properties of this type of motion as long as the definition we give to accelerated motion conforms with the essence of naturally accelerated motion. This, after repeated intellectual efforts, we are confident we have at last achieved. Our main confirmation of this is based on the principle that that which can be observed by the senses in nature must be in agreement with the properties which we can deduce [from our definition]. Finally, in the investigation of the definition of naturally accelerated motion we have been guided, as it were by the hand, by our insight into the character and properties of nature's other works, in which nature generally employs only the least elaborate, the simplest and easiest of means.

'For I do not believe that anybody would imagine that swimming or flying could be accomplished in a simpler or easier way than that which fish and birds actually use by natural instinct.

'Therefore, when I observe that a stone which falls starting from rest continually acquires additional increases in speed, why should I not believe that these increases take place in the simplest and most obvious manner? The moving body remains the same, as does also the principle of the motion. Why then would not everything else equally remain constant? So you will say: thus the speed remains the same [uniform]. Not at all. In fact what is constant is that the speed is not the same and that the motion is not uniform. Therefore we must look for the identity, or the uniformity and simplicity, not of the speed but of the increase in speed, i.e., of the acceleration. If we examine the matter carefully we find no addition or increment more simple than that which repeats itself always in the same manner. Now just what this manner is we will easily understand if we fix our attention on *the intimate relationship between time and motion*[133]. For just as uniformity and constancy of motion is defined by and conceived through equality of times and distances (thus we call a motion uniform when equal distances are traversed during equal time intervals) so also we may, in a similar manner, conceive of *equal increases of speed taking place during these same time intervals*;[134] thus we may grasp in our minds that a motion is uniformly and continuously accelerated when, during any *equal intervals of time* whatever, equal increments of speed are added to it. Thus if any equal intervals of time whatever have elapsed, counting from the time at which the moving body left its position of rest and began to descend, the degree of speed acquired during the first two time intervals will be double that acquired during the first time interval; also the degree of speed added during three of these time intervals will be treble; and that in four, quadruple that of the first time interval; in such a way that if a body were to continue its motion with the same degree of speed, or moment, which it had acquired during the first time interval and were to retain this same uniform speed, then its motion would be twice as slow as that which it would have if its degree of speed had been acquired during two time intervals.

'And thus, it seems, we shall not depart from right reason if we state that the intension of the speed[136] increases with the extension of the time.'[137]

Galileo's definition of uniformly accelerated motion explicitly postulates a *continuous* increase in speed and, in particular, its *continuous* increase starting from rest.[138] This implies, as Galileo puts it, that the body 'passes through *all* the degrees of speed and of slowness', i.e., that at the beginning of its motion it moves with infinite slowness. Galileo had accepted this idea even when he was still at Pisa, but it seemed, with good reason, odd and implausible to the best minds of the age.[139] For how could one accept a motion of infinite slowness? How could one conceive of a *continuous* transition from rest to motion, i.e., from nothing to something? Should one not, on the contrary, postulate that in physical reality motion has a minimum, and that this corresponds to the minimum of action?[140] Even Cavallieri was hesitant and requested an explanation.[141]

But Galileo was prepared for Cavallieri's question. He had already anticipated the objection in the fragment from which we quoted above:[142] 'If there is a continual addition of new speed from the first instant of motion starting from rest, and if this occurs according to the same proportions and following the same law as those whereby time, as it flows from the first instant, continually receives new additions, then there is reason to think that just after the first instant one cannot specify a time so short that there are no others and again others even shorter between it and the first instant, so similarly once the body is no longer at rest one cannot specify a degree of speed so small, or of slowness so great, that the moving body had not already before this possessed another even more slow; and since slowness can increase, or speed decrease, to infinity, we must accept that the moving body will have possessed at a given instant a moment of slowness so large that moving thus for years and years it would not travel an inch'. No doubt this seems astonishing, even absurd. However, 'even though this is astonishing at first sight it is nonetheless not false; experiment, hardly inferior to demonstrative argument, can show this to anybody'.

The experiment[143]—it is scarcely necessary to say that, as is almost always the case with Galileo, it is a matter of a thought-experiment—consists of imagining a weight dropped onto a stake driven into the earth. It is noted that the downwards motion of the stake depends on the speed with which the weight strikes it. From the fact that the weight has no, or almost no, effect when it falls from a very small height it can be concluded that its motion is almost infinitely slow.

This argument from experiment pleases Galileo greatly, and he repeats it in slightly modified form in the *Discourses* (from which we quote a long passage below); but he is well aware that it carries less weight than a demonstrative argument. Therefore he supplements his 'experiment' with the following considerations:[144] 'We must bear in mind that any given degree of speed can be acquired in greater or lesser amounts of time, and this as a result of various factors of which one, with which we are particularly concerned, is the length of the space through which the motion takes place. Actually heavy bodies not only tend to move vertically towards the centre of all heavy things, but they also move on planes inclined to the horizontal, and this the more slowly the smaller the inclination; the slowest, thus, on those with the minimum elevation, above the horizontal; and infinite slowness, i.e., rest,[145] occurs on the horizontal plane itself. Now the difference in the degrees of speed thus acquired is so great that the degree acquired by the heavy body falling vertically for one minute is only acquired on an inclined plane in an hour, a day, a month, a whole year, and this even though the heavy body descends with a continual acceleration'. That such phenomena or 'accidents' are non-contradictory and even have a very high probability can be elucidated 'by a geometrical example in which speeds are represented by lines and the continuous passage of time by the uniform motion of another line, which shows us that the degrees of speed are actually infinite in number'.

This strange argument clearly presupposes what it is supposed to prove. Moreover, it takes for granted that bodies falling from a given height always acquire the same degree of speed regardless of whether the path taken is vertical or on an inclined plane.[146]

The *Dialogue Concerning the Two Chief World Systems*, which is only a semi-scientific work,[147] skilfully evades the problem of continuity. But the *Discourses Concerning Two New Sciences* returns to the attack. At the beginning of the second section of the 'Third Day', a section concerned with the investigation of accelerated motion, Galileo, through his friend Sagredo, puts the following objection:[148] 'Although I can offer no rational objection to this or indeed to any other definition that any author whatsoever might wish to postulate, since all definitions are arbitrary, I believe that one might, without offence, raise the doubt whether such a definition as this, which is conceivable and acceptable *in abstracto*, conforms to or is confirmed by that kind of natural motion which we find in the case of heavy falling bodies. And since it is apparantly maintained that the natural motion of heavy bodies is as is given in the definition, I would like to clear my mind of certain difficulties in order that I may go on to apply myself more earnestly to the propositions and their demonstrations.'

Clearly what is at stake here is the legitimacy of mathematisation in physics. Sagredo is well aware that in pure geometry, or in pure kinematics, it is legitimate to speak of an infinite series of magnitudes, or fractions, interposed between zero and any particular value; and even that it would not be legitimate to do anything else. But what gives one the right to carry over these abstract considerations from the realm of mathematics to that of reality? Thus he continues:[149] 'I think of a heavy body falling from rest, that is, starting with a privation of all speed, and accelerating in proportion to the time from the beginning of the motion; such a motion as would, for instance, in eight beats of the pulse acquire eight degrees of speed; having at the end of the fourth beat acquired four degrees; at the end of the second, two; at the end of the first, one; and since time is divisible without limit, it follows from all these considerations that if the earlier speed of a body is less than its present speed in a constant ratio, then there is no degree of speed however small (or, one may say, no degree of slowness however great) with which we may not find this body travelling after starting from infinite slowness, i.e., from rest. So that if that degree of speed which it had at the end of the fourth beat was such that, if kept uniform, the body would traverse two miles in an hour, and if keeping the degree of speed which it had at the end of the second beat, it would traverse one mile an hour, we must infer that, as the instant of starting is more and more nearly approached, the body moves so slowly that, if it kept on moving at this rate, it would not traverse a mile in an hour, or in a day, or in a year or in a thousand years; indeed, it would not traverse a hand's width in an even greater time; this is something which *our imagination finds it very difficult to accept*,[150] all the more so since our senses show us that a heavy body falling immediately acquires a great speed'.

Sagredo invokes experience as his witness against the abstract reasoning of kinematics. Galileo replies to him by also invoking experience, or more accurately, by proposing *to perform an experiment*:[151] 'This is one of the difficulties which I also experienced at the beginning, but which I shortly afterwards removed; and the removal was effected by the very observation which creates the difficulty for you. You say that observation appears to show that immediately after a heavy body starts from rest it acquires a very considerable speed: and I say that the same experiment makes clear the fact that the initial *impetuses* of a falling body, no matter how heavy, are very slow and very weak. Place a heavy body upon a yielding material, and leave it there without any pressure except that of its own gravity; it is clear that if one lifts this body a cubit or two and allows it to fall upon the same material, it will, with the impact, exert a new and greater pressure than that caused by its weight alone; and this effect is brought about by [a combination of the weight of] the falling body and the speed which it has acquired during the fall,[152] an effect which will be greater and greater according to the height of the fall, that is according as the speed of the falling body is greater. From the quality and intensity of the impact we can thus accurately estimate the speed of a falling body. But is it not true that if a block be allowed to fall upon a stake from a height of four cubits and drive it into the earth, say, four finger-breadths, that coming from a height of two cubits will drive the stake a much smaller distance, and from the height of one cubit an even smaller distance; and finally if the block is lifted only one finger-breadth how much more will it accomplish than if merely laid on top of the stake with no impact? Certainly very little. And if it is lifted only the thickness of a leaf the effect will be altogether imperceptible. And since the effect of the impact depends upon the speed of the striking body, can anyone doubt that, when this effect is imperceptible, the speed must be even less than tiny and the motion even less than slow? Such is the force of truth, that the very same experiment which seemed at first glance to show one thing, when more carefully examined, assures us of the contrary'.

However, in Galileo's opinion a problem of this importance, the problem of the very foundations on which the science is to be built, cannot be resolved by recourse to experiment. The experiment supports or weakens an argument. It does not replace it. Thus he tells us:[153] 'But without depending upon the above experiment, which is certainly very convincing, it seems to me that it ought not to be difficult to establish this truth by reasoning alone. Imagine a heavy stone held in the air at rest; the support is removed and the stone set free; then since it is heavier than the air it begins to fall, and not with uniform motion but slowly at the beginning and with a continuously accelerated motion. Now given that speed can be increased and diminished without limit, what reason is there to believe that such a moving body starting with infinite slowness, that is from rest, immediately acquires a speed of ten degrees rather than one of four, or of two, or of one, or of a half, or of a hundredth; or, indeed, of any of the infinite number of even smaller values? Pray listen. I hardly think you will refuse to grant that the gain of degrees of speed of the stone when it is falling from rest occurs in the same manner as the diminution and loss of these same degrees of speed when, by some impelling force, the stone is thrown upwards to the same height: but in this case I do not see how you can doubt that the ascending stone, diminishing in speed, must before coming to rest pass through every possible degree of slowness.'

But the Aristotelian, Simplicio, does have an objection: 'But if the number of degrees of greater and greater slowness is infinite, they will never be all exhausted, and therefore such an ascending heavy body will never reach rest, but will continue to move without limit always at a slower rate;[154] but this is not what we observe in fact'.

As Galileo himself says, it is easy to see what it is that makes his position so difficult to grasp: it is that, in order to understand it, it is necessary to conceive the idea of passing through an infinite number of degrees of speed in a finite amount of time; and for this it is necessary to form the concept (and here *imagination* is not enough) of speed at an instant, i.e., in a manner of speaking, the concept of an immobile motion, a motion which, so to speak, apparently disowns its affinity with time:[155] in other words, the concept of the differential of motion. Thus he continues:[156] 'This would happen, Simplicio, if the moving body were to stay at each degree of speed for some time; but it merely passes through [each degree] without remaining at it for more than an instant; and since each *quantum* of time, however small, may be divided into an infinite number of instants, there is therefore a sufficiently large number of them to correspond to the infinite number of degrees of diminishing speed. That the heavy body, as it rises, does not remain for any interval of time at the same degree of speed is evident from the following: if, for a given time interval, the moving body had the same degree of speed at the last instant of this interval as at the first, then it would be able, with this degree of speed at the last instant, to rise through a height equal to that which it had risen through between the first and last instants. It would pass in the same way from the second to the third, and consequently it would continue its uniform motion forever'.

Leaving aside objections arising from the idea of infinitesimals, we can henceforth confidently accept the definition of uniformly accelerated motion:

I call uniformly or constantly accelerated motion that motion of which the moments of its speed [i moment delle sua velocità] *increase, starting from rest, with the increase of time starting from the first instant of the motion.*

It must be admitted that Galileo's account of things did not satisfy everybody. In particular it did not satisfy Descartes, who had at first accepted continuity (speed is a magnitude, and is not continuity the *proprium* of magnitudes?) but who began to raise doubts once he had substituted his concrete physics of motion in a plenum for the abstract physics of motion in a vacuum. So he wrote to Mersenne:[157] 'As for what Galileo says, that falling bodies pass through all degrees of speed, I do not at all believe that this is what usually happens, although it is not impossible that it happens sometimes. And I am not happy with the argument used by M.F.[158] to refute this, in that he says that *acquiritur celeritas, vel in primo instanti, vel in tempore aliquo determinato*; because neither of these is true. . .'

Descartes' hesitation is clear. And it is easy to understand the reason for this. On the one hand he is inclined, because of his mathematicism, to accept the continuity of acceleration, or at least its possibility; on the other hand, because of the intemporalism, or at least the instantaneism, of his physics, his tendency is to accept the possibility of discontinuous variation. For Descartes understands perfectly well that continuity means temporality, that this means the impossibility of a finite instantaneous action, and that Galileo's reasoning amounts basically to the reaffirmation of the intimate relationship between time and motion. Descartes, therefore, makes up his mind:[159]

'I have just looked over my notes on Galileo again: it is true that I did not say there that falling bodies do not pass through all the degrees of slowness; but I did say that this cannot be decided without knowing what gravity is, and this comes to the same thing. Your case of the inclined plane does prove that any speed is infinitely divisible, and I agree with this, but not that when a body begins to fall it passes through all of these divisions. And I do not believe that you think that when one hits a ball with a hammer the ball, as it begins to move, goes less rapidly than the hammer; nor that any body which is impelled by another fails to move from the first instant with a speed of the same proportions as that of the body which moves it.[160] Now my view is that gravity is just this, that terrestrial bodies really are impelled towards the centre of the earth by the subtle matter; you will see clearly what follows from this. But we must not think that because of this bodies move initially as fast as this subtle matter; because its pressure on them is oblique and they are much hindered by the air, the lightest bodies especially.'[161]

The motion of fall of heavy [*graves*] bodies is uniformly accelerated motion. But what is the cause of this motion? What is gravity [*la gravité*]? Descartes asserts that to know this is indispensable. But Galileo refuses to answer.[162] He does not even accept the question. No doubt he is privately convinced that Gilbert is right in saying that the force of gravity is something like magnetic attraction, and that the earth is a huge magnet.[163] But being convinced of something is not at all the same as having proved it. Neither Gilbert himself, nor Kepler, nor anybody else had yet provided a proof. For nobody (including Galileo, in spite of his prolonged study of the magnet and of magnetic force) was yet able to provide a rational (i.e., *mathematical*) theory of attraction and of magnetism. So they had to make do without one. Besides, it can be accepted that whatever this cause is, its action is continuous and that as a result the motion which is caused by it must be of a very specific kind. This is why, he says,[164] 'The present does not seem to be the proper time to investigate the cause of the acceleration of natural motion concerning which various opinions have been expressed by various philosophers, some explaining it by attraction to the centre, others by the fact that there is increasingly less of the medium to be passed through, and yet others by some kind of expansion of the surrounding medium which, when it reaches round behind the falling body pushes it and exerts a continuous pressure on it: all these fantasies, and others too, ought to be examined; but it is not really worthwhile. At present we need only . . . investigate and demonstrate certain of the properties of motion which is accelerated (whatever the cause of this acceleration may be) in such a way that the moments of its speed increase, after its departure from rest, in the simple proportion with which time increases, which is the same as saying that in equal times there are equal increments of speed. And if it turns out that the accidents of motion deduced from this are realised in the accelerated free fall of heavy bodies then we may conclude that the assumed definition does represent the motion of heavy bodies and that it is true that their acceleration increases with the increase in time and with the duration of the motion.'

The deduction of the 'accidents' of uniformly accelerated motion (i.e., of the relations between the time elapsed, the speed and the distance covered) was given by Galileo in two different versions: it is worth examining them both.

The demonstration in the *Dialogue* is based on the continuity of the acceleration and involves the ideas of 'instantaneous speed', of 'moment' and of 'the sum of the speeds' (this being identified with the distance covered).[165] In accelerated motion, says Galileo,[166] 'the increases in the motion are continuous . . . and changing from moment to moment . . . they are always infinite. Hence we may better exemplify our meaning by imagining a triangle, which shall be this one, ABC. Taking in the side AC any number of equal parts AD, DE, EF, and FG, and drawing through the points D, E, F, and G straight lines parallel to the base BC, I want you to imagine the sections marked along the side AC to be equal times. Then the parallels drawn through the points D, E, F, and G are to represent the degrees of speed, accelerated and increasing equally in equal times. Now A represents the state of rest from which the moving body, departing, has acquired in the time AD the velocity DH, and in the next period the speed will have increased from the degree DH to the degree EI, and will progressively become greater in the succeeding times, according to the growth of the lines FK, GL, etc. But since the acceleration is made continuously from moment to moment, and not discretely from one time to another, and the point A is assumed as the instant of minimum speed (that is, the state of rest and the first instant of the subsequent time AD), it is obvious that before the degree of speed DH was acquired in the time AD, infinite others of lesser and lesser degree have been passed through. These were achieved during the infinite instants that there are in the time DA corresponding to the infinite points on the line DA. Therefore to represent the infinite degrees of speed which come before the degree DH, there must be understood to be infinite lines, always shorter and shorter, drawn through the infinity of points of the line DA, parallel to DH. This infinity of lines is ultimately represented here by the surface of the triangle AHD. Thus we may understand that whatever space is traversed by the moving body with a motion which begins from rest and continues uniformly accelerating, it has consumed and made use of infinite degrees of increasing speed corresponding to the infinite lines which, starting from the point A, are understood as drawn parallel to the line HD and to IE, KF, LG and BC, the motion being continued as long as you please.

Now let us complete the parallelogram AMBC and extend to its side BM not only the parallels marked in the triangle, but the infinity of those which are assumed to be produced from all the points on the side AC. And just as BC was the maximum of all the infinitude in the triangle, representing the highest degree of speed acquired by the moving body in its accelerated motion, while the whole surface of the triangle was the sum total of all the speeds with which such a distance was traversed in the time AC, so the parallelogram becomes the total and aggregate of just as many degrees of speed but with each one of them equal to the maximum BC. This total of speeds is double that of the total of the increasing speeds in the triangle, just as the parallelogram is double the triangle. And therefore if the falling body makes use of the accelerated degrees of speed conforming to the triangle ABC and has passed over a certain space in a certain time, it is indeed reasonable and probable that by making use of the uniform velocities corresponding to the parallelogram it would pass with uniform motion during the same time through double the space which it passed with the accelerated motion.

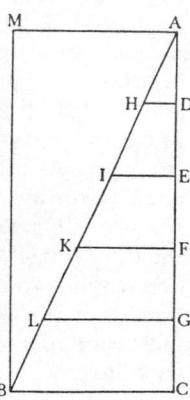

It is certainly somewhat surprising to observe Galileo calling the conclusion of his argument 'reasonable and probable'. But, we believe, the responses that he attributes to the other two participants in the discussion, Sagredo and Simplicio, elucidate for us the meaning of this expression. For Sagredo proclaims that Galileo's demonstration is a perfect mathematical proof. Simplicio, the Aristotelian, agrees with this; but, he says in effect,[167] 'mathematical rigour is out of place in the natural sciences'. And this is the major problem facing Galilean science (we will return to this problem later, dealing with it at the greater length which it calls for), the problem of the legitimacy of the mathematisation of reality.[168] For Simplicio—i.e., for Aristotle—is not altogether wrong. Reality is complex; it does not conform to the simple diagrams of geometry, or even of kinematics. Real bodies falling in real space are not the same thing as abstract bodies in a geometrical space. Galileo knows this perfectly well. And this is why, since it is a matter of *real* bodies, he says that it is 'probable' that they move according to the kinematic law that he has deduced.

In the *Discourses* the situation is not quite the same. The aim here, of course, even more than in the *Dialogue*, is the discovery of the real laws of the real world. But now there is an awareness that this investigation takes place in two stages: first the purely geometrical investigation of the 'abstract' or 'simple' case; then this is compared with the concrete case. The uniformly accelerated motion of which Galileo investigates the 'accidents' is not in the first instance thought of as a real motion of a real body on the earth. It is the motion of an 'abstract', Archimedean body, in a geometrical space. Therefore the argument is not merely probable; the conclusion is given to us as *demonstrated*. This is his argument:[169]

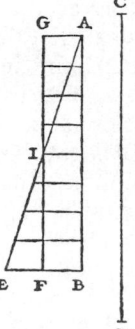

'*The time in which a given space is traversed by a body with a uniformly accelerating motion starting from rest is equal to the time in which this same space would be traversed by the same body moving at uniform speed and of which the degree of speed is the mean of the maximum and minimum degrees of the speed of the uniformly accelerated motion.*

'Let the extension AB represent the time in which the space CD is traversed by a moving body accelerated from rest at point C. And let the final and maximum degree of speed acquired in the instants of time AB be represented by EB, constructed on AB. Draw the line AE; all of the lines drawn from each of the points of line AB parallel to BE represent the increasing degrees of speed after the instant A. Then the line BE being bisected at F a parallelogram AGFB is formed of the parallel lines FG, BA and AG, BF. It is equal to the triangle AEB and the side GF divides AE at I into equal parts. If the parallel lines of triangle AEB are extended up to IG, the aggregate of all the parallel lines contained in the quadrilateral will be equal to the aggregate of those contained in triangle AEB; for those in the triangle IEF are equal to those contained in the triangle GIA, while those in the trapezium AIFB are common. Since each and every point on line AB corresponds to each and every instant of time AB and since the parallel lines drawn from these points and included in triangle AEB represent the increasing degrees of the accelerating ˌspeed, while the parallel lines contained within the rectangle represent in the same way just as many degrees of non-increasing, but uniform, speed, it is clear that there are assumed to be just as many moments of speed [*momenta velocitatis*] in the accelerated motion represented by the growing parallel lines of triangle AEB as there are in the uniform motion represented by the parallel lines of GB. For the deficiency of moments of speed in the first half of the accelerated motion (the deficient moments being represented by the parallel lines of triangle AGI) is compensated for by the moments represented by the parallel lines of triangle IEF. It is obvious therefore that equal spaces will be traversed in the same time by two moving bodies, one of which is moved with a motion uniformly accelerated from rest, while the other is moved with a uniform motion having a moment half of the moment of the maximum velocity of the accelerated motion. Q.E.D.'

It can be seen that this demonstration from the *Discourses* employs the same concepts and methods as that in the *Dialogue;* moment, instantaneous speed, sum or aggregate of moments or of speeds. However, it is more direct, more straightforward. The motion is no longer subdivided into fragments but is, as it were, considered as a whole. Thus the calculation of the space traversed no longer involves the idea of the potential motion, of the uniform motion which the body would have been able to achieve after the completion of its accelerated motion; the accelerated motion, or more accurately the aggregate of its speeds or moments, is now directly equated with that of a uniform motion with a speed of one half of the maximum speed attained in the accelerated motion. There may be some advantage to be gained from proceeding in this way. But this is by and large outweighed by the fact that, even more than in the demonstration in the *Dialogue*, Galileo's argument is applied to a completed and finished motion. No doubt the procedure as conceived is absolutely general and could be applied to any accelerated motion (as long as the acceleration is uniform) of any duration and distance. But any such motion is only grasped by this procedure as a completed motion and what is missing in Galileo's demonstration is precisely the 'intimate relationship between time and motion' and the preponderant role of time. This is why, in the

Discourses, a second theorem is added to this first one (the only one which had been proved in the *Dialogue*):[170]

'*If any moving body descends from rest with a uniformly accelerated motion, the spaces traversed in any times whatsoever by that body are related to each other in the double ratio of these times, i.e., as the squares of these times.*

'Let the flow [*fluxus*] of time beginning from some first instant A, be represented by extension AB, on which are taken any two time intervals AD and AE. Let HI be the line through which the moving body—beginning its motion at point H—descends with uniform acceleration. Let HL be the space traversed in the first time period AD, and let HM be the space through which it will have descended in time AE. I say that the space MH is to the space LH in a ratio which is the double [i.e., square] of the ratio of the time AE to the time AD; or we may say that the spaces MH and HL are in the same ratio as are the squares of EA and AD.

'Draw the line AC making any angle at all with the line AB, and from points D and E draw parallel lines DO and EP; of these lines DO represents the maximum degree of speed acquired in instant D of the time interval AD, and PE represents the maximum degree of speed acquired in instant E of time interval AE. Now it has been demonstrated above that, in regard to the spaces traversed, they are equal when one is traversed by a body moving

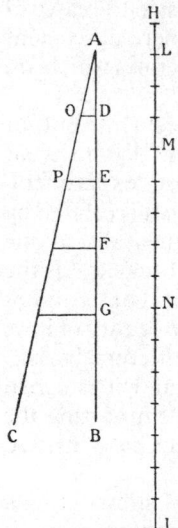

with unifrom acceleration from rest and the other is traversed in the same time by a body moving with a uniform motion whose speed is one-half the maximum speed acquired in the accelerated motion. It is therefore clear that the spaces MH and LH would be the same as those traversed in time intervals EA and DA by uniform motions whose speeds were one-half of PE and one-half of OD. Accordingly, if it is demonstrated that spaces MH and LH are in the duplicate ratio of the times EA and DA, our proposition will be proved. But in the fourth proposition of the first book it was demonstrated that the spaces traversed by two bodies moving with uniform motions are to each other as the product of the ratio of the speeds and the ratio of the times. *Now in this case the ratio of the speeds is the same as the ratio of the times* (for ½PE is to ½OD—or PE is to OD—as AE is to AD); therefore the ratio of the spaces traversed is as the square of the ratio of the times.'

The ratio of the spaces is the square of the ratio of the times. Here at last we have a formula which gives us directly the spaces traversed as a function of times elapsed and which, as it were, follows the motion without bringing it to a standstill: and which, moreover, will enable us to formulate a corollary which represents Galileo's greatest intellectual victory, since it submits motion, and therefore time, to the law of whole numbers.

'Hence it is clear that if we take any number of equal time intervals starting from the first instant or the beginning of the motion, for example AD, DE, EF, and FG, in which the spaces HL, LM, MN, and NI are traversed, these spaces will be related to each other as the odd numbers beginning with unity, i.e., 1, 3, 5, 7; for this is the ratio of the differences of the squares of the lines—which lines are uniformly increasing by an amount equivalent to the smallest of these lines [i.e., to the line representing the speed at the end of the first time interval]; or we may say [that this is the ratio of the differences] of the squares of the numbers beginning with unity. While, therefore, during equal time intervals the degrees of speed increase as the natural numbers, the increments in the distances traversed in these same times are to one another as the series of odd numbers beginning with unity.'[171]

So the deduction of the 'accidents' of-uniformly accelerated motion is accomplished. But, the question remains, is it true that 'this acceleration is that which one meets in nature in the case of falling bodies?'. It is possible to doubt this. And the Aristotelian who raises these doubts asks that at least he be shown 'one of those experiments—and there are, I understand, several of them—which in various cases are in agreement with the demonstrated conclusions.'[172]

This request, thinks Galileo, 'is a very reasonable one; for this is the custom—and properly so—in those sciences where mathematical demonstrations are applied to natural phenomena, as is seen in the cases of perspective, astronomy, mechanics, music, and others, where agreement with experiment is demanded as confirmation of those principles which are to become the foundations of the whole edifice'.[173]

Galileo and the Aristotelian seem to be in complete agreement. But, in fact, the same words are being used to mean fundamentally different things. Aristotelian empiricism demands 'observation' or 'experiment' which can function as the basis and foundation of theory; what is offered by Galilean epistemology, which is both *a priorist* and experimentalist at one and the same time (one could even say that it is the latter because it is the former), are experiments which are designed on a theoretical basis and of which the function is to confirm or refute the application to reality of laws deduced from principles which themselves have a quite different basis.

So the experiment which Galileo proceeds to relate—and it is a real experiment this time—would be completely incapable of supporting the body of classical physics, though many historians of science have insisted on loading this weight upon it.

Galileo's experiment is beautifully conceived; the idea of substituting a body rolling down an inclined plane for a body in free fall is truly a mark of genius. But, we are obliged to note, its execution is not of the same order as its conception.

It is as follows:[174] 'A piece of wooden moulding or scantling, about 12 cubits long, half a cubit wide, and three finger-breadths thick, was taken; on its edge was cut a channel a little more than one finger in breadth; having made this groove very straight, smooth, and polished, and having lined it with parchment, also as smooth and polished as possible, we rolled along it a hard, smooth, and very round bronze ball. Having placed this board in a sloping position, by lifting one end some one or two cubits above the other, we rolled the ball, as I was just saying, along the channel, noting, in a manner presently to be described, the time required to make the descent. We repeated this experiment many times in order to measure the time with an accuracy such that the deviation between two observations never exceeded one-tenth of a pulse-beat. Having performed this operation and having assured ourselves of its reliability, we now rolled the ball only one-quarter the length of the channel; and having measured the time of its descent, we found it precisely one-half of the former'. It is fortunate that Galileo tells us that 'in such experiments, repeated a full hundred times, we always found that the spaces traversed were to each other as the squares of the times, and this was true for all inclinations of the plane, i.e., of the channel, along which we rolled the ball'; and that 'we also observed that the times of descent, for various inclinations of the plane, bore to one another precisely that ratio which had been predicted and demonstrated for them'. It is very fortunate; because otherwise nobody could have believed that there could be such exact agreement between the experiment and the predictions! In fact, in spite of Galileo's assertion, one is tempted to doubt this. And this for the simple reason that a *strict* agreement like this is *strictly impossible*. Perhaps in this case it can be explained by the manner in which Galileo measured time intervals:[175] 'For the measurement of time we employed a large vessel of water placed in an elevated position; to the bottom of this vessel was soldered a pipe of small diameter giving a thin jet of water, which we collected in a small glass during the time of each descent, whether for the whole length of the channel or for part of its length; the water thus collected was weighed, after each descent, on a very accurate balance; the differences and ratios of these weights gave us the differences and ratios of the times, and this with such accuracy that although the operation was repeated many, many times, there was no appreciable discrepancy in the results'.

It is very easy to understand Descartes, who 'denied'[176] all of Galileo's experiments! How right he was! Because all of Galileo's experiments, at least all of the real ones, those which resulted in measurements, in precise values, were falsified by his contemporaries.[177]

In spite of this it is Galileo who is in the right. For, as we have seen above, he was not at all looking to found his theory on facts gained in the realm of experience: he knew perfectly well that this is impossible. And he was also aware that concrete observation—even experiment—performed, for example, in the air and not in a vacuum, or on a smooth plane and not on a geometrical plane, and so on, *cannot* produce results as predicted by the analysis of the abstract case. Therefore he does not expect this of them. The abstract case is an assumption. Experiment can confirm that it is a good assumption. It can do this within its limited means; or rather, within the limits of our means. For, as has recently been said: 'What is the point of trying to measure to five decimal places, when even the second has no significance?'[178]

CONCLUSION

We can see, then, that during the course of its development, Galileo's thought remained true to itself. It is the same in the *Dialogue* and the *Discourses* as it was in the letter to Paulo Sarpi with which we began this study. In the latter as in the former it is, one might say, regressive, 'resolvent', analytical in the deepest sense of this term. From phenomena, from experimental observations, from the 'symptoms' of accelerated motion, Galileo ascends—or descends perhaps—to its essential definition. In both the letter to Sarpi and the later works he looks for the principle, i.e., the essence, of this motion; and this, translated into a definition, will enable him to deduce and to demonstrate its 'accidents' and 'symptoms'. A comparison of his two attempts at this deduction—that which fails and that which succeeds—illuminated by the analysis which we have performed of texts by Descartes, enables us to understand the reasons for the failure and for the success.

It is obvious that Galileo's thought, or his mental attitude, differs from that of Descartes. It is not purely mathematical; it is *physico-mathematical.* Galileo does not put forward hypotheses about possible forms of accelerated motion: he looks for the real form, the form to be found in nature. Galileo, unlike Descartes, does not start from a causal mechanism so as then to translate it into purely geometrical relations: nor even so as to substitute such a relation for it. He starts from the idea—preconceived no doubt, but it constitutes the basis of his philosophy of nature—that the laws of nature are mathematical laws. *Reality is the incarnation of mathematics.* For Galileo, therefore, there is no gap between theory and observation; the theory, its formulae, are not applied to the phenomena from the outside; they do not 'save the phenomena', but express their essence. Nature replies to questions posed in mathematical language because nature is the domain of measure and order. And if experiment thus shows the way to reasoning, or as it were 'takes it by the hand', this is because in a properly performed experiment, i.e., to a well-formulated question, nature reveals its fundamental essence which, however, only the intellect is able to grasp.

Galileo tells us to start from experience. But this 'experience' is not the raw experience of the senses. The 'facts' with which the definition he is looking for must be in agreement, or with which it must be compatible, are precisely those two descriptive laws of fall (laws of the 'symptoms') which he already knows.

Galileo also tells us that he is guided by the idea of simplicity. Not merely formal simplicity, but something different; something analogous, certainly, but still different—a real simplicity, one could say, an internal conformity with the essential nature of the phenomenon under investigation.

This real phenomenon is motion. Galileo does not know how it is produced, nor how, under the influence of what force, acceleration is produced. He cannot, any more than could Descartes, make profitable use of Gilbert's work; he cannot employ an obscure idea (that of attraction) which he does not know how to mathematise. Nevertheless, it is a real phenomenon, a phenomenon really produced by nature, and this means that it is something produced *in time*.[179]

This intuition, this constant, sustained attention to the *real* character of the phenomenon, is what enabled Galileo to avoid Descartes's error, and indeed his own. Motion is, above all else, a temporal phenomenon. It takes place *in time*. Therefore Galileo tries to define the essence of accelerated motion as a function of time and no longer as a function of the distance covered. Distance is only a consequence, an accident or symptom, of an essentially temporal reality.

It is true that time cannot be *imagined*. Every graphic representation of it runs the risk of slipping into thorough-going geometrisation. But with sustained effort the intellect, thought, *conceiving* and *understanding* the continuous character of time, can symbolise it spatially without risk. Uniformly accelerated motion, therefore, will be that which is accelerated in relation to *time*.

Thus the idea of time has the role in Galileo's thought that the idea of real causality had in the thought of Beeckman and Descartes. But it was, precisely, the fact that he was able, or knew how, to do without any concrete representation of the way in which motion and acceleration are produced (force, attraction, etc.) that enabled him to keep his balance on this knife-edged interface called motion where the real and the mathematical coincide.

Galileo succeeded where Descartes failed. He knew how to grasp and how to retain in and by thought the paradoxical idea of motion, and to make of it the foundation of his reasoning. Descartes, at least in his early work, did not manage to achieve this.

Should we, then, call him to account for this? Or is there not, in Descartes' resistance, a symptom of something important and profound? We are quite willing to accept this. The classical idea of motion (that which Descartes himself later took up, and which enabled him to formulate the principle of inertia, thereby, so to speak, getting his revenge on Galileo) is not as clear and distinct as has been claimed, as Descartes himself claimed. A change which is also a state, the same and yet other—these concepts can only be brought to coincide, as they had been earlier by Plato's demiurge, by 'force'.

NOTES
THE LAW OF FALLING BODIES:
DESCARTES AND GALILEO

Introduction

1. In a letter to Paolo Sarpi, October 16 1604; see *Opere*, vol. X, p. 115.
2. See Descartes and Beeckman, *Varia* in *Oeuvres de Descartes*, ed. C. Adam and P. Tannery, vol. X, p. 58 ff; *Physico-mathematica, ibid.*, p. 75 ff. Long passages from these texts are given below.
3. Beeckman found the method of infinitesimals and the idea of a continuous variation unacceptable, and he tried to arrive at Descartes' results using a finitist conception and calculation. Cf. *ibid.*, p. 61 ff.
4. See E. Mach, *Mechanik*, Leipzig, 1921, p. 125; P. Duhem, *Etudes sur Léonard de Vinci*, vol. III, *Les précurseurs parisiens de Galilée*, Paris, 1913, p. 566 ff.
5. See Duhem, « De l'accélération produite par une force constante », *Congrès International de Philosophie*, 2nd session, Geneva, 1905, p. 859: « Aristotle formulated the law that *a constant force produces a uniform motion of which the speed is proportional to the force which produces it*. For nearly two thousand years this law dominated mechanics. Today we assert a different law: *a constant force gives rise to a uniformly accelerated motion and the acceleration of this motion is proportional to the force acting on the moving body*. This law is the basis of modern dynamics. » Duhem's way of putting this seems to us to be inaccurate. Aristotle did not have the modern idea of force and he therefore spoke (as did the Scholastics) of *cause* and not of *force*. These are not the same thing.
6. The complete statement of the law of falling bodies in fact contains *two*, distinct propositions: (a) the speed of a falling body increases in direct proportion to the time elapsed, and (b) the acceleration of fall is the same for all bodies. The merit of having formulated the second of these propositions is sometimes attributed to Benedetti, but this is incorrect (cf. above, Part I, pp. 25, 31), for Benedetti only accepted this proposition as true for bodies of different weights but of 'the same nature'. As for bodies with different 'natures', he thought that they fall with speeds proportional to their specific weights. It was Galileo who (using an argument analogous to Benedetti's) first established that heavy bodies all fall with the same speed regardless of their weights and their 'natures'. Cf. *Discorsi e dimostrazioni matematiche intorno a due nuove scienze*, 'Giornata prima', *Opere*, vol. VIII, p. 128 ff.
7. See Duhem, *Etudes sur Léonard de Vinci*, vol. III, p. 570.
8. See above, Part 1, pp. 1, 3.

Galileo

9. *Galileo a Paolo Sarpi in Venezia*, Padova, 16 ottobre 1604, *Opere*, vol. X, p. 115: « Ripensando circa le cose del moto, nelle quali, per dimostrare li accidenti da me osservati, mi mancava principio totalmente indubitabile da poter porlo per assioma, mi son ridotto ad una proposizione la quale ha molto del naturale et dell'evidente; et questa

supposta, dimostro poi il resto, cioè gli spazzii passati dal moto naturale esser in proporzione doppia dei tempi, et per conseguenza gli spazii passati in tempi eguali esser come i numeri impari *ab unitate*, et le altre cose. Et il principio è questo: che il mobile naturale vadia crescendo di velocità con quella proportione che si discosta dal principio del suo moto; come, v. g., cadendo il grave dal termine *a* per la linea *abcd*, suppongo che il grado di velocità che ha in *c* al grado di velocità che hebbe in *b* esser come la distanza *ca* alla distanca *ba*, et così conseguentemente in *d* haver grado di velocità maggiore che in *c* secondo che la distanza *da* è maggiore della *ca*. » [English translation is based on that which can be found in Marshall Clagett, *The Science of Mechanics in the Middle Ages*, Madison, 1961, p. 580.]

[10] On the history, or the prehistory, of the law of falling bodies, see P. Duhem, *Etudes sur Léonard de Vinci*, vol. III, *Les Précurseurs Parisiens de Galilée*, Paris, 1913; E. J. Dijksterhuis, *Val en Worp*, Groningen, 1924; E. Borchert, *Die Lehre von der Bewegung bei N. Oresme*, Beiträge zur Geschichte der Philosophie und Theologie des Mittelalters, Bd. XXXI, fasc. 3, Münster, 1934.

[11] The 'positivist' interpretation of Galilean epistemology was developed above all by E. Mach (see his *Mechanik*, chap. 8, p. 122 ff.). It is just as false as the analogous interpretation of Newton's epistemology.

[12] Galileo *knows* that he has no knowledge of this cause. He knows that he does not know what gravity is, or at least that he cannot make use of his hypotheses and his beliefs on the matter; cf. above, p. 101.

[13] In 1600 Gilbert, soon to be followed by Kepler, identified gravity with attraction. And there is no doubt at all that Galileo shared this belief (see below, Part 3, p. 187 ff.). But Gilbert's attraction was a soul, and Kepler's, although no longer a soul, was still a force *directed* towards the object, i.e., something even more mysterious. On Gilbert's physics see E. A. Burtt, *The Metaphysical Foundations of Modern Physical Science*, London, 1925; on Kepler see E. Cassirer, *Das Erkenntnisproblem in der Philosophie und Wissenschaft der neueren Zeit*, vol. I, Berlin, 1911, p. 328 ff.

[14] It could of course be argued that Newton himself does not explain it and that his attraction is just as mysterious as the tendency of like things to come together, which Plato and 'the ancients' spoke of and from which it in fact derives. This is certainly true. There have always been attempts to explain it (see E. Meyerson, *Identité et Réalité*, chapter 3. Paris, 1926). But Newtonian attraction is a non-directed force appropriate to geometrical space; and this is enough.

[15] The correct formula for the law '*the speed of the moving body is proportional to the distance covered*' would be an exponential function. See P. Tannery, *Mémoires scientifiques*, vol. VI, p. 441 ff.

[16] See E. Wohlwill, 'Die Entdeckung des Baharrungsgesetzes', *Zeitschrift für Völkerpsychologie und Sprachwissenschaft*, vol. XIV and XV.

[17] Duhem, 'De l'accélération produite par une force constante'. *Congrès International de Philosophie*, 2nd session, Geneva, 1905; *Etudes sur Léonard de Vinci*, vol. III, *Les Précurseurs Parisiens de Galilée*, 1913. To the texts cited by Wohlwill and Duhem should be added that of Michel Varro: cf. M. Varronis Genevensis I.C. et cos. ord. *De Motu tractatus*, Genevae, Ex officina Jacobi Stoer, MDLXXXIV. p. 12 sq.: « Vis . . . naturalis, qua resquaelibet ad locum suum naturalem tendit, subjectum suum, motu continue et ordinatim crescente, movet. Illius autem motus causa est quod facilius id moveatur, quod in motu est, quam quod quiescit. Vis igitur eadem, subjectum quod iam in motu est premens, illud magis movebit, quam si quiescat, et magis motum, magis etiam movebit : ita ut eadem vis motione major fiat, quam per se sit. Et haec est causa cur ictus, quo magis ab altero venit, eo vehementior sit. Motus autem huius spatia hanc celeritatis proportionem servant, ut quae est ratio totius spatij, per quod fit ille motus ad partem ispsius (utriusque initio inde sumpto, ubi est motus initium), eadem sit celeritas ad celeritatem. Exempli gratia, si vis aliqua per lineam ABE moverit, sitque AB illius lineae pars, quae erit ratio AE ad AB, eadem erit celeritas motus in puncto E ad celeritatem motus in puncto B.

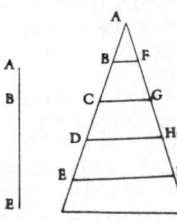

Cujusmodi proportio observatur in parallelis triangulum secantibus. Ut enim se habet AC ad AB, sic CG ad BF, et ut AD ad AC, sic DH ad CG. Itaque si in spatia aliquot dividatur totius motus spatium, finis secundi duplo citius ferretur quam finis primi : finis vero tertij citius quam finis primi et sic deinceps.

[18] See G. B. Benedetti, *Diversarum speculationum mathematicarum et physicarum liber*, Taurini, MDLXXXV, « *Disputationes de quibusdam placitis Aristotelis*, Cap. XXIV, p. 184 :« Aristot. 8. cap. primi libri de coelo, dicere non deberet quod quanto propius accedit corpus ad terminum ad quem, tanto magis sit velox, sed potius, quod quanto longius distat a termino à quo, tanto velocius existit. Quia tanto major sit semper impressio, quanto magis movetur naturaliter corpus, et continuo novum impetum recipit, cum in se motus causam contineat, quae est inclinatio ad locum suum eundi, extra quem per vim consistit. Neque etiam recte scripsit Aristo. 9. cap. lib. 8 physicorum et. 2. lib. primi de coelo esse aliquem motum ex recto et circulari mixtum, quod omnino impossible est. » Cf Duhem, *De l'accélération*, etc., p. 885 et Wohlwill, *op. cit.*, vol. XV, p. 394.

[19] Nicolo Tartaglia, *La Nuova scientia inventa da Nicolo Tartaglia*, Book I, proposition 1 (cited in P. Duhem, *op. cit.*, p. 875).

[20] His reproach is, in fact, quite unfair, for in his *Physics* Aristotle does not at all ignore the starting point: cf. *Physics*, VIII, 9, 265 b.

[21] G. B. Benedetti, *op. cit.*, p. 184: « causam moventem, id est propensionem eundi ad locum ei a natura assignatum ».

[22] G. B. Benedetti, *op. cit.*, p. 184: « tanto major sit semper impressio, quanto magis movetur naturaliter corpus, et continuo novum impetum recipit, cum in se motus causam contineat, quae est inclinatio ad locum suum eundi, extra quem per vim consistit. »

[23] See above, Part 1, note 147.

[24] On *impetus* physics see the works by Duhem and Wohlwill already cited and also E. J. Dijksterhuis, *Val en Worp,* Groningen, 1924; R. Marcolongo, « La meccanica di Leonardo da Vinci », *Atti della Reale Accademia delle Scienze fisiche e matematiche di Napoli*, vol. XIX, 1932; E. Borchert, *Die Lehre von der Bewegung bei Nicolaus Oresme* (Beiträge zur Geschichte der Philosophie und Theologie des Mittelalters, Band XXXI, fasc. 3), Münster, 1934.

[25] To equal distances covered, *whatever the starting point,* there will therefore always correspond equal accelerations; this is a conception which we find again in Galileo, and which led him to accept the mistaken belief that the value of g is always and everywhere the same. Cf. below, Part 3, p. 199.

[26] The idea of *impetus*, which was developed in order to explain violent motion, allows, or even implies, the partial elimination of teleology: for it makes it possible to think of motion as produced by an internal cause which is no longer determined by a goal.

[27] See above, Part 1, p. 14 ff.

[28] See the works cited above, notes 16, 17, 24.

[29] Alexandri Piccolominei, *In mechanicas questiones Aristotelis paraphrasis paulo quidem plenior,* a Nicolaum Ardinghellum Cardinalem amplissimum. Excussum Romae, apud Antonium Bladium Asulanum, MDXLVII cap. XXXVIII, quaestio trigesimatertia (Duhem, « De l'accélération, etc. », p. 882 ff.): 'It should be noted that there are two kinds of gravity or heaviness; one which has its source in the very nature of the body, the other a superficial one, which the Greeks called ἐπιπόλαιαν. This latter is nothing else than a certain temporary *impetus* which is either acquired in the thing itself when it is moved by its own tendency [*qui vel acquiritur in re ipsa ex suo nutu mota*] or is impressed by a motor which moves it by violence. For when a stone moves downwards it becomes ever faster because it continually, as a result of the motion, acquires a greater gravity (I am speaking now of superficial gravity) . . . Similarly, when a stone is thrown violently, it receives a certain heaviness or superficial lightness impressed by the projector. This is simply an adventitious *impetus* which moves the stone violently and which makes it able to move by itself until this *impetus* weakens and disappears. This heaviness or superficial lightness cannot become persistent or perfect, for the substantial form of the body on which it acts, namely the heaviness or lightness which is natural to the body, is opposed to its being impressed perfectly and deeply . . . The impelling force comes to an end either as the result of the resistance of some object which repels the moving body, or by the moving body's

own tendency, by an effort which results from its own nature and which becomes more powerful than the heaviness or superficial lightness . . . As soon as the real heaviness overcomes, by the power of its effort, the *impetus* impressed on the stone by the motor, it ceases to move violently and, by its own motion, it tends downwards.' Cf. *Etudes sur Léonard de Vinci*, vol. III, p. 197.

30 Julii Cearii Scaligeri, *Exotericarum exercitationum liber XV, De Subtilitate ad Hieronimium Cardanum*, Lutetiae apud Vascosanum MDLVII, cited by Duhem, 'De l'accélération etc.', p. 884. Exerc. LXXVII. *Quamobrem mota rota facilius movetur postea*: 'Heavy bodies, a stone for example, have nothing in them which incites their being set in motion; on the contrary, they are completely opposed to this. A stone which is set in motion on a horizontal plane does not move with a natural motion . . . Why, then, is the stone moved more easily after the motion has started? Because, in accordance with what we said above on the subject of projectile motion, the stone has already received the impression of the motion. After a first part of the motion there follows a second; but the first still remains. Such that even though a single motor is acting, the motions which it impresses in this succession are multiple. For the first impulsion is retained by the second, and the second by the third.' Cf. *Etudes sur Léonard de Vinci*, vol. III, p. 201.

31 See above, p. 19.

32 *Les Manuscrits de Léonard de Vinci*, edited by Ch. Ravaisson-Mollien, MS of the Bibliothèque de l'Institut, fol. 44, *verso*. Paris, 1890, cited by Duhem, 'De l'accélération etc.', p. 870 ff.: 'The heavy body, in falling, acquires at each degree of time a degree of motion more than in the previous degree of time, and similarly a degree of speed more than in the previous degree of time. Thus for each doubled quantity of time the length of the descent is doubled, just as is the speed of the motion'; *ibid.*, fol. 45, *recto*: 'The heavy body in free fall acquires a degree of motion with each degree of time, and a degree of speed with each degree of motion. Let us say that at the first degree of time the heavy body acquires a degree of motion and a degree of speed; in the second degree of time it will acquire two degrees of motion and two degrees of speed, and so on, as is said above'. Cf. *Etudes sur Léonard de Vinci*, vol. III, p. 514 ff. On Leonardo's physics see R. Marcolongo, 'La meccanica di Leonardo da Vinci', *Atti della Reale Accademia delle scienze fisiche e matematische di Napoli*, vol. XIX, 1932.

33 P. Duhem, 'De l'accélération, etc.', p. 872.

34 Space is rational—at least in form—whereas time is dialectical. Cf. E. Meyerson, *Identité et Réalité*, Paris, 1926, 3rd edition, p. 27 ff., 276 ff, 280 ff; *De l'explication dans les sciences*, Paris, 1921, vol. I, p. 151 ff., 261 ff., vol. II, p. 204 ff., 377 ff., 380.

35 This is where the error lies. For it is perfectly correct to say that the speed depends on the height, and even that it depends only on the height: this is the postulate of Galilean dynamics. Cf. below, Part 3, p. 182 ff.

36 Galileo is perfectly well aware of this. Thus when, in the *Discourses*, he puts forward his definition of uniformly accelerated motion—uniformly accelerated in relation to time— he attributes to Sagredo the following objection (*Discorsi, Opere*, vol. VIII, p. 203): '*Sagredo*: As far as I understand it at the minute it seems to me that it would be much clearer if we could define it, without changing the basic idea, as follows: uniformly accelerated motion is that in which the speed increases as the distance covered increases. In such a way, for example, that the speed acquired by a moving body in falling four cubits would be double that acquired in falling two cubits, and this latter would be double that which the body would have after the first cubit. Because there is no doubt that a heavy body falling from a height of six cubits would have a force of impact double what it would have if it fell three cubits, triple what it would have if it fell two, and sextuple what it would have if it fell one.

Salviati: It is comforting to have such a companion in error. And I can tell you that your argument is so plausible and has such high probability that our Author himself, when I put this proposition to him, did not deny that he also had for some time held this mistaken belief. But then what really amazed me was to see demonstrated in a few simple words not only the falsity but the impossibility . . . of two propositions which are so extremely plausible that when I have put them forward to many people for their opinion I have never come across anyone who did not agree with them.' Cf. below, note 51, for Galileo's counter-objection.

[37]   Historians of scientific thought in general, and of Galileo in particular, have rarely given adequate consideration to his deep and conscious Platonism. Even those who have noted it (E. Strauss, cf. his introduction to his translation of the *Dialogo, Dialoge über die beiden hauptsächlichsten Weltsysteme*, Leipzig, 1891, p. xlix; E. Cassirer, *Das Erkenntnisproblem in der Philosophie und Wissenschaft der neueren Zeit*, vol. I, Berlin, 1911, p. 389; E. A. Burtt, *The Metaphysical Foundations of Modern Physical Science*, London, 1925, p. 71; L. Olschki, *Galilei und seine Zeit*, Halle, 1927, pp. 164-174) seem to us to have underestimated both the real significance of this Platonism and the extent to which Galileo's adoption of it was conscious. Cf. below, Part 3, note 123 and p. 201 ff.

[38]   It could be objected that a formula for this 'weakening'—analogous to Fourrier's formulae—is perfectly conceivable and could have found a place in a physics such as Kepler's. This is quite true, but only on condition, precisely, that *impetus*, instead of being simply some lasting effect of *elan*, of muscular effort, is taken to be a *magnitude*.

[39]   It is interesting to note that even in theories which involve a mutual attraction between bodies (Kepler's and Newton's theories), the reciprocal relation is subdivided into, or replaced by, *two* unilateral relations.

[40]   See above, Part 1, p. 33. Galileo (*De Motu, Opere*, vol. I, p. 321; cf. Duhem, 'De l'accélération etc.', p. 892) says that he read Alexander's account of Hipparchus' theory only after he had himself developed his own idea. This may be so. All the same, an exposition of it was given by Bonamico; see above, p. 16.

[41]   It is possible that Galileo, even though he does not say so, did in fact discover the contradictory character of Hipparchus' theory.

[42]   See above, Part 1, p. 33 ff.

[43]   The word is still used by Newton.

[44]   See above Part 1, note 155.

[45]   Rest and motion thereby become *states*, and have an equal ontological rank. Now, for Aristotle and the Scholastics rest is only a privation, whereas motion is a process. Consequently rest persists without a cause (a privation has no need of a cause for it to persist), whereas motion exists only as the effect of the cause which sustains it. Therefore the principle *cessante causa cessat effectus* is applicable to motion. See above, p.91.

[46]   [On *momento* see the note by Stillman Drake (eds. Drake and Drabkin, *Galileo On Motion and On Mechanics*, p. 144): '. . .momento is a technical term of Galileo's for which no single modern equivalent exists. Galileo did not adopt the traditional mediaeval concept of "positional weight", but utilized the word *momento* to combine the notions of weight or force and the effective distance at which this acted. At the same time he recognized that other factors, especially velocity, could enter into the effective action of a weight, and thus he came to apply this word also to the product of weight and velocity. Hence the word *moment* here means sometimes 'static moment' and sometimes "momentum"'. Trans.]

[47]   Pseudo-Aristotle, *Quaestiones Mechanicae*, II, 24.

[48]   The persistence of the terminology (Newton still speaks of *impetus*) misled Duhem: he did not notice the radical transformation which the idea underwent with Galileo. This failure to understand explains, but does not justify, judgements such as the following, which is false through and through (Duhem, 'De l'accélération etc.', p. 888): '. . .at the risk of upsetting received opinion and of going against the legend, we must assert the following propositions: the opinions held by Galileo on the subject of dynamics were deeply marked by Peripatetic principles; they differed very little from doctrines accepted by a large number of 16th century physicists; they were notably less advanced than the insights of several of his predecessors'. Similar judgements, equally without foundation, can be found in *Les Origines de la Statique*, vol. I, Paris, 1905, p. 260 ff., and *Etudes sur Léonard de Vinci*, vol. III, p. 560 ff.

[49]   Galileo, *Frammenti attenenti ai Discorsi, Opere*, vol. VIII, p. 373: « Io suppongo (e forse potrò dimostrarlo) che il grave cadente naturalmente vada continuamente accrescendo la sua velocità secondo che accresce la distanza dal termine onde si partì: come, v. g., partendosi il grave dal punto *a* e cadendo per la linea *ab*, suppongo che il grado di velocità nel punto *d* sia tanto maggiore che il grado di velocità in *c*, quanto la distanza *da* è

maggiore della *ca*, e così il grado di velocità in *e* esser al grado di velocità in *d* come *ea* a *da*, e così in ogni punto della linea *ab* trovarsi con gradi di velocità proporzionali alle distanze de i medesimi punti dal termine *a*. Questo principio mi par molto naturale, e che risponda a tutte le esperienze che veggiamo negli strumenti e machine che operano percotendo, dove il percuziente fa tanto maggiore effetto, quanto da più grande altezza casca : e supposto questo principio, dimostrerò il resto.

Faccia la linea *ak* qualunque angolo con la *af*, e per li punti *c, d, e, f* siano tirate le parallele *cg, dh, ei, fk :* e perchè le linee *fk, ei, dh,cg* sono tra di loro come le *fa, ea, da, ca*, adunque le velocità ne i punti *f, e, d, c* sono come le linee *fk, ei, dh,cg*. Vanno dunque continuatamente crescendo i gradi di velocità in tutti i punti della linea *af* secondo l'incremento delle parallele tirate da tutti i medesimi punti. In oltre, perchè la velocità con la quale il mobile è venuto da *a* in *d* è composta di tutti i gradi di velocità auti in tutti i punti della linea *ad*, e la velocità con che ha passata la linea *ac* è composta di tutti i gradi di velocità che ha auti in tutti i punti della linea *ac*, adunque la velocità con che ha passata la linea *ad*, alla velocità con che ha passata la linea *ac*, ha quella proporzione che hanno tutte le linee parallele tirate da tutti i punti della linea *ad* sino alla *ah*, a tutte le parallele tirate da tutti i punti della linea *ac* sino alla *ag ;* e questa proporzione è quella che ha il triangolo *adh* al triangolo *acg*, ciò è il □° *ad* al □° *ac*. Adunque la velocità con che si è passata la linea *ad*, alla velocità con che si è passata la linea *ac*, ha doppia proporzione di quella che ha *da* a *ca*. E perchè la velocità alla velocità ha contraria proporzione di quella che ha il temp al tempo (imperò che il medesimo è crescere la velocità che sciemare il tempo), adunque il tempo del moto in *ad* al tempo del moto in *ac* ha subduplicata proporzione di quella che ha la distanza *ad* alla distanza *ac*. Le distanze dunque dal principio del moto sono come i quadrati de i tempi, e, dividendo, gli spazii passati in tempi eguali sono come i numeri impari *ab unitate:* che risponde a quello che ho sempre detto e con esperienze osservato; e così tutti i veri si rispondono.

E se queste cose son vere, io dimostro che la velocità nel moto violento va decrescendo con la medesima proporzione con la quale, nella medesima linea retta, cresce nel moto naturale. Imperò che sia il principio del moto violento il punto *b*, ed il fine il termine *a*. E perchè il proietto non passa il termine *a*, adunque l'impeto che ha auto in *b* fu tanto, quanto poteva cacciarlo sino al termine *a;* e l'impeto che il medesimo proietto ha in *f* è tanto, quanto può cacciarlo al medesimo termine *a;* e sendo il medesimo proietto in *e, d, c,* si trova congiunto con impeti potenti a spingerlo al medesimo termine *a*, nè più nè meno: adunque l'impeto va giustamente calando secondo che sciema la distanza del mobile dal termine *a*. Ma secondo la medesima proporzione delle distanze dal termine *a* va crescendo la velocità quando il medesimo grave caderà dal punto *a*, come di sopra si è supposto e confrontato con le altre prime nostre osservazioni e dimostrazioni: adunque è manifesto quello che volevamo provare. »

Cf. *ibid.*, p. 380 and p. 383: « Assumo, eam esse cadentis mobilis per lineam *al* accelerationem, ut pro ratione spacii peracti crescat velocitas ita, ut velocitas in *c* ad velocitatem in *b* sit ut spacium *ca* ad spacium *ba*, etc.

Cum autem haec ita se habeant, ponatur *ax* cum *al* angulum continens, sumptisque partibus *ab, bc, cd, de* etc. aequalibus, protrahantur *bm, cn, do, ep* etc. Si itaque cadentis per *al* velocitates in *b, c, d, e* locis se habent ut distantiae *ab, ac, ad, ae* etc., ergo se quoque habebunt ut lineae *bm, cn, do, ep*.

Quia vero velocitas augetur consequenter in omnibus punctis lineae *ae*, et non tantum in adnotatis *b, c, d,* ergo velocitates illae omnes sese respicient ut lineae quae ab omnibus dictis punctis lineae *ae* ipsis *bm, cn, do* aequidistanter producuntur. Istae autem infinitae sunt, et constituunt triangulum *aep:* ergo velocitates in omnibus punctis lineae *ab* ad velocitates in omnibus punctis lineae *ac* ita se habent ut triangulus *abm* ad triangulum *acn*, et sic de reliquis, hoc est in duplicata proportione linearum *ab, ac*.

Quia, vero pro ratione incrementi accelerationis tempora quibus motus ipsi fiunt debent imminui, ergo tempus quo mobile permeat *ab* ad tempus quo permeat *ac* erit ut *ab* linea ad eam quae inter *ab, ac* media proportionalis existit. »

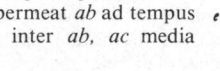

50 Cf. Duhem, *Etudes sur Léonard de Vinci*, vol. III, p. 570 ff.

[51]   Galileo's argument, which Duhem (*op. cit.*, p. 578) and Caverni (*Storia del metodo sperimentale in Italia*, vol. IV, Bologna, 1895, p. 295) both find conclusive, is as follows (*Discorsi, Opere*, vol. VIII, p. 204): 'When the speed is proportional to the spaces traversed, or to be traversed, then these spaces are traversed in equal times. For if the speed with which the body traverses a space of four cubits is double the speed with which it has traversed the first two (the former space being double the latter) then the times of these motions will be equal. But one and the same body cannot traverse the four cubits and the two cubits in the same period of time, unless the motion is instantaneous; but we have seen that a falling body's motion takes place in time, and that it passes two cubits in less time than four. Therefore it is false that the speed increases in proportion to the space'. The argument contains a similar error to that which we found in the argument discussed above: Galileo applies to motion of which the increase of speed is proportional to distance covered a calculation which is only applicable to uniformly (in relation to time) accelerated motion. See E. Mach, *Mechanik*, Leipzig, 1921, p. 245, and P. Tannery, *Mémoires Scientifiques*, vol. VI, p. 400 ff.

[52]   A correct deduction would have led Galileo to the same formulae that Descartes arrived at; see above, p.86.

[53]   The formula would actually be an exponential function.

[54]   Galileo's expression; see above, p.96.

[55]   Cf. P. Duhem, 'De l'accélération etc.', p. 907.

[56]   See below, Part 3, note 249.

[57]   However, Galileo never adopted Gilbert's *theories*, and he never attempted to use Gilbert's idea of attraction in working out a theory of fall. This is easily understood; Gilbert's physics was animist, and nobody before Newton, not even Kepler, was able to give mathematical form to the idea of attraction.

Descartes

[58]   Descartes was not one to hide his light under a bushel: cf. *Journal de Beeckman*, in *Oeuvres de Descartes*, eds. C. Adam and P. Tannery, vol. X, p. 331: « Is dicebat mihi se in arithmeticis et geometricis nihil amplius optare: id est se tantum in iis his novem annis profecisse quantum humanum ingenium capere possit ».

[59]   See especially P. Duhem, *Etudes sur Léonard de Vinci*, vol. III, *Les Précurseurs parisiens de Galilée*, Paris, 1913, p. 566 ff., and G. Milhaud, *Descartes savant*, Paris, 1920, p. 25 ff. Cf. also J. Sirven, *Les années d'apprentissage de Descartes*, Paris, 1928.

[60]   Gilbert's book (Guilielmi Gilberti Colchestrensis *De Magnete* . . . Londini, MDC; English translation by P. Fleury Mottelay, *De Magnete*, New York, 1958) in which the earth is taken to be a magnet and fall is explained in terms of terrestrial attraction, created a great stir and was of the first importance in the development and transformation of the concepts of physics. It was highly praised by Galileo, and Kepler, Gassendi and Newton were all influenced by it. Gilbert's conception of attraction (a mysterious force, rather like a soul) was, of course, contrary to the spirit of the new science and was, consequently, of no use to Galileo and Descartes. But it was precisely on this point that Gassendi and Newton concentrated their efforts in an attempt to transform Gilbert's idea of attraction into a force not directed towards its object.

[61]   *Journal de Beeckman*, 1613, *Oeuvres de Descartes*, vol. X, p. 60, note f: « Mota semel nunquam quiescunt, nisi impediantur. Omnis res semel mota nunquam quiescit, nisi propter externum impedimentum. Quoque impedimentum est imbecillius, eo diutius mota movetur : si enim aliquid in altum projiciatur simulque circulariter moveatur, ad sensum non quiescet ante reditum in terram ; et si quiescat tandem id non fit propter impedimentum aequabile, sed propter impedimentum inaequabile, quia alia atque alia pars aeris vicissim rem motam tangit. » However, we should not confuse the law of conservation of motion with the law of inertia. This mistake is still made too often; it is made, for example, by Duhem (cf. 'De l'accélération etc.', p. 904) and was made earlier by Wohlwill. The law of inertia asserts the persistence of motion *in a straight line;* the law of conservation of motion has no such implication. Thus Beeckman believed in the persistence of circular motion and he explained the persistence of the circular motion of the planets by reference to that of a candelabra suspended by a wire (this being easily

observed), for he thought that the conservation law was universally applicable (*Oeuvres de Descartes*, vol. X, p. 225): « eo modo quo in recto motu valeat hoc theorema : *quod semel movetur semper eo modo movetur dum ab extrinseco impediatur.* In vacuo vero nulla tales consideratio habenda ; magnum enim corpus, parvum, grave, leve, magna aut parva superficie, hac sive illa figura, etc. semper eo modo quo semel motum est, pergit moveri, his accidentibus nihil impedimenti afferentibus. Praeterea cum candelabra eo modo moventur quo dico annuum motum terrae fieri, si abscisso fune fieri posse, ut candelabra in aere elevata manerent neque deciderent, sed ut astra in caelo, sic haec in aere vagarentur, nulla ratio videtur esse cur non pergerent circulariter moveri, usque dum saepius aeri occursando impedita ». Beeckman's case was not unique: Hobbes also believed in the persistence of circular motion. Therefore the merit of having been the first to have clearly conceived and stated the law of inertia belongs not to Beeckman, nor even to Galileo, but to Descartes alone.

This book was already being printed when the publication by Cornélis de Waard of new fragments of Beeckman's *Journal* (*Correspondance du Père Marin Mersenne*, vol. II, Paris, 1936, pp. 118 ff., 123 ff., 235 ff., 280 ff., etc.) appreciably altered the image that people have had, in as much as they have had any image at all, of the Dutch physicist, and makes us very strongly regret that his invaluable diary has remained unpublished. For we can now appreciate that Beeckman fully deserved the title of *vir ingeniosissimus* that Descartes bestowed upon him; and more importantly, he will henceforth be considered a link of the greatest importance in the history of the development of scientific ideas: and finally, we can now see that his influence on Descartes was far deeper than has hitherto been supposed. In particular, some of the laws of motion and of impact stated by Descartes in his *Principes* are modelled on laws formulated by Beeckman (cf. *Correspondance du Père Mersenne*, vol. II, p. 633 ff.); moreover, being very well-read in the contemporary scientific literature Beeckman certainly drew Descartes' attention to books which he had not read.

Beeckman belonged to that tradition of thought that could be called the Bruno-Gilbert tradition. Like Bruno he asserted that the Universe is infinite and that there is an infinite number of fixed stars: he was also like Bruno in defending the idea of the void, which he identified with the ether and subtle matter. Like Gilbert and Kepler he saw the source and agent of attraction in this ether. Well before Descartes and Pascal he gave an explanation of the rise of liquids in closed tubes in terms of atmospheric pressure. Of particular importance in relation to our present concerns in this book is the fact that, before Descartes, he stated the principle of the conservation of motion, rejected the idea of *impetus*, and gave a correct solution to the problem of projectiles. Thus in 1620 he declared that *Motus a Deo semel creatus non minus quam corporeitas ipsa in aeternum conservatur*; and although, not knowing how to explain within this perspective the undeniable fact that in the impact of non-rigid bodies these bodies come to a halt (so that motion is lost), he wrote (*ibid.,* vol. II, p. 123): « His its positis, nunquam motus in vacuo potest intelligi ad celeriorem motum vergere, sed omnia tandem spectare ad quietem propter aequales occursus. Unde sequitur Deum opt. max. solum potuisse motum conservare movendo semel maxima corpora minima celeritate, quae deinceps reliqua ad quietem semper spectantia perpetuo resuscitant et vivificant », in 1629 he stated that this loss of motion is only apparent and that the motion is in fact conserved and is distributed to the parts and the atoms of which the bodies are composed (cf. *ibid.*, p. 259 ff.).

From 1614 he opposed the idea of *impetus* (*ibid.,* p. 236): « Lapis, projectus in vacuo, perpetuo movetur ; obstat autem ei aer, qui novus semper ei occurrit atque ita efficit ut motus ejus minuatur. Quod vero philosophi dicunt vim lapidi imprimi, absque ratione videtur ; quis nempe posse concipere, quid sit illa aut quomodo lapidem in motu contineat, quave in parte lapidis sedem figat ? Facillime autem mente quis concipiat in vacuo motum nunquam quiescere, quia nulla causa mutans motum, occurrit ; nihil enim mutatur absque aliqua causa mutationis », and in 1618, in one of the corollaries to the thesis which he had presented at Caen, he stated (*ibid.*, p. 237): « Lapis e manu emissus pergit moveri non propter vim aliquam ipsi accedentem, nec ob fugam vacui, sed quia non potest non perseverare in eo motu, quo in ipsa manu existens movebatur ».

Beeckman's achievements were clearly enormous. Nevertheless they should not be exaggerated and we should not attribute to him the discovery of the principle of inertia, as does his distinguished editor (cf. *ibid.*, pp. 122, 236, 272). For when Cornélis de Waard says (p. 236), concerning Beeckman's principle that *Omnis res semel mota, nunquam quiescit nisi propter externum impedimentum: quoque impedimentum est imbecillius, eo diutius mota movetur,* 'in the first of these notes he wrongly applies this only to celestial and circular motion, but soon afterwards (July 1613—April 1614) he extended it to cover rectilinear motion', he makes exactly the same mistake as Beeckman himself, when he says (p. 360): « *Id quod semel movetur in vacuo, semper movetur, sive secundum lineam rectam seu circularem tam super centro suo, qualis est motus diurnus Terrae et annuus* » without noticing (and this can certainly not be held against him) that the conservation of circular motion and that of rectilinear motion are *strictly incompatible.*

[62]   Not until at least thirty years later, in Gassendi's *De motu impresso a motore translato* (Paris, 1643), do we find a conception of the mechanism of fall as clear as Beeckman's. We are emphasising Beeckman's achievements because they seem to us to have been somewhat underestimated.

[63]   This, incidentally, refutes Duhem's view that there was a widespread diffusion of Oresme's formula or law in the sixteenth and seventeenth centuries (see Duhem, *Etudes sur Léonard de Vinci*, vol. III, p. 580 ff. and *passim*). Our impression, on the contrary, is that it was practically unknown.

[64]   See Descartes and Beeckman, *Physico-mathematica, Oeuvres de Descartes*, vol. X, p. 75 ff.

[65]   It is interesting to observe that Beeckman's conception, which was after all quite natural for a Copernican, and which represented a considerable advance over that of Benedetti and the young Galileo, was on the other hand similar to the traditional conception of fall as motion directed *towards an end.*

[66]   As we have seen, and will see again later, Galileo did not think of it like this. His starting point was always Benedetti's conception of an Archimedean space, and he investigated the motion of fall as *a specific type of motion.* He did not start from the concrete case.

[67]   See *Journal de Beeckman* in *Oeuvres de Descartes*, vol. X, p. 58: « Lapis cadens in vacuo cur semper celerius cadat : Moventur res deorsum ad centram terrae, vacuo intermedio spatio existente, hoc pacto ; Primo momento, tantum spacium conficit, quantum per terrae tractionem fieri potest. Secundo, in hoc motu perseverando superadditur motus novus tractionis, ita ut duplex spacium secundo momento peragretur. Tertio momento, duplex spacium perseverat, cui superadditur ex tractione terrae tertium, ut uno momento triplum spacii primi peragretur ».

[68]   This is a passage of great importance because it demonstrates quite clearly just how different are the ideas of attraction and of tendency: attraction acts *externally,* it *draws* the body towards the earth. The motion of fall is therefore—*horribile dictu*—a violent motion. No doubt Kepler was to make the situation less shocking by making the attraction mutual: but Descartes was to definitively identify the natural motion of fall with the violent motion produced by impact. See above, p. 92 ff.

[69]   'The double space is maintained'—*duplex spatium perseverat*—the double speed, i.e., that which makes the body cover a double space in one moment, is maintained.

[70]   *Journal de Beeckman, Oeuvres de Descartes*, vol. X, p. 58: « Lapis cadentis tempus supputatum : Cum autem momenta haec sint individua, habebit spacium per quod res una hora cadit ADE. Spatium per quod duabus horis cadit, duplicat proportionem temporis, id est ADE ad ACB, quae est duplicata proportio AD ad AC. Sit enim momentum spatij per quod res una hora cadit alicujus magnitudinis, videlicet ADEF. Duabus horis perficiet talia tria momenta, scilicet AFEGBHCD. Sed AFED constat ex ADE cum AFE ; atque AFEGBHCD constat ex ACB cum AFE et EGB id est cum duplo AFE.

Sic si momentum sit AIRS, erit proportio spatii ad spatium, ut ADE cum *klmn*, ad ACB cum *klmnopqt*, id est etiam duplum *klmn*. Ast *klmn* est multo minus quam AFE. Cum igitur proportio spatii peragrati ad spatium peragratum constet ex proportione trianguli ad triangulum, adjectis utrique termino aequalibus, cumque haec aequalia adjecta semper eo minora fiant quo momenta spatii minora sunt : sequitur haec adjecta nullius quantitatis fore quando momentum nullius quantitatis statuitur. Tale autem momentum est spatii per quod res cadit. Restat igitur spatium per quod res cadit una hora se habere ad spatium per quod cadit duabus horis, ut triangulum ADE ad triangulum ACB.

Haec ita demonstravit M. Perron, cum ei ansam praebuissem, rogando an possit quis scire quantum spatium res cadendo conficeret unica hora, cum scitur quantum conficiat duabus horis, secundum mea fundamenta, viz. *quod semel movetur, semper movetur, in vacuo* et supponendo inter terram et lapidem cadentem esse vacuum. Si igitur experientia compertum sit, lapidem cecidisse duabus horis per mille pedes, continebit triangulum ABC 1000 pèdes. Hujus radix est 100 pro linea AC quae respondit horis duabus. Bisecata ea in D, respondet AD uni horae. Ut igitur se habet proportio AC ad AD duplicata, id est 4 ad 1, sic 1000 ad 250, id est ACB ad ADE. »

71 Notice that Descartes, like Galileo, represents the *distance covered* by the falling body not by a line but by a *surface*. The reason for this is that neither Galileo nor Descartes consider, in the first instance, the distance covered: they consider the motion accomplished. The indivisible 'moment' that Descartes speaks of is not an 'instant'; it is exactly the same thing as Galileo's 'degree of speed'; it is an instantaneous motion or speed, the minimum or, one might say, the differential, of the motion. Therefore, since it is motion, it necessarily has *two dimensions*. So the figures (the triangle or rectangle) represent the sum of the infinite 'moments' or 'degrees of speed'. Duhem, it seems to us, did not understand this.

72 See P. Duhem, *Etudes sur Léonard de Vinci*, vol. III, p. 570 and G. Milhaud, *Descartes savant*, p. 27.

73 *Cogitationes Privatae*, in *Oeuvres de Descartes*, vol. X, p. 219 ff. « Contigit mihi ante paucos dies familiaritate uti ingeniosissimi viri, qui talem mihi quaestionem proposuit : *Lapis*, aiebat, *descendit ab A ad B una hora ; attrahitur autem a terra perpetuo eadem vit, nec quid deperdit ab illa celeritate quae illi impressa est priori attractione. Quod enim in vacuo movetur semper moveri* existimabat. Queritur *quo tempore tale spatium percurrat.* »

74 Descartes was to deny later that he had ever learned anything from Beeckman. Cf. Letter to Mersenne, 4 November 1630 (*Oeuvres*, vol. I, p. 171 ff.) and a letter to Beeckman himself (*Oeuvres*, vol. I, p. 157 ff.).

75 Gilson has pointed out this typical feature of Descartes' mind: Descartes is much less concerned to establish that a phenomenon is real than to explain it. See E. Gilson, *Etudes sur le rôle de la pensée médiévale dans la formation du système cartésien*, Paris, 1930.

76 *Cogitationes Privatae, Oeuvres*, vol. X, p. 219: « Solvi quaestionem. In triangulo isoscelo rectangulo, ABC spatium [motum] repraesentat ; inaequalitas spatii a puncto A ad basim BC, motus inaequalitatem. Igitur AD percurritur tempore, quod ADE repraesentat ; DB vero tempore quod DEBC repraesentat : ubi est notandum minus spatium tardiorem motum repraesentare. Est autem AED tertia pars DEBC : ergo triplo tardius percurret AD quam DB. Aliter autem proponi potest haec quaestio, ita ut semper vis attractiva terrae aequalis sit illi quae primo momento fuit : nova producitur, priori remanente. Tunc quaestio solvetur in pyramide. »

77 'The inequality of the motion'—*motus inaequalitatem*—i.e., the variation in the speed.

78 'The problem would be resolved by a pyramid'—*solvetur in pyramide*—i.e., the speeds would increase as the cubes and not as the squares.

79 It is interesting that for Beeckman, as for Galileo (see above, pp. 77 and 115; *Dialogo, Opere*, vol. VII, p. 251; *Discorsi e dimostrazioni, Opere*, vol. VIII, pp. 208 and 209), the flow of time is always represented by a vertical line and not, as we are used to, by a horizontal.

80 This decisive progress consisted in (a) the explicit assertion of the law of conservation of *motion*, which was thereby emancipated from the idea of *impetus*, and (b) the elimination of any *cause* within the moving body. For the first time in the history of physics a variable effect could be explained by the successive, or enduring, action of a constant force.

81 Descartes and Beeckman, *Physico-mathematica*, in *Oeuvres de Descartes*, vol. X, p. 75 ff.: « In proposita quaestione, ubi imaginatur singulis temporibus novam addi vim qua

corpus grave tendat deorsum, dico vim illam eodem pacto augeri, quo augentur lineae transversae *de, fg, hi,* et aliae infinitae transversae quac inter illas possunt imaginari. Quod ut demonstrem, assumam pro primo minimo vel puncto motus, quod causatur a primo quae imaginari potest attractiva vi terrae, quadratum *alde*. Pro secundo minimo motus, habebimus duplum, nempe *dmgf* : pergit enim ea vis quae erat in primo minimo, et alia nova accedit illi aequalis, Item in tertio minimo motus, erunt 3 vires ; nempe primi, secundi et tertii minimi temporis, etc. Hic autem numerus est triangularis, ut alias forte fusius explicabo, et apparet hunc figuram triangularem *abc* repraesentare. Immo, inquies, sunt partes protuberantes *ale, emg, goi*, etc. quae extra trianguli figuram exeunt. Ergo figura triangulari illa progressio non debet explicari. Sed respondeo illas partes protuberantes oriri ex eo quod latitudinem dederimus minimis quae indivisibilia debent imaginari et nullis partibus constantia. Quod ita demonstratur. Dividam illud minimum *ad* in duo aequalia in *q* ; iamque *arsq* est [primum] minimum motus, et *qted* secundum minimum motus, in quo erunt duo minima virium. Eodem pacto dividamus df, fh, etc. Tunc habebimus partes protuberantes *ars, ste*, etc. Minores sunt parte protuberante *ale*, ut patet. Rursum, si pro minimo assumam minorem, ut *aα*, partes protuberantes erunt adhue minores, ut *αβy*, etc. Quod si denique pro illo minimo assumam verum minimum, nempe punctum, tum illae partes protuberantes nullae erunt, quia non possunt esse totum punctum, ut patet, sed tantum media pars minimi *alde*, atqui puncti media pars nulla est. Ex quibus patet, si imaginetur, verbi gratia lapis ex *a* ad *b* trahi a terra in vacuo per vim quae aequaliter ab illa semper fluat, priori remanente, motum primum in *a* se habere ad ultimum qui est in *b*, ut punctum *a* se habet ad lineam *bc*. Mediam vero partem *gb* triplo celerius pertransiri a lapide, quam alia media pars *ag*, quia triplo majori vi a terra trahitur : spatium enim *fgbc* triplum est spatii *afg*, ut facile probatur. Et sic proportione dicendum de caeteris partibus. »

82 Note this 'at each instant' (*singulis temporibus*): when Descartes thinks about force he immediately also thinks about time.

83 'Minimum or point of motion'—*minimum vel punctum motus*—this is exactly the same thing that Descartes also called 'moment', and that was called 'degree of speed' by Galileo and his predecessors.

84 P. Duhem comments about this (*op. cit.*, p. 576): 'What Beeckman [he should, of course, have said: Descartes] said was of far greater precision and of far greater significance than the arguments of the writer on mechanics from Pisa'. The arguments of 'the writer on mechanics from Pisa' were not as bad as Duhem makes out; as we have seen (and will see again below) they involved the use of Cavalieri's geometry of indivisibles. As for Descartes' argument, it can be found, almost identical, in Gradi (see Caverni, *Storia del metodo sperimentale in Italia*, vol. IV, Bologna, 1895, p. 306 ff.).

85 This idea is perfectly correct if, like Descartes, one eliminates time and takes the *action* of the force to be timeless or instantaneous. Thus, as Newton was to say (*Philosophiae naturalis principia mathematica*, Londini, 1687, Axiomata sive leges, Lex II, p. 12), it is undeniable that 'Si vis aliqua motum quamvis generat, dupla duplum, tripla triplum generabit, sive simul et semel, sive gradatim et successive impressa fuerit.' On Descartes' 'instantaneism' see the fine book by Jean Wahl, *Le rôle de l'idée de l'instant dans la philosophie de Descartes*, Paris, 1920.

86 We have already said that Beeckman did not really understand himself, did not understand the implications of his 'principle'. This is entirely confirmed by the passages published by Cornélis de Waard. Beeckman was so far from understanding his own ideas that he denied the continuity of the acceleration that takes place during fall, and adopted the theory that the motion is discontinuous (see *Correspondance du Père Mersenne*, vol. II, p. 291 ff.). Moreover he asserted, like Aristotle, that a body thrown in the air comes to rest before beginning setting off again. However paradoxical it might seem, then, we can see that the new idea of motion was far from clear to Beeckman, and it was left for Descartes to fully elucidate it and to draw from it all its implications. But he did not accomplish this until ten or fifteen years later, at the time of writing the *Regulae* and *Le Monde*, by which time he had decided to see in motion only that which could be seen by a mathematician.

87 Descartes' physics was, unfortunately, a physics of the *imagination*, and he often took a

clear concept in physics to be one which could be clearly represented in images. Cf. L. Brunschvicg, 'Métaphysique et mathématique chez Descartes', *Revue de Métaphysique et Morale*, 1927; see also above, p. 100 ff. and note 159.

[88] He needed only to rigorously sustain the proportionality between force and speed and to continue to think causally, i.e., in relation to time.

[89] Descartes and Beeckman, *Physico-mathematica, Oeuvres de Descartes, op. cit.*, vol. X, p. 77: « Aliter vero potest haec quaestio proponi difficilius, hoc pacto. Imaginetur lapis in puncto *a* manere, spatium inter *a* et *b* vacuum ; iamque primum, verbi gratia, hodie hora nona Deus creet in *b* vim attractivam lapidis ; et singulis postea momentis novam et novam vim creet, quae aequalis sit illi quam primo momento creavit ; quae iuncta cum vi ante creata fortius lapidem trahat et fortius iterum, quia in vacuo quod semel motum est semper movetur ; tandemque lapis, qui erat in *a*, perveniat ad *b* hora decima. Si petatur quanto tempore primam mediam partem spatii confecerit, nempe *ag*, et quanto reliquam: respondeo lapidem descendisse per lineam *ag* tempore ⅛ horae; per spatium *gb*, ⅞ horae [this is obviously a mistake; the fractions should be the other way round]. Tunc enim debet fieri pyramis supra basim triangularem, cuius altitudo sit *ab*, quae quocunque pacto dividatur una cum tota pyramide per lineas transversas aeque distantes ab horizonte. Tanto celerius lapis inferiores partes lineae *ab* percurret, quanto majoribus insunt totius pyramidis sectionibus. »

[90] As always in Descartes, the line represents the trajectory.

[91] I.e., the proportion will be to the cube and not the square: this is the second of the two hypotheses discussed in the *Cogitationes Privatae*.

[92] Cf. P. Duhem, *Etudes sur Léonardo de Vinci*, vol. III, p. 570.

[93] See G. Milhaud, *op. cit.*, p. 28 ff.

[94] I.e., the situation is similar to that of Leonardo da Vinci and Benedetti, discussed above.

[95] See the letter to Mersenne of 14 August 1634 (*Oeuvres de Descartes*, vol. I, p. 303), where Descartes says that he had glanced through the copy of Galileo's *Dialogue* which Beeckman had lent him between Saturday and Monday: 'Beeckman came here Saturday evening and lent me Galileo's book, but he took it away again to Dort. this morning, so that I have only had it in my hands for 30 hours. I did not fail to glance through it and I find that he philosophises very well on the subject of motion, though there are very few things he says about it which I find entirely correct. But as far as I could see he is more deficient where he follows received opinions than where he departs from them, although an exception is what he says concerning the tides which I find rather far-fetched. I have also explained this myself, in *Le Monde*, by the earth's motion, but in a completely different way from his.

'However, I must say that in his book I did come across several of my own ideas, and among them two which I think I have written to you about before. The first is that the spaces through which heavy bodies pass when they fall are to each other as the squares of the times taken in falling, i.e., that if a ball takes 3 moments to fall from A to B, it will take only 1 to continue from B to C, etc., which is what I have said, with many qualifications, for in fact it is never completely true as he believes he has proved'. Descartes' qualification is odd, but it is perfectly comprehensible in the context of his physics. Galileo's solution assumes a vacuum and attraction, whereas Descartes now accepts neither of these. But this is not the point of interest for us here: this is the fact that Descartes believes that he has found in Galileo his own solution to the problem, although it is actually quite different.

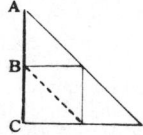

[96] It is interesting that Duhem believes that these two replies were identical (see P. Duhem, *op. cit.*, p. 569). Moreover, in order to follow Descartes' argument Duhem found it necessary to turn the accompanying diagram on its side (p. 566).

97 Letter to Mersenne of 13 November 1629, *Oeuvres de Descartes,* vol. I, p. 71.

98 Notice this detail; Beeckman had said only 'moves for ever in the same way'; Descartes specifies, 'with the same speed'. Beeckman, of course, would have taken his statement to mean just this; for him this would have gone without saying. But it is necessary to say it, for a body could perfectly well move for ever without maintaining its speed, and even move 'in the same way', for example with constant acceleration or deceleration. The law of the conservation of motion no doubt implies the conservation of speed, but it is necessary to make this explicit. It would be enough for Descartes to add to this the conservation of direction to arrive at the law of inertia: this addition is sufficient, but also absolutely indispensable. Thus, contrary to the beliefs of Duhem ('De l'acceleration etc.', p. 904) and de Waard (*Correspondance de Mersenne,* vol. II, pp. 236, 237), neither Descartes nor Beeckman, *in any of the texts cited by us thus far,* stated the principle of inertia.

99 The motion of fall as it is pictured by Descartes is faster than it is in reality. In fact the distance covered in 3 and 4 'moments' is as 3^2 to 4^2, i.e., 9 to 16. Thus it is not 'twice' as large during the fourth 'moment'. If Descartes had remembered this calculation five years later when he was looking through the *Dialogue* he would not have been able to believe in the identity of his solution with that of Galileo. For while, in Galileo, the distances covered in successive time intervals are *sicut numeris impares ab unitate,* they are not so for Descartes. But by the time Descartes read Galileo he had abandoned all hope of being able to give an exact numerical solution to the problem of real fall. And the abstract case of fall in a vacuum, which was that investigated by Galileo, and earlier by himself, no longer interested him: the concept of a vacuum was absurd, and a physics of clear ideas could have no use for it.

100 See above, p. 71 ff.

101 Gravity successively produces instantaneous forces, *impetuses,* which move the body and which are conserved during its motion. *Impetus* here, as in Cardan and as occasionally in Galileo himself, is in fact identified with the motion and the speed. This is the heritage of an older conception preserved within a new physics. As for the abandonment of the idea of attraction, this is quite typical of Cartesian thought—Descartes manifestly prefers the idea of gravity to this obscure idea of action at a distance.

102 These distances are, of course, infinitely small.

103 In a sense this is quite correct: the acceleration is in effect produced at each point on the path.

104 Letter to Mersenne of 18 December 1629, *Oeuvres de Descartes,* vol. I, p. 89. The letter is in Latin.

105 Cf. above, Part 1, p. 23 ff.

106 See the well-known works of L. Brunschvicg, *La Causalité physique et l'expérience humaine,* Paris, 1925, and *Le Progrès de la conscience dans la philosophie occidentale,* Paris, 1927.

107 See E. Bréhier, *Histoire de la philosophie,* vol. II, Paris, 1928, p. 93 ff.

108 See E. Meyerson, *Identité et Réalité,* Paris, 1926, p. 282 ff.; *La déduction relativiste,* Paris, 1925, p. 135 ff.

109 Letter to Mersenne of 12 September 1638, *Oeuvres de Descartes,* vol. II, p. 355.

110 Letter to Mersenne of 11 October 1638, *Oeuvres de Descartes,* vol. II, p. 380: 'I find, in general, that he is a philosopher quite out of the ordinary in that he abandons Scholastic error as much as he can, and tries to investigate problems in physics by mathematical reasoning. In this I am entirely in agreement with him and I hold that there is no other means for discovering truth. But he seems to me to be deficient in that he digresses continually and does not remain with anything long enough to give a thorough explanation of it; this shows that he did not examine things in an orderly way, and that, without having investigated the first causes in nature, he merely looked for the reasons for some particular effects, and thus that he build without foundations.'

111 From a certain point of view Descartes' criticism of Galileo is justified, in principle but not in fact. In effect Descartes' charge is that Galileo built a mathematical physics, which *contradicted common sense and everyday experience* (cf. *Le Monde, Oeuvres de Descartes,* vol. XI, p. 41) illegitimately, i.e., without a supporting metaphysics. In

principle Descartes is right. In fact, however, he is wrong, for Galileo was a Platonist.

112 Letter to Mersenne of October-November 1631 (*Oeuvres de Descartes*, vol. I, p. 228). In 1638 (Letter to Mersenne of 11 October 1638, cited in note 110) Descartes would write (*Oeuvres*, vol. II, p. 386): 'He assumes that the speed of falling bodies always increases uniformly, and I used at one time to believe this myself; but now I believe that I can prove that this is not true'—because Galileo's deduction is based on the idea of a vacuum and neglects, which one cannot do, the resistance and the motive force which determines the acceleration. Finally, in 1640 Descartes would write (Letter to Mersenne of 30 August 1640, *Oeuvres*, vol. III, p. 164 ff.): 'I have already written to you several times that I do not believe that the speed of falling bodies always increases *in ratione duplicata temporum*, but that it might well increase approximately in this way at the beginning of fall, though this could not continue; and even that when they reach a certain speed this cannot increase any more; and this is confirmed by what you write concerning rain drops, etc.'. Notice, incidentally, that now that he does not believe in the law Descartes no longer claims to have been its originator.

113 Cf. *Regulae ad directionem ingenii*, XII (*Oeuvres de Descartes*, vol. X, pp. 419, 420.

114 *Le Monde, Oeuvres de Descartes*, vol. XI, p. 39.

115 *Ibid.*, p. 40.

116 Emphasis added.

117 *Le Monde, op. cit.*, p. 38.

118 *Ibid.*

119 It should be noted that for Descartes and the Cartesians extension is a substance or essential attribute, whereas duration is conflated with being and time is only a mode—and even a subjective mode.

120 Letter to Mersenne of 16 October 1639 (*Oeuvres de Descartes*, vol. II, p. 593 ff.): 'To understand how the subtle matter which swirls round the earth pushes heavy bodies towards the centre, fill some round vessel with small lead shot and mix in with this lead some bits of wood, or some other material lighter than lead, bigger than the bits of lead; then rotate the vessel very fast and you will see that the shot will push all the bits of wood, or other material, towards the centre of the vessel, in the same way that the subtle matter pushes earthly bodies, etc.'. On the Cartesian theory of gravity see the excellent work by P. Mouy, *Le développement de la physique cartésienne*, Paris, 1934.

121 Letter to Mersenne of 25 December 1639 (*Oeuvres*, vol. II, p. 635).

122 Letter to Mersenne of October-November 1631 (*Oeuvres*. vol. I, p. 230).

123 See below, Part 3, p. 263. In fact Descartes does not at all misunderstand the idea of relativity; on the contrary, he uses it.

124 Letter to Mersenne of 11 March 1640 (*Oeuvres*, vol. III, p. 37 ff.). [*toise* = an old French measure, equal to 1,949 metres, and divided into six feet— *Trans*.] Cf. Letter to Mersenne of 11 June 1640 (*Oeuvres*, vol. III, p. 79): 'The reason why I say that falling bodies are pushed less by the subtle matter at the end of their motion than at the beginning is simply that there is then less difference between their speed and that of the subtle matter. For example, if the body A, being without motion, is hit by the body B which is tending to

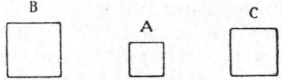

move towards C with a speed such that it could travel one league in one quarter of an hour, it would be pushed more by the body B than it would if it were itself already moving towards C with a speed such that it could travel one league in half an hour, and it would not be pushed by it at all if it were already moving just as fast [as B], i.e., such that it could travel a league in a quarter of an hour.'

125 Letter to Mersenne of November-December 1632 (*Oeuvres*, vol. I, p. 260).

126 *Ibid.*, p. 231.

127 Galileo had said that a cannon-ball fired horizontally from the top of a tower would reach the ground at the same instant as another one which had been dropped vertically. See

Dialogo, Opere, vol. VII, p. 181; Letter to Mersenne of 19 August 1634 (*Oeuvres*, vol. I, p. 305).

[128]   Letter to Mersenne of 22 June 1637 (*Oeuvres*, vol. I, p. 392). Cf. Letter to Mersenne of 12 September 1638 (*Oeuvres*, vol. II, p. 355): 'For it is impossible to say anything definite concerning the speed unless one has a true explanation of what gravity is and altogether the whole world system. Now, because I did not wish to undertake this I have found a way of omitting a consideration of it, and of so separating it from all other considerations that I can explain them without it. For while there is no motion which does not have a speed, yet it is only the increase and decrease in speed which can be considered, and when people speak about a body's motion postulating that it takes place with a speed which is the most natural to it, this is just as if they had not said anything about it at all.'

[129]   Cf. E. Bréhier, *Histoire de la philosophie*, Paris, 1928, vol. II, p. 97 ff.

[130]   Cf. G. Bachelard, *La Valeur inductive de la relativité*, Paris, 1929.

[131]   It is the awareness of this failure which gives Cartesian physics the pragmatic quality which it has in the *Principes*.

Galileo again

[132]   Galileo, *Opere*, vol. II, p. 261 ff. Cf. *Discorsi, Opere*, vol. VIII, p. 197.

[133]   Emphasis added.

[134]   Emphasis added.

[135]   Emphasis added.

[136]   'The intension of the speed' or the degree of speed is the instantaneous speed of the moving body. Descartes called it 'moment' or 'point' of speed.

[137]   This admirable fragment, which was published by Alberi with the works of Galiléo's Pisa period, i.e., as one of his very earliest works, is attributed by Favaro to the later, Padua period. Favaro dates it as having been written in 1604. We cannot accept this date. Because (i) the letter to Paolo Sarpi dates from 16 October 1604; now, Galileo tells us that the discovery of the correct definition of accelerated motion required 'prolonged mental effort', and this is confirmed by the number of fragments, published by Favaro as an Appendix to the *Discorsi* (*Opere*, vol. VIII, p. 370 ff.) which give the incorrect deduction of the law, based on the incorrect definition: all this would be incomprehensible if Galileo already had the correct law as early as 1604; (ii) Galileo's use of the methods of the calculus of indivisibles would require us to accept that he had developed this twenty years before Cavalieri.

Therefore it seems to us that, while we have no need to adopt Caverni's hypothesis (*Storia del metodo sperimentale in Italia*, Bologna, 1895, vol. IV, p. 307 ff.) which dates the discovery as late as the years 1622-23, we must accept that of Wohlwill, which locates it in 1609; we must therefore assign this fragment a date later than that attributed to it by Favaro.

[138]   This was not unnecessary: the proof of this is that Descartes, who accepted only instantaneous actions, doubted it. See Letter to Mersenne of 11 October 1638 (*Oeuvres de Descartes*, vol. I, p. 399) and above, p. 100 ff.

[139]   Cf. Mersenne, *L'Harmonie universelle*, vol. I, Paris, 1636, p. 74: 'The human mind is not capable of understanding how it could be possible that one continuous motion could be slower than another, and this has led the Spanish philospher Arriaga and a number of others to say that slowness of motion is nothing else than an intervention of some [periods of] rest, though these are not perceptible by the senses, and that they are longer and greater in number the slower the motion . . . And they postulate this also for the natural motion of stones and heavy bodies falling towards the centre of the earth . . .' Cf. *Correspondance du P. Marin Mersenne*, vol. II, p. 291 ff.

[140]   Beeckman, among others, was of this opinion: cf. *Correspondance du P. Marin Mersenne*, vol. II, pp. 260, 400. This is not at all an absurd idea; it is, precisely, the idea at the centre of quantum theory.

[141]   Letter to Galileo of 21 March 1626 (*Opere di Galileo Galilei*, vol. XIII, p. 312).

[142]   *Opere*, vol. II, p. 262.

[143] Cf. *Opere*, vol. II, p. 263.

[144] *Ibid.*, p. 264.

[145] The identification of rest with 'infinite slowness' appears to re-establish the continuity between 'rest' and 'motion'. But this is actually an illusion: the transition from the infinite to the finite is no easier than that between nothing and something.

[146] That the speed of the body in its descent depends only on the *height* of fall is put forward by Galileo as a 'postulate' or axiom. Cf. below, Part 3, p. 182.

[147] On the literary and intellectual structure of the *Dialogue* and the *Discourses*, and on the roles assigned to the interlocutors, see below, Part 3, p. 158 ff.

[148] *Discorsi, Opere*, vol. VIII, p. 198.

[149] *Ibid.*, p. 198.

[150] Emphasis added. Sagredo is absolutely right: the *imagination* refuses to accept mathematical *reasoning*. This is precisely why it is necessary to replace the former by the latter.

[151] *Ibid.*, p. 199.

[152] The product of the speed and the weight is the *momento;* see above, p. 76 and note 46.

[153] *Ibid.*, p. 200.

[154] That it does in effect happen like this, i.e., that there is no necessary point of rest in a continuously decelerating motion, had already been shown by Benedetti (see above, Part 1, p. 26). But Galileo could have invoked the authority of Aristotle himself, for he explains in his *Physics* (Book V, chap. 6, 230 b; Book VI, chap. 8, 238 b) that there is never a first or a last moment of motion, and consequently that there is also neither a first nor a last moment of rest.

[155] In fact it does not do so at all. On the contrary, it is only the idea of motion at an instant, the idea of a moment or element of motion and of speed, which makes it possible to resolve the difficulties brought to light by the arguments of Zeno.

[156] *Discorsi, Opere*, vol. VIII, p. 202. Cf. p. 62.

[157] Letter to Mersenne of 11 October 1638 (*Oeuvres de Descartes*, vol. II, p. 399 ff.

[158] Probably Frenicle.

[159] Descartes' decision is in effect in favour of the imagination.

[160] Cf. Galileo's handwritten comment on his copy of the *Dialogo* (*Opere*, vol. VII, p. 48): 'If a body however heavy, moving with a speed however great, hits some other body which is at rest, however weak this latter body and however small its resistance, the moving body will never, by hitting it, confer on it its own speed: this follows self-evidently from the fact that one hears the sound of impact, which one would not hear, or rather which would not exist, if the body at rest were to receive, immediately on the arrival of the moving body, the same speed as the latter.'

[161] Letter to Mersenne of 22 January 1640 (*Oeuvres de Descartes*, vol. III, p. 9 ff.).

[162] See above, p. 68 and below, Part 3, p. 187.

[163] *Dialogo*, p. 426 ff. (*Dialogue*, p. 399).

[164] *Discorsi*, p. 202.

[165] The proof in the *Dialogue* begins with the analysis of a concrete example: a body which is thrown upwards and which, after having risen, falls down again to its starting point, moves with a continually decreasing speed in the first (ascending) part of its motion, and continually increasing speed in the second (descending) part while covering the same distance in the same amount of time in each of these two motions, and while passing (in reverse order, of course) through the same series of speeds. Now, these two series are manifestly complementary: if we add the speed which the ascending body has at an instant n after the beginning of the motion to the speed which it has at n during the motion of descent we will always obtain the same number, which will clearly be equal to the maximum speed. Therefore in its total motion the body will have covered the same distance as it would have covered if it had moved for the same amount of time with the maximum speed. But since it has performed a double motion (up and then down) each of these parts will be one half of the total motion, i.e., one half of the motion which would be effected (and hence one half of the distance which would be covered) by the body if it moved for the same amount of time with the maximum speed. Cf. *Dialogo*, p. 254 (*Dialogue*, p. 227).

166 *Dialogo*, p. 255 ff. (*Dialogue*, p. 228 ff.).

167 *Dialogo*, p. 256 (*Dialogue*, p. 230).

168 See below, Part 3, p. 201 ff.

169 *Discorsi*, p. 208 ('Third Day', Theorem I, proposition 1) [The English translation is based on that given in M. Clagett, *The Science of Mechanics in the Middle Ages*, Madison, Wisconsin, 1961, p. 410]. The diagrams which Galileo gives with his proof are extremely interesting. He seems to be aware of the fact that his way of representing the space covered, the trajectory of the motion, (he represents a *line* by a *surface*) is hardly natural, and of how easily this method of representation could lead one into the error of thorough-going geometrisation, an error which he had once committed himself. It must be possible to represent the trajectory by a line. But Galileo does not know how to do this. Therefore he merely draws a line at the side of the diagram, a line which has no relation whatever with the diagram itself.

170 *Discorsi*, p. 209 ('Third Day', Theorem I, proposition 2) [Clagett, *op. cit.*, p. 410].

171 *Ibid.*, p. 210.

172 *Ibid.*, p. 212.

173 *Ibid.*, p. 212.

174 *Ibid.*, p. 212 ff.

175 *Ibid.*, p. 213.

176 Letter to Mersenne of April 1634 (*Oeuvres de Descartes*, vol. I, p. 287).

177 Mersenne, *L'Harmonie Universelle*, vol. I, p. 112: 'Corollary I. I doubt whether Galileo actually performed the experiments of fall down inclined planes, since he does not speak of them, and since the ratio he gives is often contradicted by experiment: I wish that a number of people would try the same thing on different planes, taking all possible precautions, so as to see whether their experiments are in agreement with ours. . . Corollary II. Those who have seen our experiments and who have helped with them, know that they could not be performed more carefully, both in relation to the plane, which is very straight and very smooth and which compels the body to descend in a straight line, and in relation to the roundness and heaviness of the balls, and in relation to the falls; from which it could be concluded that experiment is not capable of giving rise to a science, and that one should not place too much trust in reasoning alone because it does not always agree with the truth of appearances, from which it often diverges: this will not prevent me from speaking about the uniformly inclined plane, that is a plane such that heavy bodies will press and weigh equally at each of its points.' Cf. the present author's article 'Galilée et l'expérience de Pise'. *Annales de l'Université de Paris*, 1936.

178 Cf. R. Poirier, *Remarques sur la probabilité des inductions*, Paris, 1931.

179 On the other hand, as we have seen, Descartes succeeded at the point where Galileo met with failure; for it was Descartes and not Galileo who explicitly stated the *principle of inertia*, whereas Galileo remained all his life obstructed by the phenomena. See below, Part 3, p. 200.

PART III

GALILEO AND THE LAW
OF INERTIA

INTRODUCTION

Descartes-the-physicist's greatest claim to fame is certainly that he gave a 'clear and distinct' formulation to the principle of inertia, and that he realised its importance.

Of course it could be argued that by the time he did this, i.e., by the time his *Principes* was published (twelve years after Galileo's *Dialogue* and six years after his *Discourses*) it was no longer a particularly notable achievement nor a very difficult one. For by 1644 the law of inertia was no longer an unheard-of or novel idea. On the contrary, thanks to the work and writings of Gassendi, Torricelli and Cavalieri it was beginning to have the status of a generally accepted truth. Moreover, it could be pointed out that even though Galileo himself did not state the law *expressis verbis*, or at least did not present it as a fundamental law of motion, his physics was nevertheless so impregnated with it that Baliani, a thinker in no way comparable to those we have just mentioned, was able to extract it quite spontaneously therefrom.[1]

One could point to Newton's judgement; he attributed all the credit for the discovery to Galileo, with no mention of Descartes at all. And if, in defence of Descartes' claims in the matter, one were to point out the fact that he had already stated the law of inertia in his *Le Monde*, it could be replied that, on our own account above, it was Beeckman whom Descartes had to thank for the principle of the conservation of motion.[2]

All this is no doubt perfectly correct: and it is not our intention to minimise in any degree the merits of those who, together with Descartes and Galileo, laid the foundations of classical science. Above all, we do not have the slightest wish to detract from the role and credit of Galileo himself: quite the contrary, as will become clear below.[3] Nevertheless when, after the subtly hesitant and cautious texts of Galileo, after the contorted explanations of Gassendi, and after the admirably clear but barren mathematical formulations of Torricelli, we arrive at the lapidary statements of Descartes, it seems to us to be impossible not to accept the evidence of a decisive advance in clarity of thought and understanding. This is so striking that one could, *mutatis mutandis*, characterise the difference between Galileo and Descartes by invoking Pascal's well-known statement about the difference between coming across a concept by luck and proceeding to use it without prolonged reflection, and perceiving in a concept a whole, marvellous train of consequences and so using it as a solid and well-established principle on which to base an entire physics.[4]

The law of inertia is simplicity itself. It merely states that a body, left to itself, remains in its *state* of rest or motion so long as nothing intervenes to change it.[5] At the same time it is a law of capital importance: for it implies a concept of motion which conditions the general interpretation of nature; it implies a wholly new conception of physical reality itself. This new concept of motion asserts that it is a *state*, and although it rigorously contrasts it with rest it locates them both on the same ontological level.[6] It asserts, implicitly or explicitly, that the body—moving or at rest—is completely *unaffected* by which of these two contrasting states it is in, and that the fact of being in one or the other in no way changes it. In other words, neither of these two *states* brings about in the bodies of which they are the *states* any modification or change, and that the transition from one of these *states* to the other one is, for the moving body itself, absolutely without consequence. It implies, therefore, that it is impossible to attribute the state of rest (or of motion) to a given body except *in relation to another one* which is taken to be in motion (or at rest), and that it is purely and completely arbitrary which of these two *states* is attributed to which of the two bodies.[7] Motion is thus conceived as a *state*; but not as a state just like any other, for it is a *relational state*.[8]

The classical concept of motion implies not only that the moving body is unaffected by motion but also that one motion is unaffected by another: two motions never interfere with one another.[9] It is of this strange entity, this substantial relation, an entity no less paradoxical that the famous substantial qualities of mediaeval physics, that the principle of inertia asserts the eternal persistence.

However, the principle of inertia does not assert the eternal persistence of *all* motions, as we have just stated, but only of *uniform rectilinear* motion. The principle does not hold for circular motion. Nor does it hold for rotational motion.[10] It could be said that while ancient and mediaeval physics contrasted natural circular motion with violent rectilinear motion, this contrast is inverted by classical physics, for which it is rectilinear motion which is natural, and circular motion which henceforth is considered to be violent.[11] But this would be inaccurate. For in classical physics there is no *natural* motion, and nor is there, strictly speaking, any *violent* motion. No motion is the consequence of the 'nature' of the moving body, and no such 'nature' could bring about its rest. It follows, obviously, that 'violence' is never done to the nature of the moving body: it is, as we have pointed out, completely unaffected by the state it is put in. On the other hand it follows from this that it is only by 'force'—although no longer by 'violence'—that the *state* of the body can be changed from one to the other. Since all motion (or at least all putting in motion) and all rest (or at least all cessation of motion), all acceleration and deceleration, involve a cause or, more accurately, a *force*,[12] this must necessarily be conceived as external to, as not belonging to, the moving body which is in itself *inert*.[13]

The classical concept of motion—that of Galileo, Descartes and Newton—seems to us today not only obvious but even 'natural'. However, this 'obviousness' is still very recent; it is scarcely three hundred years old.

And it is to Descartes and to Galileo that we owe it.

The principle of inertia did not emerge straight away in final form, like Athena from the head of Zeus, from the thought of Descartes or Galileo. The new concept of motion, with its implication of a new conception of physical reality, of which the principle of inertia is both the basis and the expression, was elaborated increasingly accurately as the result of a long and difficult mental labour. The Galilean and Cartesian revolution—for it was nonetheless a revolution—had a prolonged preparation. It is the history of this preparation which we intend to investigate here,[14] a history which constitutes a necessary prerequisite for any understanding of Galileo's achievements, a history in which the human mind can be observed obstinately grappling over and over again with the same problems, untiringly coming up against the same objections, the same difficulties, and slowly and laboriously forging the instrument which would enable it to overcome them.

Classical physics investigates in the first instance the motion of heavy bodies, i.e., in the first place the heavy bodies with which we are surrounded. Thus it was the attempt to explain the facts of everyday experience (the fact of free fall, the fact of projectiles, etc.) that gave rise to the development of ideas which resulted in the discovery of the law of inertia. It is interesting, however, that this attempt did not give rise to it directly. Nor was this its main source. The new physics was not only born on earth; it was born equally in the heavens. And it was in the heavens also that it achieved its maturity.

This fact, that classical physics had a celestial 'prologue' and 'epilogue', or to put it less figuratively, the fact that classical physics was born as a dependent of astronomy, and that it remained tied to it throughout its whole history, is very significant and is rich with implications. It is an expression of the replacement of the idea or concept of the Cosmos—a closed whole with a hierarchical order—by that of the Universe—an open ensemble interconnected by the unity of its laws.[15] It implies the impossibility of initiating and elaborating a terrestrial mechanics without having, or at least without initiating and elaborating simultaneously, a celestial mechanics. It explains the partial failure of Galileo and of Descartes.

1

THE PHYSICAL PROBLEM
OF COPERNICANISM

Let us now proceed to the facts. The new physics developed, we have just said, as a dependent of astronomy. More precisely, it developed as a result of the problems posed by Copernican astronomy, and in particular of the need to reply to the *physical* arguments put forward by Aristotle and Ptolemy against the possibility of the motion of the earth.

1. Copernicus

In fact it was not hard to reply to the 'geometrical' arguments in favour of geocentrism. Those who, from the fact that all circular motion implies an axis or a fixed point around which the motion takes place, deduce that the earth is stationary, obviously confuse geometry and physics.[16] So, having shown the vacuity of this argument, Copernicus goes on:[17] 'This is why the ancient philosophers attempted to find other reasons for assigning to the earth a fixed position at the centre of the world. And as the main consideration they put forward heaviness and lightness. For the element earth is the heaviest, and all things which have weight are carried towards it and strive towards its centre. Now since the earth, towards which all heavy bodies from all directions are carried perpendicular to its surface by virtue of their own nature, is round, they would all meet at its centre if they were not stopped at its surface. . . But it appears that things which are carried towards its centre try to get there in order to rest at the centre. All the more reason then that the earth should be at rest at the centre, and receiving all falling things, should remain there at rest because of its weight. They also attempt to prove the same thing by means of an argument based on motion and its nature. For Aristotle says that the motion of a simple body is simple:[18] now, of simple motions, one is rectilinear, the other is circular: and as for rectilinear motions, one is upwards and the other downwards. Consequently all simple motions are either towards the centre, i.e., downwards, or away from the centre, i.e., upwards, or around the centre in the case of circular motion. To be carried downwards, i.e., to tend towards the centre, is normal only for earth and for water, which are regarded as heavy; in contrast to this, air and fire, which are endowed with lightness, tend upwards, to move away from the centre. It appears that rectilinear motion should be attributed to the four elements, whereas to celestial bodies should be attributed motion around the centre. This is what Aristotle teaches.

'So, according to Ptolemy of Alexandria,[19] if the earth were to revolve, at least with a daily revolution, this would have to happen contrary to what has just been said. For this motion, which would complete the whole circuit of the earth in twenty four hours, would have to be extremely vigorous and of an unsurpassable speed. Now things which move with a violent rotation seem to be completely incapable of coming together again, but must rather be dispersed, *unless they are held together by some force*.[20] So, he says, the earth would long ago have been scattered to the heavens (nothing could be more ridiculous): and this would be even more true of living things and of other loose masses which would in no way have been able to remain stable. Furthermore, freely falling bodies would not travel vertically, to their predestined place, a place which would meanwhile have been so swiftly removed from beneath them. And we would observe clouds and everything else suspended in the air always carried towards the west.'

It would be wrong to underestimate these objections. Of course one could reply, as Copernicus did,[21] and as his defenders did after him, that gravity is nothing else than the natural tendency of the parts of a Whole to collect together, and that terrestrial heavy bodies do not at all seek to arrive at the centre of the world in order to rest there, but merely tend to move towards their Whole, the earth. This, however, would leave it unexplained why it is that they move towards the centre of the earth, and it is not easy to explain this. Furthermore, one would still have to reply to the arguments concerning the rectilinear motion of fall. Now Copernicus' reply to these latter arguments is in fact quite superficial, even mere quibbling.

In fact Copernicus does not see the weakness of the 'centrifuge' argument. He takes it seriously and deals with it just like the others. Thus there is one and only one objection which he puts to his opponents. Extending to the earth an idea which was accepted for the heavens,[22] he asserts of it the *natural* character of its motion. Now, since it is *natural*, this motion on the one hand cannot give rise to the disastrous effects mentioned by Ptolemy (a *natural* motion which destroyed the very *nature* of the moving body would be a contradiction in terms), and on the other hand, being *natural* to the *earth*, all bodies of terrestrial nature and origin are naturally impelled by it, even those which are not in direct contact with it; for they are nonetheless *physically attached* to it.

In Copernicus' judgement, and no doubt he is right, the Aristotelian argues on the basis of his own system of physics: he takes this physics for granted. He quite naturally applies his own categories, and considers the motion of the earth to be 'violent'. This is the implicit premise on which his objections are based. Therefore, Copernicus replies:[23] 'But if someone thinks that the earth moves then he would certainly say that this motion is natural, and not violent. Now things which happen in conformity with nature produce different effects from those which happen by violence. For things which are subjected to force or violence must inevitably be destroyed and cannot last for long; but objects which exist naturally are in an appropriate state, and remain at their best. So Ptolemy has no need to fear that the earth and all terrestrial things would be destroyed by a rotation resulting from natural agency, an agency which is quite different from an artificial one or from one resulting from human ingenuity. But why does he not have this fear in relation to the whole world, for its motion would have to be as much faster as the heavens are larger than the earth? Have the heavens become so vast because their motion, by its unimaginable rapidity, has moved them away from the centre, and would they collapse if this motion were to cease?'

We have shown elsewhere[24] just how weak this counterargument by Copernicus would seem from an Aristotelian point of view. Let us now consider his reply to the last argument, which is based on the motion of bodies separated from the earth, i.e., the flight of birds, the movement of clouds, and the vertical fall of heavy bodies. This is the Aristotelian's strongest argument. For, according to his physics, motion is a process which has an effect upon the moving body, which expresses its nature, which 'exists' in the moving body. The heavy body, as it falls, goes from A to B, from a certain point above the surface of the earth towards the latter, or more precisely *towards its centre*. It follows the straight line which connects these two points. If, during its fall, the earth were to move, neither this line (the line connecting point A to the centre of the earth) nor this point, nor the body, would in any way participate in this motion: the motion of the earth would not affect the body which is separated from it. If the earth were to slip away from underneath it, too bad! The heavy body could only follow its path. It could not chase after the earth. It follows that if the earth were in motion a body thrown from the top of a tower could never fall at its foot; equally, a body thrown vertically into the air could not fall back to the place from which it started. It follows *a fortiori* that a cannon-ball dropped from the top of a ship's mast would never fall to the foot of the mast.[25]

This is Copernicus' reply:[26] 'But what do we say about the clouds and other things suspended in the air, and about things which fall or those which, on the contrary, tend upwards? Quite simply that it is not only the earth, together with the element of water which is connected with it, which move thus [i.e., naturally] but also a considerable part of the air and everything else which has the same relation as it has to the earth. Either the air which is close to the earth, mixed with earthy and watery matter, has the same nature as the earth, or the air's motion is acquired and shared without resistance because of the contiguity and the perpetual motion of the earth . . . This is why the air which is closest to the earth appears to be at rest, as do things suspended in it, unless they are moved about by the wind or by some other force . . .

'As for falling and rising things, we believe that their motion must be double in relation to the world, and must in general be composed of both rectilinear and circular motion.[27] Because things which are drawn downwards by their weight are the most earthy; now it is certain that the parts keep the same nature as the whole. And it is for the same reason that this happens to things which are drawn upwards by the force of their fiery nature. For terrestrial fire is above all nourished by earthy matter;

therefore it is said that flame is nothing but burning smoke . . .'

Copernicus' reply, while it is strong enough if it is taken to be an argument *ad hominem*, is in itself very weak. For how is it possible to assert that—given that circular motion from west to east is *natural* for all terrestrial bodies—this *natural* tendency which impels them (and which explains, no doubt, why clouds, the air, birds and falling bodies or bodies thrown in the air, follow the motion of the terrestrial sphere and do not get left behind) does not in any way interfere with their motion from east to west? Heavy bodies are impelled by a natural downwards motion. Therefore it is very difficult to impress on them an upwards motion; if terrestrial bodies were impelled by a *natural* motion towards the right it would be practically impossible to make them move towards the left.

Yet there is already, in a manner of speaking, hidden beneath Copernicus' reasoning, the germ of a new concept which is yet to be developed. Copernicus' argument applies to terrestrial phenomena the laws of 'celestial mechanics'. By this very fact the division of the Cosmos into sub- and super-lunary regions is implicitly abandoned. Moreover, Copernicus' argument puts forward an explanation of the fact that bodies 'do not get left behind', and of the fact that a falling body follows a line which is *vertical for us* and falls at the foot of the tower from which it is thrown: this explanation is found in the fact that bodies *participate in the earth's motion.*[28]

What would have to be changed in Copernicus' arguments for them to be acceptable rather than absurd? Rather a lot: it would be necessary to replace the mythical explanation of the participation of heavy bodies in the earth's motion (their participation in the earth's 'nature') with a physical explanation, or more accurately a mechanical one; i.e., it would be necessary to make explicit the ideas which underlie the argument, and in particular the idea that for a group of bodies *which are impelled by the same motion*, this motion in which they all take part can be discounted. In other words it would be necessary to formulate the idea of a physical system, and to recognise the *physical*, and not merely the *optical* (as does Copernicus) relativity of motion. But in order to really work this through one would have to give up the Aristotelian idea of motion and to replace it with a different one: and this in its turn implies giving up Aristotelian philosophy and replacing it with a different philosophy. For, as will become increasingly clear in what follows, it is not a matter merely of a scientific problem; throughout the whole debate it is a philosophical problem that is involved.

2. Bruno

It was Bruno's achievement to make explicit the idea of a physical system. His work was very uneven, of course; it was tumultuous and chaotic. And it was vitiated, from our scientific point of view, by the deep animism of his thought. Nevertheless, his obscure and muddled thought played an important role in the history of science.[29] This was a positive role because by a brilliant intuition Bruno understood that the new astronomy involved the idea of *infinity*. Therefore, with indomitable courage he challenged the mediaeval vision of the well-ordered and finite Cosmos with his own vision of an infinite Universe. His role was also negative, however, because in connecting his metaphysical and cosmological views (plurality of worlds, even of inhabited worlds) with the new astronomy, and therefore also with the new physics, he made it seem in the eyes of the Church that the former were necessarily associated with the latter, and he thereby became the occult, but real, cause of the condemnation of both Copernicus and Galileo.[30]

In his defence of Copernican astronomy Bruno came up against the same physical objections as had Copernicus. In replying to them it was, of course, the ideas that had been sketched out by the master that he developed. In his development of them, however, he changed them, as a result of his remarkably intelligent use of *impetus* physics.

Against the possibility of the earth's motion Aristotelians had pointed to the winds, the clouds and birds. Bruno replied that since the air surrounding the earth is drawn along by the earth's motion, the motions of winds, clouds and birds take place in exactly the same way as if the air were stationary. As for the argument from vertical fall, this is also dealt with by the same consideration. Bruno writes:[31] 'Your reply to the argument concerning winds and clouds also provides a reply to another argument, that which Aristotle puts forward in the second book of *De Caelo*[32] where he says that it would be impossible for a stone thrown upwards to fall down again along the same vertical, but that the very fast motion of the earth would necessarily leave it well behind, towards the west'.

This famous argument is worthless, in Bruno's opinion, because it fails to take into account an extremely important fact, the fact that the experiment in question (throwing a stone vertically upwards) is performed *on the earth*. Consequently, 'all the relations of straightness and obliqueness are necessarily modified with [the earth's] motion'.[33]

In contrast to Copernicus, who distinguished between the 'natural' motion of the earth and the 'violent' motions of everything on it, Bruno explicitly assimilates them. What happens on the moving earth is precisely equivalent to what happens on a ship moving on the water; for in this case too the overall motion of the ship has no effect on the motions of its 'parts', *'for there is a difference between the ship's motion and the motion of things which are on the ship.*[34] If this were not so then it would follow that when the ship is moving on the sea it would be impossible for anyone to ever throw anything in a straight line from one side to the other: and it would be impossible for anyone to be able to jump up and to fall with his feet in the same place from which he had jumped'.[35] This is precisely what follows from the Aristotelian conception of the matter, although Aristotelians refuse to draw this conclusion and even deny it. As for Bruno, he goes further with the analogy between the motions taking place on the ship and those which take place on the earth. The latter are produced in a quite different way from that envisaged by Aristotle because 'everything which is on the earth moves together with the earth'.[36] The hypothetical phenomena predicted by Aristotle (falling to the west etc.) would only take place *if the origin of the (stone's) motion were external to the earth.*

Of course, if 'something were thrown to earth from a place external to the earth the earth's motion would cause [its motion] to lose its straightness. Similarly it can be seen on a ship, for if it is moving down a river and someone on the bank throws a stone straight towards it, the stone will miss its target, and this to an extent which depends on the ship's speed. But if someone is on the ship's mast then, however fast the ship is moving, he will not miss his target by an inch: so that the stone or any other heavy object thrown towards a point at the foot of the mast, or towards any other point on the ship's deck or its hold, will travel there in a straight line. Similarly, if someone on the ship throws a stone straight upwards towards the top of the mast, or towards the crow's-nest, the stone will fall down again along the same path, regardless of how fast the ship is moving, as long as it is not rolling.'[37]

The originality of Bruno's thought compared with that of Copernicus is quite clear. Bodies which are 'on the earth' participate in the motion of the earth not at all because they participate in its 'nature', but simply because they are 'on it', in precisely the same way in which bodies which are 'on the ship' participate in its motion. What this means, and moreover Bruno actually says this, is that it is not a matter of participation in a 'natural' motion, but in motion *tout court*, of the fact that the moving body is part of a mechanical system. This idea underlying Bruno's argument, of a mechanical system, of a set of bodies interconnected by virtue of their participation in a common motion, has no place in Aristotle's physics.

Aristotle sees motion as a function or expression of the 'nature' of the moving body. He sees it as a journey from a place A to another place B, where these 'places' are seen as specified in relation to the centre and the circumference of the Cosmos. It follows that from a given place there can only be, for a given body, one 'natural' motion. This, in Bruno's view, amounts to saying that Aristotle conceives of 'places' as external to the physical system of the earth. For Bruno 'places' are not specified by their relation to the Cosmos; they are specified by their relation to some or other mechanical system. Thus a given 'place' can belong to different mechanical systems, and bodies moving from this place can perform quite different motions depending on which system they are a part of. This conclusion, from which any Aristotelian would shrink in horror, is drawn explicitly by Bruno.[38]

'Consider two men, one on the moving ship and the other not on it; suppose they both hold a hand in the same point in the air and that from this point they simultaneously each drop a stone, without giving the stones any impulse. The stone of the first man would not lag behind at all and would not deviate at all from its [vertical] path and would arrive at its predetermined place. That of the second would find itself carried towards the rear. And this results simply from the fact that the stone dropped from the hand of the man on the ship, the man who consequently is moving with the ship's motion, has a certain impressed virtue which the other does not have, i.e., that which comes from the hand of the man who is not on the ship: and this is so even though the stones are equally heavy and that, since they leave as near as possible from the same point, they have traversed the same amount of air, and have started with the same impulse. We can give no reason for this difference except that things which are connected with the ship, or are a part of it in this sense, move together with it; and that one stone, that which moved together with the ship, carried with it the motor's virtue, whereas the other did not have this participation with it. From which it is obvious that whatever moves does not receive the virtue to go in a straight line from the point from which it starts, nor from the point towards which it goes, . . . but from the efficacy of the virtue already impressed on it; and it is from this that the whole difference derives. And this seems to me to be an adequate reply to the above mentioned argument.'

Bruno is certainly not wrong, or at least not completely. The idea of *impetus*, a virtue or force which *impels* the moving body, which produces the motion, (the *impetus* or 'the impressed virtue impels as long as it lasts' he says;[39] when we throw something upwards we impress on it a relative lightness,[40] and the medium plays no role in the motion even though it is a necessary condition of it, since if there were no space[41] in which it could take place no translation would be possible) is in fact enough to overthrow the system of Aristotelian physics. It is enough, in particular, in order to establish the idea of a physical system of bodies, to explain their unity, their continuing interconnectedness even in the absence of contact. However, it is far from enough for the establishment of the new physics, or even to serve as the foundation of Copernican astronomy. It is not even enough as the basis for Bruno's own physics. For while *impetus* physics is certainly compatible with the distinction which he makes between the motion of the ship and that of bodies on the ship, it is not at all equivalent to this. In fact none of the defenders of this famous theory before Bruno had had the idea of drawing from it the consequences, in relation to the ship, which Bruno himself drew.

This distinction, which is broadly speaking equivalent to the principle of the relativity of motion, implies, as we have seen, the formal negation of the Aristotelian theory of place. It would actually be more accurate to say that it is this negation which gives rise to the distinction.

It is worth repeating that the Aristotelian theory of place is *metaphysically* based on the idea of the Cosmos, a well-ordered system of objects each of which has its own nature: a system in which the geometrical (or spatial) arrangement (or distribution) expresses, and is explained in terms of the differences between these 'natures'. It is supported, as far as physics is concerned, by the theory of the 'natural' motion of bodies, i.e., by the fact of the 'downwards' motion of 'heavy' bodies and the 'upwards' motion of 'light' bodies.[42]

Now it is both this support which Aristotelian doctrine derives from physics, as well as its metaphysical foundation, which Bruno explicitly rejects.

First the supporting physics: Bruno says[43] 'the theory of heaviness and lightness to be found in Aristotle is completely false. Concerning this we assert the following most true propositions: heaviness and lightness are not said of natural bodies, naturally constituted, neither of their whole spheres nor of their parts, if it is suited to the terrestrial sphere and any fixed star to have parts in one and the same place'.

This is, as we have seen, the theory already proposed by Copernicus. Bruno continues:[44] 'Heaviness and lightness are nothing other than the impulse of the parts towards their place, in which they either move or remain at rest . . .; this is why any particular part is taken to be at one time heavy at another time light; however, where it is born and where it should be it is neither heavy nor light. Consequently 'heavy' and 'light' is only a relative difference, and in relation to absolute differences of the local world it is nothing. Therefore Plato was right to say in the *Timaeus* that in the heavens there is nothing which is above and nothing which is below because all of the parts thereof are the same'. And to make it clear that in the great debate between Aristotle and Plato (and this is a very significant piece of information) he adopts the position of the latter against the former, Bruno adds: 'It is useless for Aristotle to attempt to raise objections against this'. It could not be otherwise because it is usually Plato who is right on the theory of gravity; it was Plato who said, again in the *Timaeus*, that heaviness and lightness do not exist *qua* qualities of bodies, but that there is only more or less heavy or light: 'heavier is that which is [made up of] more, and lighter is that which is [made up of] less [parts]'.

As for the metaphysical or cosmological foundation, it is sufficiently well-known as to need no emphasis that Bruno, as we have mentioned, was one of the first, if not the first, to assert the infinity of space and to oppose to the traditional finite Cosmos his infinitely infinite Universe, and to take to its logical conclusion what Copernicus had only hesitantly hinted at, the identification of the earth with the heavens.

'The world' he says,[45] 'which the ancient philosophers said was once created and thereafter eternal . . . is not the Universe, but this machine and other similar machines. . .'

It is no longer only the earth which is identified with the planets in an enlarged but still limited 'world'. The sun itself, which in Copernicus occupied the centre of the Universe, has lost its privileged position. No doubt it retains its central location in *our* world; but our world, the solar system, is only one 'machine' among the infinite number of other 'machines' which fill Bruno's infinite Universe. Therefore the sun is not at the 'centre' of the Universe, because in this infinite Universe, where an infinite number of stars, other suns, move according to eternally fixed laws, there is no more a centre than there is a circumference. There is no limit to the infinity of space.[46] Therefore nothing could be more absurd than Aristotle's effort to base his cosmological finitism on an alleged analysis or classification of motions. Upwards motion! Downwards motion! For Bruno 'up' and 'down' are purely relative concepts, just as relative as 'left' and 'right'. Everything is to the left or to the right of something, and everything is up or down as one likes. As for circular motion 'around the centre', any point in space can be taken as the 'centre' because none is the centre in reality. Each point in infinite space is equivalent, and each inhabitant of each of the stars can believe himself to be at the centre of the Universe and, consequently, stationary.

Each inhabitant of each of the stars! This is a dangerous idea, and one for which Bruno, and Galileo, would pay dearly.

Each inhabitant of each of the stars can believe himself to be stationary. But none has the right to do so. For the infinity of Bruno's Universe implies the complete geometrisation of space: no more 'places' or directions with special status.[47] This in turn implies that space and bodies are not in themselves affected by motion[48] or by rest.

Space does not resist the motion of bodies. Why would it do so? The journey of a body from one place to another is not its passage from 'its' place to another which is not 'its' place. All 'places' are places, because they are all equivalent. For precisely the same reason a body never resists motion, for it always moves from its place to its place. Thus all bodies have the same capacity for motion; and for rest, because being at their places they have no tendency to go anywhere else.[49]

It is clear that space is the real 'place' of bodies. This is where Aristotle's 'places' really are. For these (the surrounding surfaces of bodies) are themselves in Bruno's space. The Universe itself is located in space: an immense, infinite void which under-lies and accommodates the real.[50]

Bruno rejects[51] the Aristotelians' objections concerning the impossibility, logical as much as metaphysical, of infinity, and concerning the physical impossibility of a void. Quite the contrary, it is Aristotle's finiteness (the bounded Cosmos) which is unknowable, false and impossible; it is the infinite which is known, true and even necessary;[52] and note that this is the infinite not potentially but actually, because according to Bruno matter itself is everywhere and always actuality. As for the void it is explicitly identified with space, which contains all bodies. The void is infinite and its parts are everywhere *beneath* bodies. No doubt empty space does not in fact exist, except at the point of contact between bodies, for it is the air or the ether which fill space. But this in no way goes against the fact that the void, metaphysically and in itself, is something different from the bodies which fill it. It is the necessary support and receptacle[53] of that which fills it. Now, this is the true meaning of the term 'place', the correct reply to the question 'Where?'. Where are bodies? In the void, replies Bruno, in space, which is their common receptacle and which is the 'fixed place' in which everything is contained. It is a 'fixed place' because it is infinite, and because the infinite as such cannot be moved.[54] All finite things, on the contrary, can be moved. Aristotle asserts that motion presupposes a 'place', and that this would be made impossible in a vacuum (motion in a vacuum would be instantaneous and would be at infinite speed). Not at all!, says Bruno. It is not a 'place' which is presupposed by motion, but space. The void, far from making it impossible, is on the contrary its necessary condition. All motion takes place in the void, towards the void, and even because of the void.[55] Moreover, motion in the void never takes place instantaneously with an infinite speed.[56] Aristotle's argument is unacceptable.

One can only be amazed at how daring and radical was Bruno's thought. It accomplished a transformation, a revolution, in the traditional picture of the world and of physical reality. The infinity of the Universe, the unity of nature, the geometrisation of space, the rejection of place, the relativity of motion: we are so close to Newton. The mediaeval Cosmos is destroyed. One could say that it has disappeared in the void dragging Aristotle's physics with it and opening up space for a 'new science' that Bruno, however, could not himself construct.

What was it, then, that prevented him from making this advance? No doubt it was in the first place the very exuberance of his thought, its religious inspiration, its animist character, the emotional value for him of 'the Universe', the great chain of being. But it was also the facts, experiments, phenomena.

Bodies fall, the earth spins round, the planets trace out circles around the sun. Aristotle explains these things; Bruno, fundamentally, does not know how to do so.[57] It is here that Bruno is at his weakest. Because it is not enough to confront Aristotle's physics with a *metaphysics*. It is another *physics* that is needed. No doubt it was only from a *new metaphysics* that the new physics could emerge. But Bruno's animist and anti-mathematical *metaphysics* could not give birth to it; it could only encourage him to stick with the old Parisian physics (*impetus* dynamics), that of Copernicus. So we observe this strange drama of a man who had been lifted so high and carried so far by his deep metaphysical intuition, but who turns back, stumbles and falls. *Impetus*, the force-cause of motion, the tendency of the parts of a Whole to come together, the natural circular motion of Wholes, the natural circular motion of the spheres, the stars guided by souls.[58]

But we should not be too harsh. Thought abhors a vacuum; a scientific theory only disappears when it is replaced by another. And this other theory, only Newton achieved that.

3. Tycho Brahe

To us the arguments with which Bruno battled against the positions of Aristotle seem reasonably convincing. It must be said, however, that in his own day they convinced nobody. Not Tycho Brahe who, in his polemic with Rothmann, calmly put forward the old Aristotelian arguments (although dressing them up a bit to make them more modern[59]); not even Kepler who, although he was influenced by Bruno, felt himself obliged when fighting the Aristotelians, to go back to Copernicus' arguments, although he strengthened them with a new concept, or substituted for Copernicus' mythic conception of the community of nature the physical conception of attractive force.

Tycho Brahe's objections against the earth's motion—and against Copernicus' arguments—are not entirely without merit. Basically he is quite right in saying that the idea of 'natural' (as opposed to 'violent') motion scarcely allows that a body can be impelled by *two* motions of this kind, and in adding that these motions would necessarily impede each other. He is equally right when he estimates that the explanation given by Copernicus for the motion of bodies separated from the earth (clouds, air, etc.) is 'hardly probable'. It is strange, however, that without noting that this is a view held by both Copernicus and Aristotle, he goes so far as to formally deny the principle-axiom on which this is based: it is false, he says, that a part separated from a Whole conserves its 'virtue'. Quite the contrary, it could be said that this is never the case.[60]

Tycho was less fortunate in taking up the classical objection of the body falling from a high tower and of the body thrown vertically into the air.[61] But the argument appealed to him. He correctly saw it as the strongest of the Aristotelians, objections (and Tycho, in spite of his innovations in astronomy, remained an Aristotelian in physics). Thus he refused to accept Bruno's reasoning. He wrote to Rothmann:[62] 'Those who believe that a cannon-ball thrown upwards on a moving ship will come back down at the same place as it would if the ship were at rest are seriously mistaken. In fact the cannon-ball will be left behind to an extent that depends on the speed of the ship'.

Perhaps, for us, Tycho's obstinacy puts him in a bad light. But we should be fair. We should try to understand what it is about Bruno's statement of the case that makes it so implausible from an Aristotelian point of view.[63]

It was not enough, however, simply to reject Bruno's arguments or to appeal to experience. Tycho was concerned to modernise the classical debate by introducing into it a recent invention—the cannon.[64] 'Now, I ask you what would happen', he wrote, 'if one were to fire a cannon-ball from a large cannon towards the east . . .; and if then, from the same cannon and from the same place, one fired another one . . . towards the west? Are we to believe that both of them . . . will travel equal distances over the earth?'

What Tycho is hinting at is that in order to perform this feat the cannon-balls would somehow have to know what they had to do, in particular they would have to know that they must conform to the (Copernican) theory according to which all 'terrestrial' objects must follow the earth as it moves. Tycho did not accept this theory. Moreover, he believed that even if one were to accept it for the case of bodies at rest in relation to the earth, it would be impossible to apply it to the case of the cannon-balls, or even to that of freely falling bodies. From Tycho's point of view this cannon-ball case is quite different from the former. The cannon-ball is in violent and extremely rapid motion; how could this motion coexist with the 'natural' motion of rotation without impeding it, and without being impeded by it? For, as we have pointed out several times above, from the point of view of pre-Galilean physics, nothing could be less plausible than the mutual independence of motions. To accept this would be to accept a cause which has no effect. Thus Tycho Brahe goes on to explain[65] just how odd it would be if 'the extremely violent motion caused by the gun-powder in opposition to nature' and the natural and extremely rapid motion of the earth could combine without resistance. For, according to Copernicus and Bruno[66] 'the ball fired from the cannon would have three motions: one by which it would, as a result of its heaviness, tend to move in a straight line towards the centre of the earth, another by which, because of its common nature with the earth, it would untiringly imitate its rotation, and a third produced . . . by the violence . . . of the gun-powder explosion . . . which would impel the ball to move very rapidly where, by its own nature, it has no wish to go. Now this extremely violent motion impedes the other, i.e., that by which heavy bodies necessarily and naturally descend in a straight line; therefore it is only after having travelled a great distance, and only after this violence is exhausted and has gradually died down that the cannon-ball can come down to earth. I ask, therefore, what will be the effect of this second motion (namely the circular rotation) . . . and how is it that it is not completely impeded by this very violent concitation, produced against nature? Experience shows that a ball of a given size and weight fired in opposite directions by the same amount and strength of powder, covers approximately the same distance over the earth's surface, the same towards the east as towards the west, as long as it is fired, as I have said, with the cannon at the same inclination and as long as the air remains calm and there is no accidental cause promoting or impeding the impulsion. Yet, in following the extremely rapid diurnal motion of the earth (if this existed) a ball, when fired towards the east, would never be able to cover as much distance over the surface of the earth, the earth by its own motion going ahead of it, as a ball fired in the same way towards the west. . .'

Tycho Brahe believes, then, as a good Aristotelian, that the ball's violent motion prevents it from falling to earth. He thinks that this motion overrides the motion of fall, and this not because it carries the ball upwards but simply because it is there in the ball and is more rapid, stronger, than that which would carry it towards the centre of the earth. Therefore he cannot understand how it could happen that a ball impelled, according to Copernicus, by an *extremely rapid* natural motion could come under the influence, as if nothing were happening, of the violent motion. Tycho believes that if terrestrial bodies were actually impelled by this motion, more rapid than that of a ball fired from a cannon, then the two motions would impede each other. One would override the other, and all cannon-balls would always fly in the same direction. So if the earth were turning they would cover, *in relation to the moving earth*, different distances when fired in different directions. Now this does not happen, and this is because the natural motion in which the balls are supposed to be participating does not exist. The cannon-ball is only impelled by the violent motion.

We should not take Tycho lightly. Basically, when he says that it is impossible to accept the earth's motion as long as nobody can produce new and stronger arguments to show clearly that the violent motion is not prevented, nor even at all influenced, by the natural motions (of fall and of the earth's rotation), he is perfectly right.[67] This is why, as we shall see, Galileo, following Kepler, spent so much time on this problem.

4. Kepler

Kepler's counter-arguments are of very great interest. Not that they contribute a definitive solution to the problem. But they demonstrate for us, once again, just how original and uncommon was the thought of Bruno and Galileo. They demonstrate the difficulty of the obstacles which had to be overcome. Finally, they reveal the fundamental—philosophical—origin of these difficulties.

For it was, in fact, philosophy, ontology, metaphysics, that were involved throughout this debate: not pure science. It was for philosophical reasons, much more than for purely scientific reasons, that Kepler was held back—and to which we owe the term *inertia*[68] itself—and which prevented him from laying the foundations of the new dynamics.

From a strictly scientific point of view Kepler's was certainly the foremost intellect of his age. Did he not combine mathematical genius of the very highest order with an unparalleled boldness of thought, boldness which enabled him to free astronomy, and thereby physics and mechanics, from its obsession with circularity? Did he not write a *Physics of the Heavens*[69]—a conjunction of terms as astonishing for his time as that of *Creative Evolution* has been for ours—and declare, following Plato, the reign of geometry in the material world?[70] Yet philosophically he was closer to Aristotle than he was to Descartes or Galileo. Philosophically he was still of the Middle Ages. For him motion and rest are opposed in the same way as are light and darkness, as being and lack of being.[71] Therefore he required a cause to explain the existence and the persistence of motion; he did not need one to explain rest, nor coming to rest.[72]

Kepler abandoned, of course, the classical conception of the 'natural places' of bodies. The 'natural place' of a Keplerian body is space; and Keplerian space, like that of Bruno, is sufficiently homogeneous for each 'place' in it to become for each body a 'natural place'. Therefore a body will remain in this natural 'place' as long as it is not removed from it by a force. It will not leave it by itself because it is, on Kepler's view, *inert* and without natural tendencies. Furthermore it will *come to rest* by itself in each place unless a force pushes it or drags it somewhere else. This is also an aspect of *inertia*. Thus for clouds, birds, or stones which have been dropped or thrown, to follow the earth in its diurnal motion they must, for Kepler, be drawn along with it or by it or, in other words, they must form together with the earth a real unity or system.[73] Now, thanks to the magnetic force of attraction this physical unity does actually exist. A stone, the clouds, the air, are attached to the earth, connected to it as if by straps or chains. This is the explanation of the vertical projectile, and of other phenomena which Tycho and his defenders, not knowing about this force, could not understand or explain.[74] For they did not take into account the fact that the earth and everything which is on it or part of it form a real unity. They imagined that things fly around in the air isolated from the (magnetic) action which the earth exerts on them. To put it another way, they imagined that the physical situation of things which are in close proximity to each other is the same as that of objects which are at a great distance from one another. In fact[75] 'while the attractive force of the earth, as was said before, does extend a long way upwards, it is nevertheless true that if a stone were separated from [the earth] by a distance of the order of the earth's diameter, the stone would not keep up with the moving earth; its resisting force would combine with the earth's attractive forces, and thus it would to some extent shield itself from the sway [raptus] of the earth'. In Kepler's opinion, 'violent motion releases projectiles to some extent from the earth's raptus, so that those projected eastwards run ahead of it and those projected westwards are caught up by it; therefore under the compulsion of this force they move away from the place from which they are projected: the earth's raptus cannot impede this violence in solids as long as the violent motion still retains its strength. But since no projectile becomes separated from the earth's surface by as much as a hundred thousandth of the earth's diameter, and since even clouds and smoke which contain a minimal amount of terrestrial matter do not reach as high as one thousandth of the [earth's] radius, the resistance to motion and the inclination to rest of clouds, smoke and things which are thrown vertically into the air, can achieve nothing. I mean that it can do nothing to impede this raptus because the strength of this resistance is in no way comparable to that of the raptus.[76] Therefore anything thrown vertically into the air will fall back to its place, and is not at all impeded in this by the earth's motion, which cannot step aside beneath it but pulls along with it all things flying in the air, these being no less attached to it by the magnetic force than if they were joined to it.

'If these propositions are grasped and diligently considered by the mind, not only will the absurdity and the falsely imagined physical impossibility of the earth's motion disappear, but also it will be clear how to reply to the physical objection however it is formulated'.

In Kepler's opinion his own views were no different, except in form, from those of Copernicus.[77] 'Copernicus was content to endow the earth and all earthly things, in that they come from the earth, with the same motive soul which, in turning the body of the earth turns also all particles which come from the earth's body. . .' But according to Kepler, Copernicus only attributes a soul to the earth so as to enable 'the force of this soul, distributed throughout all the particles, . . . to be the agent of violent motion'. Thus this doctrine is, according to Kepler, 'excessive'. It is unnecessary to postulate an animal faculty where a corporeal faculty would suffice. Kepler says, therefore, that 'the force of the corporeal faculty (which we call gravitational or magnetic force) makes itself felt in corporeal motions in its drawing along of bodies attracted by the earth and in thus causing them to participate in its motion'.

No doubt this corporeal faculty is more acceptable than the animal one, about which, in fact, Copernicus never actually spoke a single word. It was Kepler himself who had earlier believed in the souls of the planets and it is fascinating to see that, for him, Copernicus' 'nature' and soul are one and the same thing. There were, however, plenty of people who thought that the one was no better than the other and who were unable to understand or to swallow Kepler's views. He therefore found himself obliged to return to the attack and to deal explicitly with Tycho's objections.

Kepler's friend Fabricius,[78] in a letter of 26 January 1605, put the following question, concerning a passage from Tycho's *Letters*[79] in which Tycho had given an account of the arguments he had employed in refuting Rothmann, a defender of Copernicus. Fabricius says: 'What is your reply, as a Copernican, to Tycho's argument of the cannon shot? It is clear that if the cannon is fired towards the east the ball, because of the more rapid motion of the earth, will come to rest rather towards the west, and would not be able at all to move towards the east. This is clearly an argument of Herculean strength against the diurnal motion of the earth, and once this is destroyed everything else falls easily'. No doubt. But, Kepler replies:[80] '. . . as for Tycho's argument, using the cannon to throw doubt on the earth's motion, what you ask is the same as that recently asked by the Chancellor of Bavaria. Therefore I give the same reply; the motions combine and the one does not hinder nor abolish the other. The earth moves from west to east, and with it all the air with which it is surrounded, and all heavy bodies, whether they be on the earth's surface or in the air. For why not heavy bodies in the air? What would prevent them? Gravity? But this tends towards the centre of the earth, towards the centre of the surface of the earth which faces the stone; and the earth, by a magnetic force, attracts the stone more strongly than if it were attached to it by a hundred chains and most strongly tensed ligaments. Could the air which it must travel through prevent it? But this also follows the earth, at least when it is this close. So what would prevent it? You cannot point to anything. So I will point out what it is that prevents [the stone from following the earth's motion], but at the same time I will reply [to these objections].

'All material bodies, in themselves and by nature, are stationary in whatever place they are at. For rest, just like darkness, is a kind of privation which does not need to be created but adheres in created things as a kind of nothingness. Motion, on the contrary, is something positive like light. Therefore if a stone is in local motion this is not because it is material but because something external pushes it or pulls it or because of some intrinsic faculty to tend towards something. The Aristotelians say that it tends towards the centre of the world. This I deny, for if it were so it really would be prevented from following the motion of the earth. They bring forward, I know, futile arguments based on the contrary nature of fire, but this is a *petitio principii*. For fire does not seek for the heavens but flees from the earth.[81] This is why I only define gravity, i.e., that force which moves the stone intrinsically, as a magnetic force which brings together similar things, which is numerically the same in large and in small bodies, and which is divided according to the mass of the bodies and takes on the same dimensions as the bodies. Thus if a stone were located close to the earth, a stone of which the mass was of the same order as that of the earth, and if they were both free of all other motions, then, I say it would happen that not only would the stone move towards the earth but also the earth towards the stone, and they would divide the space between them in inverse proportion to their weights. If the masses are as A to B, then the ratio BC to CA would be the same, and C would be the place where they would come together; this is clearly the same ratio that the arms of a balance would have.'

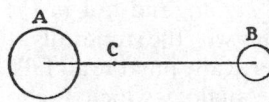

But let us go back to the question:[82] 'I have said that if you were, in thought, to withdraw from the stone that faculty which brings similar things together there would remain in the stone only a complete impotence to change place. Then only an external force and exertion could remove it [from the place where it happened to be]. Now let us imagine a stone suspended in the air, and let us withhold from it the force which brings it together with similar things, i.e., gravity, but we will leave this force of the earth acting on the stone. Let it be thus, even though this might be absurd in reality, so as to argue our case. This stone suspended in the air will then have the force to stay in its place and thereby to oppose the earth's force of rotation. The struggle of these material and corporeal properties will produce a combination such that each force will win or be beaten in proportion to the bodies.

'Thus is produced that which I promised to explain, namely the impediment whereby such a stone suspended in the air is unable to perfectly follow the rotation of the earth. Now this impediment is a very real one. So let us destroy our imaginary case and postulate that the links which are stretched between the surface of the earth and the stone are not like strings but really as is indicated to us by nature, namely like highly tensed ligaments such that the stone would actually descend towards the surface and the centre of the earth. I say that because of its innate powerlessness to move it would necessarily happen that in its descent the stone would be displaced somewhat from the vertical line which runs from the centre of the earth, through its surface, to the centre of the stone; therefore, since the earth moves from west to east, the vertical from the stone would be displaced somewhat towards the western parts of the surface. It would not altogether follow the earth but would fail to keep up with it. This then is the cause whereby the stone would not follow the earth, which you needed in order to confirm your view but were unable to point to.

'Listen now to the solution. It is true that if the stone were at a significant distance from the earth then things would happen thus. But it is 860 miles from the centre [of the earth] to its surface, and in fact no bird flies so high that it is half a mile from the earth; for a bird is no more capable of flying in the ether than we are of flying in the air, or than the stone is of swimming in water.' Kepler draws the conclusion from this that the lag due to the stone's resistance to being drawn along[83] would be minimal and in practice imperceptible. Tycho's error is thus in his failing to understand the nature of gravity, and consequently in his being mistaken as to its action. For Tycho, along with all the Aristotelians, believed that gravity is a tendency belonging *to the stone*, a tendency which carries it *towards a particular place*; whereas in fact it is a force which resides in the particles of matter which make up bodies, a force which is in the stone but also in the earth. In fact for all practical purposes it resides in the earth and *attracts* the stone externally.[84] Therefore the moving earth draws the stone with it and makes it follow its motion. As for the stone's resistance, in practice this plays no part, because the ratio of the forces is equal to that of the masses of the bodies. Thus it is clear that it is *a real physical action*, and not just *a mechanical state* (the state of motion), which explains why the stone falls at the foot of the tower, and that a cannon-ball fired vertically into the air falls back to the place from which it started. As for the analogy which Tycho tried to establish between the horizontal motion of the ball, a motion which prevents it from falling, and the more rapid motion which the ball would have were it to participate in the earth's circular motion, this analogy does not exist because the participation in the earth's motion, i.e., the ball's being drawn along, is a result precisely of the attraction of gravity. Moreover, here again the ratio of the forces is the same as that of the masses, and the force of the gun-powder explosion, however large, is nothing in comparison with that of the earth's attraction.

So Kepler continues:[85] 'Now we come to Tycho's cannon. Since it has been shown that the stone falling vertically downwards does not have to deviate from that line as it falls, it is easy to deal with the cannon-ball (and with the stone thrown at an angle, clouds moved by the wind, and birds flying in the air). For what I said right at the beginning is true; the two motions in the ball, both that which comes from the earth and that which comes from the cannon, combine together. Similarly the distances combine. For when the ball is fired towards the east then, of course, a greater distance, *relative to the Universe as a whole*, will be covered in a given time than when it is fired towards the west. For in the former case the earth also tends towards the east, while in the latter case the earth opposes the westward motion and draws the ball along towards the east. It is even clear that in relation to the whole Universe the ball will never tend in a direction contrary to the earth's motion, because this is far more rapid than that with which the ball is projected. But as for the distance considered in relation to the earth's surface we have already shown that a stationary stone, suspended in the air, must follow the earth; it follows that a given force will propel the stone the same distance on the earth towards the west as towards the east. For whichever way it impels the stone the stone's attractive force is the same and so the effect of the propelling force in moving the stone is the same. But if, as mentioned above, the stone's fall deviates somewhat from the vertical then it will result from this that the distance projected towards the west will be shorter than towards the east. But this will not happen for the reasons given by Tycho, but for that which I have carefully explained here.'

Kepler believed that his replies, both to the classical objections and to the more recent ones, were quite adequate. In fact, however, they leave a great deal to be desired. For, just as with Copernicus' reply, if it has been explained how the cannon-ball, the stone and the clouds, etc., are drawn along towards the east, their free movement, towards the west just as much as towards the east, has not been explained at all. It could even be said that these explanations would make it completely impossible. For how could a projectile, however violently fired or thrown, overcome the enormous power of the 'bonds' of the earth's attraction? So it is not very surprising that Fabricius, completely unconvinced, returned to the attack.

Kepler, a bit irritated, has to reply all over again. He writes to his friend:[86] 'You want me to explain to you the answer to Tycho's argument against the earth's motion. [His argument] is not as fierce as the impact of this machine [the cannon]. Clearly it coincides with the following objection: how is it possible for a ball thrown vertically upwards to fall back to the same place if the earth has moved in the meanwhile?[87] The reply is that it is not only the earth which has moved in the meanwhile but also together with it those infinite and invisible magnetic chains by which the stone is attached to the subjacent and surrounding parts of the earth, and by which it is drawn vertically closer to the earth. Now with violent upwards motion all the chains are more or less equally stretched: but when the ball is propelled by the force of the cannon towards the east it is the chains to westward which are stretched, whereas those to eastward [are stretched] when the vapours [of the cannon] propel the ball westward. Now the one is not promoted nor the other impeded by the overall motion of the earth together with all the chains. For the violence of the motion with which the ball is projected takes place at the interior of the complex of all the chains, which are so strong that even the strongest contrary wind can do nothing against them, so they remain undisturbed and turn with the earth'.

The picture of the situation as Kepler imagines it is quite clear. The stone suspended in the air is attached to the earth by an infinite number of elastic 'chains' or 'ligaments'. Together they form a cone with the stone at the apex. It is thus pulled from all sides, but the oblique 'chains' pull with equal force and thus cancel each other out. Therefore the overall resultant of all the forces is directed vertically 'downwards', i.e., towards the earth. If we try to push the stone upwards the tension in all the chains counteracts this. If we try to push it to left or right we only have to overcome the resistance of half the chains involved. But whatever the direction of the horizontal motion the number of chains, and hence the resistance, will always be the same. Of course if the earth is turning then the chains which are pulling (those on the east) will be somewhat more stretched than the others. But the difference will be very small. And the stone's own *inertia*, its resistance, is not of the same order of magnitude as the earth's attractive force. In practice it can be ignored. If there were no attraction things would be quite different; but in this case they would, however, also be quite different from what Aristotle thought; for it is clear that[88] 'if there were no such chains the stone would remain suspended in the ether while the earth moved away, and there would be no reason for it to fall'. Thus the presence of the attracting chains explains both fall and the drawing along of the projectile; and the effect of the *inertia* is very much smaller because the length of the trajectory and the speed of the cannon-ball's motion are not of the same order of magnitude as the earth's diameter and the speed of rotation of the earth and the chains of magnetic force. 'Therefore, since this is how things are, and since I am convinced in my own mind of this, do not ask me to betray the truth so as to gain the favour of the crowd.' It is too bad if the world does not accept the light of truth. Kepler is resigned to this. He knows that darkness can be victorious: did not the night of error extinguish the torch lit by Aristarchos of Samos? But Kepler is also full of confidence: for did not Aristarchos' work reappear in that of Copernicus? *Magna est veritas et prevalebit.* The truth will prevail. At least Kepler hopes that it will, even in the mind of Fabricius, who will cease to be seduced by absurd objections. So he continues, unwearied:[89]

'Your objection concerning the winds obviously imitates their nature for it carries no great weight. I readily accept that which is correct in your argument: if the air were stationary, with the earth moving through it, then your objection from the experience of the winds would be correct. But actually the vapour, the wind's matter, is contained within the complex of the earth's magnetic force; and since it is of tenuous substance it is not strongly attracted toward the earth. Therefore any force can easily displace it and detach it from the magnetic bonds of the earth. The magnetic force has its greatest power there where it resides, in the earth, the most dense body; it is weakened in an object of which the matter is rarer. It can be compared with the force which produces violent motion. A boy can throw a small stone a great distance. But even using all his strength he could never throw a pumice-stone of the same size as far.

'But let us go back to the vapours: they are drawn along by and keep up with the earth, being subjected to its magnetic force, so that they are at rest in relation to the subjacent parts of the earth as long as no cause acts to push them away from them.' Such a cause would easily move them through the magnetic chains, regardless of direction. For, as Kepler explains, if one is investigating the motion of bodies within the field of terrestrial gravity the only thing that needs to be considered is movement across the earth, the number of magnetic chains that are crossed, and not the absolute distance through the ether. Then Kepler recalls, in a way that only confuses the issue, a quite different example, that of Bruno's ship.[90] 'For in order to pass through the ether bodies do not need any effort of their own; they can rely on the force of the earth, or the ship. Thus the example of the ship, and of the movements on it of the passengers that it carries, is obviously relevant here, although the ship does not attract the things which it carries by a magnetic force but draws them along solely by contact, while the earth still attracts them with the force of gravity which it does not communicate to the ship's motion. By contrast the ether does not attract vapours and projectiles, so that these are only attracted by their ship, i.e., by the earth.' Therefore when bodies move away from the places on the earth towards which they tend by the 'moments of gravity', 'their movements are not affected by the motion of the earth as are the motions of projected bodies as a result of the ship's motion; for they do not tend towards any part of the ether, but are only attracted by the subjacent surface of the earth by means of magnetic chains'.[91]

Kepler clearly does not accept Bruno's purely *mechanical* point of view. He does not accept that motion will continue and persist in the moving body. And he does not accept that the fact of participating in a common motion creates a connection between things and that this *in itself* separates them from the rest of the Universe. The idea of a mechanical system does not exist for him; he only knows about physical systems, real unities, *real* links or connections. The ship is one thing, the earth is something quite different. In practice, no doubt, the difference is imperceptible, and Tycho is wrong about both cases. But theoretically there is a distinction. Tycho is right theoretically, philosophically. Motion and rest are not on the same ontological level. Motion is a form of being. Rest is only a privation of it.

Moreover we must admit that this is in agreement with experience and common sense. How otherwise could we explain that a force or an exertion is necessary to set bodies in motion, and that this force must be proportional to the body, to its mass? If material bodies were unaffected by motion or rest would not the fact that a *greater* exertion is needed to produce a *more rapid* motion in a body, or to set a *larger body* in motion, remain incomprehensible?

How can it be denied that motion, much more than rest, needs to be explained in terms of a cause? For nobody, except perhaps Descartes, had ever asked why it is that there is rest in the world; on the contrary, everybody had always searched for the cause or origin of *motion*. Nobody, except Descartes, had ever thought of the idea of the quantity of rest; everybody had always spoken about the quantity of motion.

Kepler never changed his mind about this. While he was quite capable of moving from a cosmic vitalism or animism to a 'physical' conception of things, and while he was quite capable of geometrising matter to the point of taking away from it any inherent propensity to motion, yet he never accepted the ontological equivalence of motion and rest, nor that matter has no inherent tendency towards the one rather than the other. Inertia always remained, for him, a force of *resistance to motion*. It never became, as it did for Galileo and Descartes, simply a persistence in a particular state, because he never came to think of motion as a *state*. Thus the account we find in the *Epitome of Copernican Astronomy* is along the same general lines as the theory developed in the *New Astronomy*.

Once again we find here the famous argument about bodies thrown into the air:[92] 'If the earth revolved on its axis then objects projected vertically upwards would not fall back to their original place, from which they were projected, because, the centre remaining stationary, the place on the earth's surface where the projector was located would in the meanwhile have moved away from the line drawn from the centre of the earth to the projectile. If heavy things tended towards the centre *per se* this argument would be conclusive. But as has been said in the above paragraph the centre is not *per se* and primarily the goal of the motion of heavy bodies; it is so only *per accidens* and secondarily; i.e., heavy bodies go towards the centre only because it is the middle and the most interior of the body which they seek primarily and *per se* and to which heavy things are attracted.

'Now, since heavy bodies seek the body of the earth *per se*, and are sought by it, they move more vigorously towards the proximate parts of the earth than towards the remote parts. Consequently, when the closest parts, located vertically underneath, are in motion, heavy bodies falling towards this moving surface follow it in its circle just as if they were attached to the place vertically beneath them, and similarly by an infinite number of less strong oblique lines or ligaments, which all gradually contract themselves.

'But it has been said that material bodies, by their natural inertia, resist motion which they receive from outside. And if this were true then heavy bodies would to some extent become detached from this vertical *raptus*, and from the other bonds. They would in fact become to some extent detached from them if they were at a distance from the earth of the same order of magnitude as its radius, or at least at a distance comparable to the visible horizon.' But since this never happens this lag will in fact be imperceptible because the attractive force of the earth is incomparably greater than that of the bodies' inertia. As for Tycho's objection, Kepler once again gives an accurate account of it:[93] 'Cannon-balls, one fired towards the east and the other fired towards the west, would fall at different distances from the original place. The distance towards the west would be greater because the parts of the earth to westward move eastwards towards the ball; the distance would be smaller towards the east because the eastward part of the earth where the ball would have fallen if the earth had remained at rest moves away from the ball towards the east'.

But, yet again, Tycho's reasoning is incorrect. Tycho argues as if the cannon-ball were at a great distance from the earth rather than on it. Taking up again the parallel case of the ship Kepler explains that, of course, relative to the earth, i.e., relative to the stationary banks of the river on which it is moving, there is a perceptible difference between when an object is projected forwards and when it is projected backwards in relation to the ship's motion. From the point of view of the earth the distance covered will not be the same and nor will the projectile's force and impact. The force of the projection and that of the ship's motion will be added or subtracted in the two cases, but all this, and all these composite causes and their effects, have no existence for someone on the ship. For him, regardless of whether the ship is stationary or moving, the object always has the same weight, and when thrown it always covers the same distance.

'We can apply the same argument, *mutatis mutandis*, to the case of the cannon. Consider a large cannon-ball, of which the flight through the air lasts for two minutes, and which travels westwards one German mile over the earth; meanwhile the earth at the equator will have travelled eight miles in the other direction. So, in relation to the space of the world the ball is carried with a violent motion in this direction, towards the east, for a distance of seven miles, and the explosion by which it was fired in the opposite direction to that of the earth's motion accomplishes no more than to subtract the eighth mile and to slow down the ball's eastward motion. For the powder cannot completely free the ball from the grip of the earth, and it always remains within the field of its attractive virtue. . . The other ball, fired towards the east at the same time, is carried eight miles by the earth's *raptus* and, since it is violently projected by the explosion towards the east, it adds a ninth mile itself. So whether it is fired to east or west it will always travel eastwards, though somewhat more in the one case than in the other. Now this composed space of the world is quite different from space on the earth, that which men are able to measure. The space covered by the balls on the earth is roughly the same in each case, since the force is the same and the magnetic bonds are the same. . . .

'Now the westward motion results from a combination of two causes. For the cannon-ball is in itself *inert* to motion, and if it were not carried towards the east it would, by itself, remain in the west, and as the place moved away towards the east a violent motion should impel it more easily to the west than in the opposite direction. For the eastwards motion has to overcome not only the earth's magnetic attraction but also the inertia of the ball's material, which holds it back . . . But this force of resistance of the ball is not measurable, and the two forces are not comparable one with the other. In fact if the ball were located outside the field of the earth's attractive virtue and were fired by the powder with the same explosive force, it would fly through the space of the world not for a distance of one mile or eight miles but for an incredible distance.

'Though it is accepted that there is a difference which is in itself perceptible, it is true nonetheless that there is no possibility of performing the experiment. For who could assure us that the powder's explosive force, and all the other circumstances, remained the same in the two cases?'

This is Kepler's last word on the subject. The philosophical, or rather metaphysical, origin of his failure is clear. It is entirely accounted for by his refusal to put rest and motion on the same ontological level.

2

THE *DIALOGUE CONCERNING THE TWO CHIEF WORLD SYSTEMS* AND THE ANTI-ARISTOTELIAN POLEMIC

It would certainly be an exaggeration to claim that Galileo's whole work issued from his concern with cosmological questions, and to present the whole of it as a struggle on behalf of the Copernican conception of the Universe (as Henri Martin[94] and, more recently, E. Wohlwill[95] have done). We should not forget the *Discorsi e dimostrazioni*. Nevertheless it is true that cosmological questions did play a role of the greatest importance in Galileo's thought and his research, and that from his youth, from the time when he drafted the treatises and dialogues on motion at Pisa, he can be seen raising problems of which the full significance can only be understood in relation to the Copernican conception of the Universe.[96] Furthermore it can be seen that even then he came up against the very same difficulties, which remained insoluble at that time, which were to be such a hindrance to the progress of thought some forty or fifty years later.

The central problem with which Galileo was concerned at Pisa was that of the persistence of motion. Now it is clear that when he investigated the case of the motion (of rotation) of a sphere located at the centre of the world, and also that of a sphere located away from this centre, he had in mind the situation raised by Copernican doctrine; without doubt the marmoreal sphere of which he analysed the motions represented the earth, and its motions were the earth's motions.[97]

But the conclusion at which he arrived, which was moreover in contradiction with the basic premises of *impetus* physics, shows us in a striking way the difficulties, and the origin of these difficulties, with which the new physics and astronomy were confronted.

For the result at which Galileo's analysis arrived was the natural persistence, or more precisely the privileged status, of *circular* motion.[98] This result, strongly confirmed by everyday observation and above all confirmed by the Copernican 'observation' of the circular motion of the earth (orbital motion and rotational motion), in corroboration of the astronomical observation of the circular motion of the planets, would later turn out to be an insurmountable obstacle which would block the progress of Galilean science.

As we have had occasion to mention above experience is hardly on the side of the new physics.[99] Bodies fall and the earth spins round—these are two facts which it is difficult to reconcile and with which the new physics at first found it very difficult to cope.

Contrary to what has often been said the law of inertia does not have its origin in common sense experience, and is neither a generalisation nor an idealisation of it. What we find in experience is circular motion, or more generally curved motion. We never see rectilinear motion, except in the untypical case of free fall, and this is precisely not a case of inertial motion. Yet it was curved motion that classical physics would struggle to explain on the basis of the latter, rectilinear motion. This is a very strange approach, for it does not involve the explanation of an observed phenomenon by reference to a hypothetical underlying reality (as was the case in astronomy, which explained phenomena, i.e., apparent motions, by reference to a combination of real motions), and nor does it involve the analysis of the phenomenon into its simple constituents and then its reconstruction in thought (the method of resolution and composition to which the novelty of Galileo's method is often incorrectly reduced); what it involves, strictly speaking, is the explanation of that which *exists* by reference to that which *does not exist*, which never exists, by reference even to that which *never could exist.*

The explanation of the real by reference to the impossible: what a strange way of proceeding! A paradox if ever there was one. We shall call this approach Archimedean, or rather, Platonist; the explanation, or rather the reconstruction, of empirical reality on the basis of an ideal reality. This was a paradoxical, difficult and hazardous venture. The examples of Galileo and Descartes allow us to identify straight away its basic contradiction: it requires a total conversion, a radical substitution, of a mathematical, Platonic world for the world of empirical reality (for it is only in the former world that the ideal laws of classical physics hold or are realised) while at the same time it renders this total substitution impossible, since instead of explaining empirical reality it does away with it, because instead of saving the phenomena it creates, between empirical and ideal realities, a fatal chasm of unexplained facts. Right from its early days in Pisa, Galileo's Archimedean thought found itself confronted by these unexplained facts.

We know that in Aristotelian physics all motions are divided into two general kinds, or rather are classified into two categories: 'natural' motions and 'violent' motions. It is this very classification which is attacked by Galileo. The division, he points out, is not coherent.[100] For the two terms are not mutually exclusive; there are motions which must be considered simultaneously both natural and violent. Moreover, and this is more serious, there are motions which belong in neither class, motions which are neither natural nor violent. An example of this is the circular (rotational) motion of a sphere located at the centre of the world. This motion is certainly not natural for the sphere, because a sphere has in itself no propensity to motion at all. However it cannot be considered violent; no violence is done to the sphere since it remains at its place and since its motion changes nothing, neither raises nor lowers any weight. This is all the more true in that the sphere, located at the centre of the world, i.e., at its natural place, has no weight.[101]

This case of the sphere located at the centre of the world is, however, far from being the only one. For in fact all circular motion (motion around the centre) is like this, neither natural nor violent. Here too the motion effects no change, i.e., neither raises nor lowers any weight. Lastly, the motion of a spherical heavy body rolling on a horizontal plane is equally neither natural nor violent. In this case, once again, the motion neither raises nor lowers any weight. It follows, says Galileo, that if we eliminate all external resistance (an absolutely smooth plane, absolutely hard and spherical bodies, etc.) then the motion of these bodies can well go on without ever stopping, can well be prolonged indefinitely.[102]

But what in fact is a horizontal plane? Or more precisely, what is a horizontal plane for a heavy body? Or still more precisely, what is a real horizontal plane for a real heavy body, on the earth? It is not at all the horizontal plane of geometry, or of an Archimedean physics. On a horizontal plane of this kind (i.e., a geometrical one), on the earth, for example on a plane tangential to the earth's surface, a heavy body would be in quite a different situation. For in moving on such a plane it would move away from the centre of the earth (or of the world) and consequently would be raising itself. Its motion would therefore be violent, and would in fact be comparable to that of a body going up an inclined plane, i.e., on an upwards sloping plane. Therefore not only would it not be prolonged indefinitely, it would, on the contrary, necessarily come to a halt. The only real motion which would be neither natural nor violent, the only motion which would neither raise nor lower any weight, the only motion which would not take the body away from or towards the centre of the earth (or of the world), is motion along the circumference. Consequently it would be a circular motion. In other words, the *real* horizontal plane is a *spherical surface.*[103]

It can be seen that this argument is in agreement with experience. Circular motion occupies an absolutely privileged position in physical reality. At the same time we must draw the following conclusion: geometrical concepts cannot be applied just as they stand to physical reality. We could put this, in non-Galilean terms of course, by saying that real space is neither Archimedean nor Euclidean; it differs from these precisely as a spherical surface differs from a geometrical plane.

Galileo's position, which remains almost identical from Pisa to Padua and Florence, is as follows: there is an undeniable and yet inexplicable phenomenon, a phenomenon which is moreover indispensable for Galilean dynamics,[104] namely the phenomenon of weight, the fact that there are heavy bodies and that they fall. There is another fact, intimately related to the first, namely that the straight line is in reality a circumference, the plane is in reality a spherical surface. It is the circle, and not the straight line, which physically has a special status.[105]

This is explicitly stated by Galileo. Rectilinear motion is impossible in reality. There can be no *natural* rectilinear motion. 'Straight motion being by nature infinite (because a straight line is infinite and indeterminate), it is impossible that anything should have by nature the principle of moving in a straight line; or, in other words, toward a place where it is impossible to arrive, there being no finite end. For nature, as Aristotle well says himself, never undertakes to do that which cannot be done, nor endeavours to move whither it is impossible to arrive.'[106]

This is a curious passage, and one to which we shall return (one which is, moreover, echoed by many others[107]), and in which can be found most of the ideas from which we are supposed to have been liberated by Galilean physics.

If this is so how could Galileo possibly have been the founder—or among the founders—of modern physics, which is based as we have said on the primacy of the straight line over the circle, on the geometrisation of space, and on the law of inertia? Or is this a mistake, a case of fruitful misinterpretation? Did the successors and disciples of Galileo—Gassendi, Toricelli, Cavalieri—all simply misunderstand him? Did they lose sight of the distinction and,.ignoring their master's repeated warnings, identify the real plane with the geometrical plane, and thereby extract from Galileo's teachings something which had in fact never been contained in them? This is the view of Wohlwill[108], but on the other hand it is strenuously denied by E. Mach[109] and above all by E. Cassirei,[110] who believe that, on the contrary, Galileo's physics is so intimately connected with the principle of inertia that it is impossible that Galileo himself was not aware of it.

So, did Galileo or did he not formulate, or at least suggest, the principle of inertia? In our view to pose the problem in these terms is to oversimplify it. History is more complicated, more subtle and richer than this. Moreover it is to evade the only problem which is truly instructive and interesting: *why* is it that in his struggle to mathematise reality, Galileo did not manage to state, at least not explicitly (and Cassirer himself would not deny this), this principle of inertia which his successors and disciples, we are assured, found it so easy to accept? For it is not simply a matter of ascertaining a fact; we must also understand it. To achieve this it is necessary to investigate the real thought of the great Florentine in its own right.

This is precisely what we propose to undertake.[111] We will observe the extremely interesting fact that if Galileo did in fact fail in his project (Wohlwill's thesis being broadly speaking correct) then this was because, in contrast to Descartes, he did not know how, or was unable, to free himself from the grip of the phenomena, nor to accept the inevitable consequence of the mathematisation of reality, the total geometrisation of space, namely the infinity of the Universe and the destruction of the Cosmos.[112]

We have already said that modern physics was born in the heavens as much as on earth,[113] and that it was an integral part of an astronomical, or rather cosmological, project. Galileo's works, the *Dialogue* as much as *The Assayer*, were in the first place Copernican works, and Galileo's physics was Copernican physics, a physics which had to defend the great astronomer's work, and in particular the movement of the earth, against both ancient objections and modern attacks. Galileo saw better than anyone else that this new physics needed to be *toto caelo* different from the old physics. In order to build it this latter had first of all to be destroyed, i.e., its very basis, the *philosophical* foundations on which it rested, had to be destroyed. Galileo knew perfectly well that for the new, mathematical, Archimedean physics to be built it would be necessary for him to reconstruct all his concepts on new foundations, and that it would have to rest as firmly as possible on a *philosophy*. Hence the subtle mixture of 'science' and 'philosophy' in Galileo's works, and the impossibility, for any historian who has not given up all effort to understand them, of divorcing these two integral aspects of Galileo's thought.

The *Dialogue Concerning the Two Chief World Systems* claims to be an account of two rival astronomical systems.[114] But it is not in fact a book about astronomy,[115] nor even about physics. It is above all a critique, a polemical and combative work; and at the same time a didactic work and a philosophical work. It is also a book about history, for it is about Galileo's own intellectual history.

A polemical and combative work: it is this which in part determines the literary form of the Dialogue.[116] It is against traditional science and philosophy that Galileo aims his artillery. But though the *Dialogue* is aimed *against* the Aristotelian tradition it is not addressed, or is scarcely addressed, to its adherents, to the philosophers of Padua and Pisa, to the authors of treatises *De Motu* and to the commentators on *De Caelo*. It is addressed 'To the Discerning Reader'.[117] Thus it is written not in Latin, the academic language of the universities and Schools, but in the vernacular, in Italian, the language of the Court and of the bourgeoisie. All the reformers adopted this practice; remember Bacon and Descartes.

It is 'the discerning reader' that Galileo wishes to win over to his cause, and he must be persuaded and convinced. He must not be overwhelmed and exhausted. Hence, in part, the dialogue form of the book, its light conversational tone, its constant digressions and recapitulations, the apparent disorder of the debate. For this is what a conversation and discussion between gentlemen, in the salons of the Venetian patricians or at the Court of the Medicis, would really be like. Hence the variety of 'weapons' used by Galileo; the calm discussion which searches out a proof and attempts to establish an argument, the eloquent speech which seeks to persuade, and lastly, the most powerful, the weapons of the polemicist, the sharp, incisive, cutting criticism, the joke which, in making fun of an opponent, makes him seem ridiculous, and thereby undermines and erodes whatever is still left of his authority.[118]

A 'didactic' work: for it is not merely a matter of convincing, of persuading, of proving; it is also a matter, and perhaps this above all, of little by little putting the gentleman reader in a position to be persuaded and convinced, of being able to understand the argument and accept a proof.[119] For this task it is necessary to perform a double labour of destruction and reconstruction; destruction of prejudices, of traditional mental habits and of common sense; and the creation in their place of new habits and of a renewed capacity for reasoned argument.

These are the reasons why the modern reader finds some passages unbearably tedious; the modern reader has the benefit of the consequences of the Galilean revolution. Hence also the repetitions, the recapitulations, the discussion over and over again of the same arguments, the multitude of examples. For it was necessary to educate the reader, to teach him to give up his confidence in authority, tradition and common sense. It was necessary to teach him to think.

A philosophical work:[120] for it is not only the traditional physics and cosmology which Galileo attacks and struggles against: it is also the whole philosophy and *Weltanschauung* of his opponents. Moreover, physics and cosmology were at that time an integral part of philosophy. In struggling against Aristotle's philosophy Galileo does so from the vantage point of another philosophy; he fights under the banner of the philosophy of Plato, or of one version of Plato's philosophy.[121]

Hence Galileo's attack, right from the beginning of the *Dialogue*, on the traditional conception of the Cosmos, with its clear separation between the heavens and the earth, the celestial world and the sublunary world,[122] an attack in which Galileo makes use of all the phenomena provided by the new astronomy, the discoveries of the *Nuntius Sidereus*, which reveal the moon to be a body strictly comparable to, and of the same nature as, the earth. Hence also the allusions to Plato which are scattered throughout the book, of which the dialogue form is without any doubt inspired by Plato, and which moreover opens with a pseudo-Platonic cosmological myth. Hence the allusions to the Socratic method, a method which is in fact employed, successfully, by Galileo's spokesman, Salviati. All of this as if to say, 'Take notice; in the age-old battle between the two great philosophers, we are on the side of Plato!'[123]

A historical work: of course Galileo does not strictly speaking tell in it the history of his own thought, but being quite aware of the titanic labour which he had himself had to perform in order to move from Aristotle's physics to *impetus* physics, and from that to the physics of the *Discourses*, he takes us with him, as it were, back over the journey which he had had to travel himself. Hence the fact that there are arguments, only a few pages apart, which belong to quite different stages and levels of his thought.[124] Hence the use of some traditional terms, which are retained even though their meaning is progressively changed,[125] and the absence of a rigorous terminology. Hence also the sense of a play of light and shadow throughout the *Dialogue*, the sense that a real progress of thought is being achieved. Hence, finally, a certain reserve, a cautiousness which deliberately leaves certain problems obscured, which avoids certain names and doctrines which are too difficult, or sometimes too dangerous.[126]

Let us now open the *Dialogue*. The roles of the interlocutors[127] in it are clearly demarcated.[128] Salviati, Galileo's spokesman, represents the mathematical intellect of the new science; Sagredo, the *bona mens*, the mind already freed from the prejudices of the Aristotelian tradition and the illusions of common sense, a mind which is therefore capable of grasping the new truth of the Galilean arguments, and even, having grasped it, of drawing out its consequences; Simplicio represents common sense, imbued with the prejudices of Scholastic philosophy, believing in the authority of Aristotle and of official science, and struggling laboriously under the burden of tradition.

In the course of the discussion it is usually Simplicio who is given the task of confronting Copernicus with both the old and the new arguments of the defenders of geocentric astronomy. Yet when we come to the *physical* arguments, the old arguments of the clouds and the birds, the argument of the heavy bodies falling vertically to the earth, Simplicio gives way to Salviati. For the physical objections, in contrast to the others, have to be taken seriously. To discuss and refute them Salviati, *explicitly* relying on Galileo's investigations in mechanics, needs all of his subtlety.

We are familiar with these objections, and also with the replies. Now Galileo's reply does not, at least at first sight, differ very much from that of Bruno. Just like him Galileo replies to the Aristotelian arguments with the principle of the relativity of motion and with *impetus* dynamics.

Simplicio[129] recites the famous passage from *De Caelo*.[130] 'Whether the earth is moved either in itself, being placed in the centre, or in a circle, being removed from the centre, it must be moved with such motion by force, for this is not its natural motion. Because if it were, it would belong also to all its particles. But every one of them is moved along a straight line toward the centre. Being thus forced and preternatural, it cannot be everlasting. But the world order is eternal; therefore, etc.

'Second, it appears that all other bodies which move circularly lag behind, and are moved with more than one motion, except the *primum mobile*.[131] Hence it would be necessary that the earth be moved also with two motions; and if that were so, there would have to be variations in the fixed stars. But such are not to be seen; rather, the same stars always rise and set in the same place without any variations.[132]

'Third, the natural motion of the parts and of the whole is toward the centre of the universe, and for that reason also it rests therein.'

Simplicio explains that Aristotle then goes on to discuss 'the question whether the motion of the parts is toward the centre of the universe or merely toward that of the earth, [and he concludes] that their own tendency is to go toward the former, and that only accidentally do they go toward the latter, which question was argued at length yesterday.

'Finally he strengthens this with a fourth argument taken from experiments with heavy bodies which, falling from a height, go perpendicularly to the surface of the earth. Similarly, projectiles thrown vertically upward come down again perpendicularly by the same line, even though they have been thrown to immense height. These arguments are necessary proofs that their motion is toward the centre of the earth, which, without moving in the least awaits and receives them.

'He then hints at the end that astronomers adduce other reasons in confirmation of the same conclusions—that the earth is in the centre of the universe and immovable. A single one of these is that all the appearances seen in the movements of the stars correspond with this central position of the earth, which correspondence they would not otherwise possess. The others, adduced by Ptolemy and other astronomers, I can give you now if you like; or after you have said as much as you want to in reply to these of Aristotle'.[133]

As we know these Aristotelian arguments are by no means negligible. Galileo will have to deal with them one by one. But before doing this, and before going on to the discussion of certain of Ptolemy's arguments, held in reserve by Simplicio,[134] Galileo thinks it necessary to develop more fully the famous argument concerning the fall of heavy bodies, which Simplicio had skated over rather too rapidly, and to round off the argument concerning the tower and bodies thrown in the air by adding the more 'modern' proofs of the moving ship and the cannon shot.[135] So Salviati, putting off to another day the discussion of the astronomical arguments, replies as follows:[136]

'As the strongest reason of all is adduced that of heavy bodies, which, falling down from on high, go by a straight and vertical line to the surface of the earth. This is considered an irrefutable argument for the earth being motionless. For if it made the diurnal rotation, a tower from whose top a rock was let fall, being carried by the whirling of the earth, would travel many hundreds of yards to the east in the time the rock would consume in its fall, and the rock ought to strike the earth that distance away from the base of the tower. This effect they support with another experiment, which is to drop a lead ball from the top of the mast of a boat at rest, noting the place where it hits, which is close to the foot of the mast; but if the same ball is dropped from the same place when the boat is moving, it will strike at that distance from the foot of the mast which the boat will have run during the time of fall of the lead, and for no other reason that that the natural movement of the ball when set free is in a straight line toward the centre of the earth. This argument is fortified with the experiment of a projectile sent a very great distance upward; this might be a ball shot from a cannon aimed perpendicular to the horizon. In its flight and return this consumes so much time that in our latitude the cannon and we would be carried together many miles eastward by the earth, so that the ball, falling, could never come back near the gun, but would fall as far to the west as the earth had run on ahead.

'They add moreover the third and very effective experiment of shooting a cannon-ball [parallel to the horizon] to the east, and then another one with equal charge at the same elevation to the west; the shot toward the west ought to range a great deal further out than the other one to the east. For when the ball goes toward the west, and the cannon, carried by the earth, goes east, the ball ought to strike the earth at a distance from the cannon equal to the sum of the two motions, one made by itself to the west, and the other by the gun, carried by the earth, toward the east. On the other hand, from the trip made by the ball shot toward the east it would be necessary to subtract that which was made by the cannon following it. Suppose, for example, that the journey made by the ball in itself was five miles and that the earth in that latitude travelled three miles during the flight of the ball; in the shot toward the west, the ball would fall to earth eight miles distant from the gun—that is, its own five toward the west and the gun's three to the east. But the shot toward the east would range no further than two miles, which is all that remains after subtracting from the five of the shot the three of the gun's motion toward the same place. Now experiment shows the shots to fall equally; therefore the cannon is motionless, and consequently the earth is, too. Not only this, but shots to the south or north likewise confirm the stability of the earth; for they would never hit the mark that one had aimed at, but would always slant toward the west because of the travel that would be made toward the east by the target, carried by the earth while the ball was in the air. And not merely shots along the meridians, but even those made to the east or west would not range truly; for the easterly shots would carry high and the westerly low. . .'

Let us now proceed to the criticism. It is at one and the same time very profound and very simple. The arguments of the Aristotelians, says Galileo, are no more than paralogisms, for they rest on presuppositions which in fact require proof. This is certainly true. But the Aristotelian could perfectly well reject this criticism, which is a repetition of the objection which had already been made by Copernicus to the effect that Aristotle does not argue, as he claims, on the basis of the facts but, on the contrary, on the basis of a theory.[137] To which the Aristotelian could quite reasonably reply, firstly that it is impossible to do otherwise, and secondly that Galileo does the same himself.

The Aristotelian argument does in fact presuppose a theory, or to put it another way, a specific idea of motion—a process which affects the moving body. Furthermore it presupposes that we are able by the perception of the senses to directly grasp physical reality,[138] and that this is in fact the only means of grasping it, and that, consequently, a physical theory can never throw doubt on the phenomena given directly in perception.

Now Galileo explicitly rejects this. His starting point is diametrically opposed to this. He asserts (a) that physical *reality* is not given in perception but is, on the contrary, grasped by reason; and (b) that motion does not affect the moving body, which remains unchanged by any motion which impels it, and that motion only affects the relations between a moving body and a stationary object.

The Aristotelian argument, while it is a paralogism from Galileo's point of view, is incontestable in itself. Dialectically, however, at least within the *Dialogue*, Galileo does of course have the right to pronounce the Aristotelian argument to be a paralogism. This is so because he has already, before giving an account of the physical and mechanical proofs of the immobility of the earth, postulated the dual principle of the mechanical as well as the optical relativity of motion.[139]

The optical relativity of motion had, of course, always been appreciated. Copernicus had already drawn from it the conclusion that there could be no purely optical discrimination between geocentric and heliocentric astronomies: for any apparent motion of the vault of the heavens could be interpreted physically in either way.[140] This is precisely the reason for the importance of the physical proofs put forward by Aristotle and by Ptolemy.

The optical relativity of motion cannot be doubted. Therefore, says Galileo, it must be put forward as a 'principle' right from the beginning of the discussion.[141] 'Then let the beginning of our reflections be the consideration that whatever motion comes to be attributed to the earth must necessarily remain imperceptible to us and as if nonexistent, so long as we look only at terrestrial objects; for as inhabitants of the earth, we consequently participate in the same motion. But on the other hand it is indeed just as necessary that it display itself very generally in all other visible bodies and objects which, being separated from the earth, do not take part in this movement. So the true method of investigating whether any motion can be attributed to the earth, and if so what it may be, is to observe and consider whether bodies separated from the earth exhibit some appearance of motion which belongs equally to all . . .' The diurnal motion which bodies separated from the earth have in common is precisely such a motion. Thus, optically or astronomically speaking, this can be attributed either to the earth or to the heavens; or as Sagredo jokingly puts it,[142] the role of *primum mobile* can be attributed either to the earth or to the heavens.

In fact, however, the 'principle' put forward by Galileo is more extensive than that of optical relativity; in postulating the impossibility of perceiving motion in which we ourselves participate, he is already postulating the *physical* relativity of motion. He even puts it forward as equivalent and equipollent to optical relativity. For if motion is absolutely imperceptible to those participating in it, it follows that the earth's motion would have no effect whatsoever on phenomena which take place on it. This, to put it in modern terms, implies the attribution to all motion, and in particular to *circular* motion, of the properties of inertial motion.

We will have an opportunity to return to this question later. In the meanwhile let us follow Galileo.[143] 'For consider: motion, in so far as it is and acts as motion, to that extent exists relatively to things that lack it; and among things which all share equally in any motion, it does not act, and is as if it did not exist.[144] Thus the goods with which a ship is laden leaving Venice, pass by Corfu, by Crete, by Cyprus and go to Aleppo. Venice, Corfu, Crete, etc. stand still and do not move with the ship; but as to the sacks, boxes, and bundles with which the boat is laden and with respect to the ship itself, the motion from Venice to Syria is as nothing, and in no way alters their relation among themselves. This is so because it is common to all of them and all share equally in it. If, from the cargo in the ship, a sack were shifted from a chest one single inch, this alone would be more of a movement for it than the two-thousand-mile journey made by all of them together.'

At first sight it seems that Galileo is saying nothing new; it seems that what he says could be accepted by an Aristotelian. But only at first sight.

For it is very important not to confuse, as happens too often, Aristotelian and Galilean relativity of motion, (the latter ought more accurately be called Cartesian or Newtonian). For Aristotle motion as such necessarily implies a point of reference. For local motion, in particular, there must be some point, not itself in motion, with which it can be compared. But since motion is not seen as a pure and simple relation between two things but, as we have said before, as a process which has a real effect on the moving body, the point of reference or comparison must be something which is in fact really stationary—the world, and as a particularly privileged case, the stationary centre of the world. There is nothing like this in Galileo's conception: motion conceived as a state-relation, which does not affect the moving body, does not at all imply the existence of a point really and absolutely at rest. It implies only the existence of a point, or more accurately of a body, which does not participate in the motion in question: for example, the sacks and boxes in relation to one another, the ship in relation to the boxes, Corfu and Crete in relation to the ship, and so on. From this Galileo correctly draws the conclusion that a motion that is common to several moving bodies is without effect and as if it did not exist as far as the relations between them are concerned, since nothing in these relations changes, and that it only has effects on the relations which these bodies have with others which do not participate in the motion.

These arguments put forward by Salviati receive the reply which we would expect. For if a stone and a tower both participate in the earth's motion then this motion, as far as they are concerned, will be as if it did not exist, and things will happen as if it really did not exist, i.e., as if the earth were at rest. We can see immediately that this would have extremely important consequences; in particular it would imply the compatibility of all the motions; even more drastically it would follow that no motion would impede any other, that in relation to one another the motions impelling a particular body would be as if they did not exist. Now it is precisely this that the Aristotelian cannot accept. For him motions express the nature of the moving body and must be described in terms of this nature. For him motion must not be treated in a way which ignores the moving body and the cause of the motion, as if it existed in itself. Whether or not various motions are compatible with each other depends on whether or not they are compatible with the nature of the moving body. So he cannot accept Galileo's position. If the earth did move then its circular motion would be of an entirely different order and nature from the rectilinear motion of fall, and there would be no reason for these motions to combine. No doubt he would, if pressed, accept that things could be so arranged that a heavy body had two motions at one and the same time, but this would be, precisely, a case of 'mechanical' motion and at least in part violent.[145]

Now, consider the stone in free fall. What, then, could make it follow the tower's motion? If, as we presume, there is nothing connecting it to the tower, it is hardly plausible that it would do so. It would be much more plausible to say, on the contrary, that the stone falling from the top of the tower would behave (if the earth were spinning round) quite differently from the way that it does in fact behave, that it would behave in the same way as a stone falling from the top of a ship's mast which, as we know from experience (according to this account), falls at the foot of the mast when the ship is at rest and at a distance from it (behind it) when the ship is moving. This, of course, is Tycho's argument. But in putting it forward Tycho went too far. For he thereby admitted that one can consider terrestrial (the ship) and cosmic (the earth) processes on the same level, and in this he went a long way toward betraying the Aristotelian position which is based entirely (as Galileo made a point of reminding us at the beginning of the *Dialogue*[146]) on the *essential* difference between terrestrial and celestial laws. Galileo, of course, will take advantage of this and will, like Bruno, draw conclusions about the earth from the ship, and about the heavens from the earth.[147]

So the argument continues:[148] '*Salviati*: You say, then, that since when the ship stands still the rock falls to the foot of the mast, and when the ship is in motion it falls apart from there, and conversely, from the falling of the rock at the foot it is inferred that the ship stands still, and from its falling away it may be deduced that the ship is moving. And since what happens on the ship must likewise happen on the land, from the falling of the rock at the foot of the tower one necessarily infers the immobility of the terrestrial globe. Is that your argument?

Simplicio: That is exactly it . . .

Salviati: Now tell me: if the stone dropped from the top of the mast when the ship was sailing rapidly fell in exactly the same place on the ship to which it fell when the ship was standing still, what use could you make of this falling with regard to determining whether the vessel stood still or moved?

Simplicio: Absolutely none . . .

Salviati: Very good. Now, have you ever made this experiment with the ship?

Simplicio: I have never made it, but I certainly believe that the authorities who adduced it had carefully observed it. Besides, the cause of the difference is so exactly known that there is no room for doubt.'

Nobody has ever performed the experiment, replies Salviati.[149] All the authors alluded to by Simplicio derived their beliefs about this from the authority of their predecessors. If they had performed the experiment they would have seen, as will anyone else who performs it, that the stone always falls at the foot of the mast and that nothing can be concluded from this either for or against the motion of the ship; and similarly, nothing follows from the fact that a stone falls at the foot of a tower either for or against the motion of the earth. But now it is Simplicio's turn to ask the question:[150] 'So you have not made a hundred tests, or even one? And yet you so freely declare it to be certain?' For only experiment can decide. 'If you had referred me to any other agency than experiment, I think that our dispute would not soon come to an end; for this appears to me to be a thing so remote from human reason that there is no place in it for credulity or probability.'

The modern reader will perhaps judge that for once the Aristotelian Simplicio is right. For how can one decide between two different and rival theories except by experiment? So, expecting that Salviati will at this point provide Simplicio with a detailed account of an experiment, he will be astonished to hear him declare that this is completely unnecessary, not only for Salviati himself but also for Simplicio who has just been demanding it. 'Without experiment, I am sure that the effect will happen as I tell you, because it must happen that way; and I might add that you yourself also know that it cannot happen otherwise, no matter how you may pretend not to know it—or give that impression. But I am *so good a midwife of the brain*[151] that I shall make you confess this in spite of yourself.'[152]

We should pause here for a moment. For this passage, which is by no means an untypical one in Galileo's works,[153] seems to us to be of cardinal importance. In our view it is a key to the whole interpretation of Galileo's work, and therefore of classical science in general.

For what is at stake here is the whole function and place of experiment in science. Now classical science is more often than not declared to be based primarily on experiment. Its wealth and fertility of experimental reasoning is contrasted with the sterile and verbose *a priorism* of Scholastic physics. For example, Galileo has usually been seen as the cautious and shrewd observer,[154] the founder of the experimental method,[155] as a man who weighed, measured and calculated and who, refusing to adopt the method of abstract, *a priori* argument from first principles, attempted on the contrary to base the new science on the solid foundations of empirical reality. Of course this is not all totally incorrect. It is clear that it was the observation of the real motions of the planets which led Kepler to reform astronomy. Similarly it is clear that it was by directing his telescope at the skies, by observing the heavens, that Galileo delivered a mortal blow to the mediaeval Cosmos. It is also certain that Galileo's works are full of appeals to experiment and observation[156] (the pendulum, the inclined plane, and so on), and of furious attacks against those who refuse to accept what they can see with their own eyes because what they see contradicts their beliefs (for example, to accept that the nerves start in the brain and not in the heart as Aristotle taught[157]), or who even refuse to look at all from fear that they might see things which their principles tell them to be impossible.[158] Galileo's works are also full of passages in which he asserts the infinite richness of nature and condemns the arrogance of those who claim to be able to say in advance what it can and cannot do. In spite of this, however, it is not Galileo's spokesman, Salviati, but the Aristotelian Simplicio who is presented as being in favour of experiment, and it is Salviati who declares that it is unnecessary.

We will come back to this problem. For the moment let us simply note the fact: good physics is done *a priori*.[159]

The proof of this, according to Salviati, is that, to Simplicio's great astonishment and even indignation, Simplicio has no need to resort to experiment in order to recognise what is true. For the matters we are concerned with are not at all 'remote from human reason', but are, on the contrary, the closest to it. They are so close that men are already in possession of the true principles of the nature of the physical world even in advance of any experiment. Men know the truth, though they may not be aware of this. Therefore it is not necessary to teach it to them (this is even impossible); it is enough, by putting suitable questions to them, to show them (and us) that they know it already.

This is why Salviati, taking up the analyses with which the young Galileo's study of motion had begun at Pisa, asks Simplicio:[160] 'Now tell me: suppose you have a plane surface as smooth as a mirror and made of some hard material like steel. This is not parallel to the horizon, but somewhat inclined, and upon it you have placed a ball which is perfectly spherical and of some hard and heavy material like bronze. What do you believe this will do when released? Do you not think, as I do, that it will remain still?

Simplicio: If that surface is tilted?

Salviati: Yes, that is what was assumed.

Simplicio: I do not believe that it would stay still at all; rather, I am sure that it would spontaneously roll down.'

Nobody has taught Simplicio this reply. It is provided by his natural reason. Therefore it is proved, for the reader,[161] that recourse to experiment, as demanded by Simplicio, is not always necessary. This use by Salviati of the Socratic method (for it is impossible not to recognise this; we cannot fail to think of the *Theaetetus* and the *Meno*) reveals to us the meaning of his apriorism; it shows us that he is taking up his position on the side of Plato. So Salviati can go on:[162] 'Now how long would the ball continue to roll, and how fast? Remember that I said a perfectly round ball and a highly polished surface, in order to remove all external and accidental impediments. Similarly I want you to take away any impediment of the air caused by its resistance to separation, and all other accidental obstacles, if there are any.[163]

Simplicio: I completely understood you, and to your question I reply that the ball would continue to move indefinitely, as far as the slope of the surface extended, and with a continually accelerated motion. For such is the nature of heavy bodies, which *vires acquirunt eundo*; and the greater the slope, the greater would be the speed.

Salviati: But if one wanted the ball to move upward on this same surface, do you think it would go?

Simplicio: Not spontaneously, no; but drawn or thrown forcibly, it would.

Salviati: And if it were thrust along with some *impetus* impressed forcibly upon it, what would its motion be, and how great?

Simplicio: The motion would constantly slow down and be retarded, being contrary to nature, and would be of longer or shorter duration according to the greater or lesser *impulse*[164] and the lesser or greater slope upward.

Salviati: Very well; up to this point you have explained to me the events of motion upon two different planes. On the downward inclined plane, the heavy moving body spontaneously descends and continually accelerates, and to keep it at rest requires the use of force. On the upward slope, force is needed to thrust it along or even to hold it still, and *motion*[165] which is impressed upon it continually diminishes until it is entirely annihilated. You say also that a difference in the two instances arises from the greater or lesser upward or downward slope of the plane, so that from a greater slope downward there follows a greater speed, while on the contrary upon the upward slope a given movable body thrown with a given force moves farther according as the slope is less. Now tell me what would happen to the same movable body placed upon a surface with no slope upward or downward.

Simplicio: I completely understood you, and to your question I reply that downward slope, there can be no natural tendency toward motion; and there being no upward slope, there can be no resistance to being moved, so there would be an indifference between the propensity and the resistance to motion. Therefore it seems to me that it ought naturally to remain stable. . .

Salviati: I believe it would do so if one set the ball down firmly. But what would happen if it were given an *impetus* in any direction?

Simplicio: It must follow that it would move in that direction.

Salviati: But with what sort of movement? One continually accelerated, as on the downward plane, or increasingly retarded as on the upward one?

Simplicio: I cannot see any cause for acceleration or deceleration, there being no slope upward or downward.

Salviati: Exactly so. But if there is no cause for the ball's retardation, there ought to be still less for its coming to rest;[166] so how far would you have the ball continue to move?

Simplicio: As far as the extension of the surface continued without rising or falling.

Salviati: Then if such a space were unbounded, the motion on it would likewise be boundless? That is, perpetual?

Simplicio: It seems so to me, if the movable body were of durable material.'

The principle of the eternal persistence of horizontal motion, and of its speed, is now established. Historians of Galileo, and of physics, have cited this passage, and other similar passages, as a statement of a limited principle of inertia.[167] In fact, however, Galileo cannot now, any more than at Pisa, abstract from gravity, a natural quality of heavy bodies; no more than at Pisa can he forget that the *real horizontal plane is the surface of a sphere*. This is shown absolutely clearly as the discussion continues.[168]

'*Salviati*: That [the movable body is of durable material] is of course assumed, since we said that all external and accidental impediments were to be removed, and any fragility on the part of the moving body would in this case be one of the accidental impediments.[169]

'Now tell me, what do you consider to be the cause of the ball moving spontaneously on the downward inclined plane, but only by force on the one tilted upward?

Simplicio: That the tendency of heavy bodies is to move toward the centre of the earth, and to move upward from its circumference only with force; now the downward surface is that which gets closer to the centre, while the upward one gets farther away.

Salviati: Then in order for a surface to be neither downward nor upward, all its parts must be equally distant from the centre. Are there any such surfaces in the world?

Simplicio: Plenty of them; such would be the surface of our terrestrial globe if it were smooth, and not rough and mountainous as it is. But there is that of the water, when it is placid and tranquil.

Salviati: Then a ship, when it moves over a calm sea, is one of these moveables which courses over a surface that is tilted neither up nor down, and if all external and accidental obstacles were removed, it would thus be disposed to move incessantly and uniformly from an impulse once received?

Simplicio: It seems that it ought to be.

Salviati: Now as to that stone which is on top of the mast; does it not move, carried by the ship, both of them going along the circumference of a circle about its centre? And consequently is there not *in it an ineradicable motion*,[170] all external impediments being removed? And is not this motion as fast as that of the ship?'

So we have returned to the classical problem of the persistence of motion in a moving body which is separated from the motor, and transposed, apparently, into the situation discussed by Bruno.[171] Must we then choose, as he did, between Aristotle's theory of the action of the medium and the Parisian doctrine of *impetus*?[172]

Yes and no. The Aristotelian doctrine is, no doubt, to be rejected straight away. But the Parisian theory is not to be adopted just as it stands. It must undergo, or rather it has already undergone, a profound change: *impetus* is no longer taken to be the *cause* of the motion, but is identified with the motion itself.

The Aristotelians' strongest objection against the *impetus* theory was an ontological objection: an accident does not pass from one body to another. Therefore, *impetus* cannot do this. This is true, replies Galileo, if *impetus* means a force which causes a motion; but the *motion itself* can be transmitted.

Galileo employs the old 'Parisian' arguments against the Aristotelian theory. The wind can easily carry away a feather or a piece of cork but cannot carry away a rock or a cannon-ball. Yet a rock or a cannon-ball, thrown by hand or fired from a cannon, travel very much further than a feather or a piece of cork. Other arguments point to the fact that a heavy pendulum keeps moving very much longer than a light one: and that an arrow can be shot against the wind, and that it flies better when it is shot point foremost than when it is shot sideways. Aristotle, who located motivity in the air, made the mistake of confusing the ease with which it is moved with the power to store and conserve motion. The former is no doubt connected with lightness, the latter, on the contrary, with heaviness. The medium can of course have an influence on the motion of the moving body; usually it is an obstacle. But this motion belongs to the moving body itself. From this we can conclude that the *impetus* with which the ship moves remains *ineradicably impressed* on the stone after it has separated from the mast, and that this motion causes no hindrance nor retardation to the 'natural' downwards motion of the stone. It follows from this that although the actual trajectory of the stone can be lengthened by any amount, with the speed of the ship, nevertheless the stone will take no longer to complete its trajectory than it does when it falls to the foot of the mast when the ship is at rest. It also follows that a cannon-ball which is fired horizontally from the top of a tower will reach the earth, regardless of whether it has travelled two, three, six or ten thousand yards, at precisely the same instant as a cannon-ball which has fallen straight down, without having been given any impulsion by the cannon.[173]

As we might expect, these amazing and paradoxical conclusions are not enough to win over Simplicio to Galileo's doctrine of the relativity of motion and the mutual independence of each *impetus*. His doubts are far from being removed. Perhaps, he politely suggests, this is his own fault since he does not grasp things as quickly as Sagredo. In any case it seems to him that 'if this *motion which the stone shares*[174] while on top of the ship's mast were . . . conserved in it also *in an ineradicable manner*[175] after it is separated from the ship, then it would likewise be necessary for a ball dropped to earth by the rider of a galloping horse to continue to follow the horse's path without lagging behind. I do not believe that this effect is seen except when the rider throws the ball forcibly in the direction in which he is riding. Outside of that, I believe that it will remain where it strikes the ground'.[176]

The modern reader is bound to feel impatient. What purpose, he will say, do these interminable repetitions serve? Is it not obvious that Simplicio's example contributes nothing new and in no way clarifies the issues? The suspicion might even cross his mind that Galileo wishes to make fun of the Aristotelian by making him out to be as dumb as an ox. But the modern reader would be wrong. The example of the horseman does contribute something new and does take us a step further. In having Simplicio introduce him into the discussion Galileo is not making fun of Simplicio. Quite the contrary, he is representing him as being very intelligent.

For us, in fact, the two cases, that of a ball falling from the top of a mast and that of a ball dropped by a horseman, are indeed identical. But they were not so for sixteenth century physics. To throw a ball and to drop it are not the same thing. Gassendi would later have a great deal to say on this topic.[177] Furthermore, a ball falling from the top of a mast is, of course, separated from it, but less radically so than that which is dropped by the horseman. For to continue to move through the air *before* reaching the ground is quite a different matter from continuing to move *after* having arrived there.

Therefore Galileo wants to demonstrate the equivalence of the two cases distinguished by Simplicio, that of 'throwing' and that of 'dropping' the ball:[178] 'When you throw it with your arm, what is it that stays with the ball when it has left your hand, except the *motion*[179] received from your arm which is conserved in it and continues to urge it on? And what difference is there whether that *impetus* is conferred upon the ball by your hand or by the horse? While you are on horseback, doesn't your hand, and consequently the ball which is it in, move as fast as the horse itself? Of course it does. Hence upon the mere opening of your hand, the ball leaves it with just that much *motion*[179] already received; not from your own *motion*[179] of your arm, but from motion dependent upon the horse, communicated first to you, then to your arm, thence to your hand, and finally to the ball.

'I should add that if the rider threw the ball in the direction opposite to his course, when it struck it would sometimes still follow the horse's route, and sometimes it would lie still on the ground; it would move away from him only if the *motion*[179] received from the arm *exceeded in speed*[179] that of the rider. And it is folly to say, as some do, that a cavalryman can cast his javelin before him, pursue it on his horse, overtake it and recapture it. I say this is folly because in order to have the projectile return to his hand he would have to throw it straight up, in the same way as if he were standing still'. This, of course, goes without saying once we have seen that an arrow or any other object projected by the rider, having shared his motion, keeps it when it is projected into the air: in other words, once we have seen that in the mechanical system of the rider and his javelin (as in the mechanical system of the ship) the common motion is 'as if it did not exist'.

It would take too long, and is anyway unnecessary, to analyse in detail the phenomena by reference to which Sagredo, leaping to Salviati's assistance, and Salviati himself, now illustrate the key principles of Galilean physics: the relativity, mutual independence and conservation of motion. These striking and seemingly paradoxical 'cases' are brought forward in order to familiarise the reader with the principles of the new physics: the case of a letter written on a moving ship; the case of a ball rolled down a tilted board fixed to the side of a moving wagon so that when it rolls off and drops to the ground it sometimes stops motionless, sometimes even rolls backwards, and sometimes goes forwards, far enough to overtake the wagon; the case of the bowls players who, by conferring a spin on the ball, can, while throwing the ball forwards, make it roll backwards; the case of the ball which moves at different speeds as it rolls and bounces in the air . . .[180] These cases are also introduced, and this is no minor consideration, in order to dissociate in the reader's mind the ideas of motion of translation and motion of rotation. For in the new physics it is no longer uniquely rotatory motion, but motion *tout court*, which is conserved.

The modern reader has, no doubt, had enough by now; the discussion has gone on long enough. This is because he is already convinced in advance. The classical idea of motion has been *commonplace* for a long time now. But for a reader contemporary with Galileo it was not a familiar idea. To him this idea, the idea of something which both is and is not, which is conserved, and which passes from one object to another, seemed, not unreasonably, far more obscure than the Aristotelian idea of motion as a process. He would not have denied the *phenomena* adduced by Sagredo, but he would still have his doubts. Through the mouth of Salviati he asks yet again for experimental confirmation.[181] 'I should like to find some way of setting up an experiment which corresponds to the motion of these projectiles', he says. Sagredo suggests taking a little carriage and placing in it a crossbow with the bolt at half-elevation (i.e., that at which the bolt when fired will cover the greatest distance). Having set the carriage in motion the crossbow should then be fired once in the direction in which the carriage is moving and then in the opposite direction, carefully noting each time the position of the carriage. Then it will be seen in which direction the bolt is carried furthest.[182]

The experiment seems well conceived to Simplicio, and he says:[183] 'I have no doubt that the shot (that is, the space between the arrow and the place where the carriage was when the arrow struck the ground) would be much less when it went in the direction of the carriage than when it went the other way. Let the shot in itself be 300 yards, for example, and the travel of the carriage while the arrow is in the air, 100 yards. Then, when the shooting is with its course, the carriage will pass 100 of the 300 yards of the shot, so that at the time the arrow strikes the ground the space between it and the carriage will be only 200 yards. But on the other hand in the shot with the carriage running opposite to the arrow, when the arrow shall have passed over its 300 yards and the carriage its 100 additional the other way, the distance between them will be found to be 400 yards.

Salviati: Would there be any way to make these two shots travel equally?
Simplicio: I don't know of any other way than to make the carriage stand still.
Salviati: That, of course; but I mean with the carriage going full speed.
Simplicio: Only by bending the bow harder with the course and more weakly against the course.
Salviati: Then there is another way, and this is it. But how much would you need to strengthen your bow, and later to weaken it?
Simplicio: In our example, in which we have assumed that the bow would shoot 300 yards, it would be required for the shot along the course to strengthen the bow so as to shoot 400 yards, and the other way to weaken it so as to shoot no more than 200. Thus each shot would go out 300 yards with respect to the carriage, which, with its travel of 100 yards which is to be subtracted from the shot of 400 and added to that of 200, would reduce both to 300'.

Simplicio's argument, yet again, is by no means absurd. In Aristotelian physics the motion of a projectile is produced by the action of the surrounding medium. Consequently this motion is completely independent of that of the source in exactly the same way as the propagation of a light-wave is for us.

Sagredo's imaginary experiment has the same relation to the argument of the cannon shot as does the case of the moving ship to that of the heavy body falling from the top of a tower. And in each case we draw conclusions concerning phenomena of celestial physics from phenomena of terrestrial physics. In each case we eliminate the 'natural' character of the motion.

We have arrived, in fact, at one of the decisive points of the *Dialogue*.

Salviati asks:[184] 'What effect does the greater or lesser strength of the bow have upon the arrow?', to which Simplicio replies, 'The strong bow shoots it with *greater speed*,[185] the weaker with less. The same arrow goes as much farther one time than the other as the speed with which its nock goes forth is greater at one time than the other.

Salviati: So that to shoot the arrow in one direction as well as the other and have it depart equally from the moving carriage, it is necessary that if on the first shot of the given example it leaves with, say, four *degrees of speed*, then on the other shot it must leave with only two. But if the same bowing is employed, *three degrees* will always be received from that.

Simplicio: That is it. That is why the shots cannot go forth equally if shot with the same bowing while the carriage is running. . .

Salviati: Yes, this makes the accounts balance. But, tell me, when the carriage is running, don't all the things in the carriage move with the same speed?

Simplicio: No doubt about it.

Salviati: Also the bolt, and the bow, and the string with which this is strung?

Simplicio: That is right.

Salviati. Then when the bolt is discharged in the direction of the carriage, the bow impresses its three *degrees of speed* upon a bolt which already possesses one degree, thanks to the carriage which carries it at that speed in that direction. Thus when the nock leaves the string it does so with four *degrees of speed*. And on the other hand, shooting the other way, the same bow confers its three degrees upon a bolt moving with one degree in the opposite direction, so that at its separation from the string only two *degrees of speed* remain with it. But you yourself have already declared that in order to make the shots equal it is required that the bolt leave with four degrees in one case, and with two in the other. Hence, without changing the bow, the course of the carriage itself regulates the flights, and this experiment clinches the matter for those who would not or could not be convinced of it by reason.[186]

'Now apply this argument to the cannon, and you will find that whether the earth moves or whether it stands still, shots made with the same force must always carry equally no matter in what direction they are sent.'

We should pause here for a moment.

The results we have arrived at (the law of the conservation of motion, the uniformity and the indefinite persistence of circular motion) were in fact stated at the beginning of the *Dialogue*:[187] and the principle of the relativity of motion, which dominates the whole of the later discussion, is postulated, as we have seen, at the beginning of the 'Second Day'. However, these principles, while they may be self-evident, while they may be (to use a term which is not Galilean, though it could be) *innate* in our reason,[188] are nevertheless so strange and have such apparently astonishing consequences, that Simplicio, even though he had agreed to them, basically does not accept them. He rebels at the first opportunity. For his mind, the mind of a cultivated man, is so cluttered with old habits, with second-hand ideas (ideas acquired during his education), that it is at first impossible for him to think without relying on these traditional concepts. He has certainly agreed to the law of the conservation of motion (being unable to do otherwise, and having in a manner of speaking deduced it himself), but because he carries on thinking about motion in Aristotelian categories, and because the new concept of motion is neither familiar nor clear to him, he straight away relapses into his old habits and brings up all over again objections which have in theory already been dealt with. He must, therefore, be brought to think *habitually* with the newly acquired ideas.[189]

Now, how is Galileo to make these ideas penetrate into the reader's mind? Will he do it like Descartes by simply rejecting the Scholastic definition of motion in order to replace it with another one, his own? Not at all. Galileo wants the transition to be gradual and smooth. His argument recapitulates the historical development, and from this point of view the progress achieved was by no means negligible. The discussion of Aristotle's arguments picks up at the point where it had been left by Copernicus: namely, with a *qualitative* distinction between natural and violent motion as the explanation for the difference between their effects. Now there is a subtle modification, and the earth's *natural* motion (which, logically, is explained by its 'nature' or 'form') comes to be attributed to bodies which are on the earth, no longer as a result of a commonness of nature but solely because of the fact that they participate in this motion. Another subtle change and now the earth's motion is no longer seen as having any special status over and above the fact that it is circular, and this property, by yet another shift, is attributed by extension to the motion of a ship moving across the sea. The special status of *natural* motion has now completely vanished. Henceforth motion is conserved not because it is *natural* but simply because it is *motion*. It is motion as such which is conserved and which is *ineradicably impressed* on the moving body. Simplicio himself had understood and accepted this. For he does not look for a cause for the persistence of the motion of an absolutely round ball rolling on a horizontal plane. It is enough that there is no cause which would act so as to bring it to a halt.

The same tactics are applied to the transformation of the idea of *impetus*. Galileo opens his attack on Aristotelian physics with the help of the objections and ideas accumulated and developed by 'Parisian' physics. The time comes, however, when, being convinced of its hybrid and muddled character, Galileo abandons the concept of *impetus*, seen as the origin and cause of motion. So as the *Dialogue* progresses *impetus* can be found identified with moment, with motion, with speed—successive subtle modifications which imperceptibly guide the reader towards the conception of the paradox of motion which is conserved by itself in the moving body, and of speed which is 'ineradicably impressed' on bodies in motion.

In theory the special status of circular motion is now ready for destruction. It is motion as such which is conserved and not circular motion. But this is in theory. In practice the *Dialogue* does not take this step. Regardless of what others have claimed, this move is not in fact taken, and nor is the move to the principle of inertia. Neither in the *Discourses* nor in the *Dialogue* does Galileo anywhere assert the eternal conservation of rectilinear motion. This is for a simple reason, namely that for *heavy bodies* rectilinear motion is impossible, and that, for Galileo, non-heavy bodies are not bodies at all and are incapable of motion.[190]

3

GALILEO'S PHYSICS

Galileo's physics is a physics of heavy bodies. A physics of bodies which fall, which move downwards. This is why the motion of fall has such an important role to play in it, a role which is such that one could characterise it as 'a physics of fall'. For the motion of fall is not only seen by Galileo as a natural motion but is, in fact, the only motion which he accepts as natural.

So it is obvious that the term 'natural motion' does not have for Galileo, or does not have within Galilean physics, the same meaning that it had for Aristotle. The latter identified several distinct kinds of natural motion, each of which expressed in their various specific characters the differences between the natures of different kinds of moving bodies. Of these only one remains in Galilean physics. Moreover, this motion is common to all bodies. This indicates, of course, that all bodies are identical in nature,[191] although it does not tell us what this nature is.

In Galilean physics a motion never reveals nor expresses the nature of a body. We have already had occasion to mention the extent to which motion is extrinsic to the moving body. It does not affect the moving body in itself—in itself motion is as nothing and non-existent[192]—and is only possessed by the moving body in relation to something other than itself. Motion and rest are pure accidents. Therefore, in the strict, Aristotelian sense of the term there are for Galileo no natural motions, and equally there are no violent motions. This Aristotelian distinction is unacceptable to Galileo, and many years earlier Galileo had already objected to it on the grounds that it is neither exhaustive nor absolute, and that it is not applicable to any motion considered in itself.[193] So-called natural and violent motions can in fact change from one into the other. A ball thrown into the air falls back down again, and one which rolls down a slope rolls up again on the other side; the tip of a pendulum does not stop at the lowest point in its swing but goes back up again and then down yet again; and if there were a hole right through the earth a stone thrown into it would not stop at the earth's centre but would come up again to the opposite side:[194] these are classical examples from the *impetus* theorists,[195] examples which had become very well known and which Galileo makes a point of reproducing.

But if the terms 'natural' and 'violent' as applied to motion no longer have any theoretical meaning in Galilean physics, what then is their use? Simply that they refer to the common sense distinction between those motions which happen by themselves (fall, downwards motion) and those which are only performed by a body as a result of external agency (projectiles, upwards motion). Now this fact, that Galileo retains within his physics a distinction deriving from common sense, seems to us to be highly significant.

We will return to this problem later. In the meanwhile let us go back to that of free fall. As everybody knows, and as Galileo explicitly says, fall is a natural motion *of heavy bodies*.[196] Now in Galilean physics all bodies are 'heavy'. There are none which are without weight. *A fortiori* none is 'light'. In contrast to Aristotle, Galileo does not accept the existence in bodies of a distinct property called 'lightness'. This is also the reason why, for him, upwards motion is not a natural, i.e., a spontaneous, motion. No body moves upwards *of its own accord*. If a body does so it is because it is pushed and squeezed out of the place it occupies by other, heavier, bodies. All ascending motion is motion of extrusion.

It is known that these ideas, which Galileo adopted from the time of his earliest works on physics,[197] are neither very original nor very recent. They had been extensively developed long before Galileo by the Parisian nominalists. They had been taught by Copernicus, and later by Benedetti. Bonamico had written about them, and this is certainly where Galileo had come across them.[198]

Galileo, moreover, never claims that these are his ideas. While it is true that he mentions neither Benedetti nor Copernicus in this connection he does state that these are very old ideas, and that his theory of heaviness, or gravity [*la pesanteur*] as a universal quality or property of bodies, is none other than that of the ancient philosophers, and in particular that of Plato.[199]

For the young Galileo gravity or heaviness was a natural property of bodies. It was even their *only* natural property. This clearly explains both why the motion of fall is a natural motion and also why, as a natural motion, it is universal.

In the physics of the young Galileo heaviness or gravity was a source of motion. Since it is the only natural property of bodies, it is also the *only* natural source of motion: and since it is a *universal* natural property of *all* bodies, it produces in all bodies a *natural* motion in a 'downwards' direction.

Now we have seen that in the physics of the *Dialogue* (and it is no different in that of the *Discourses*), all bodies are heavy, and that any body placed on an inclined plane, or more simply any unsupported body, 'descends' and naturally moves in a downwards direction.[200]

One might be tempted, therefore, to characterise Galileo's physics as a physics of heaviness or gravity, as Descartes' has been characterised as a physics of impact and that of Newton as a physics of force. But this would, strictly speaking, be incorrect. For Galileo actually refuses to see gravity as a *natural quality* of bodies, and he equally refuses to see it as a *source* or a *cause* of downwards motion. This is for the simple reason that he is perfectly well aware that he does not know what it is. The fact is that for Galileo heaviness or gravity is not a theoretical property of bodies. It is an empirical property; it is a quality derived from common sense. This is the explanation of the curious fact that Galileo, in both the *Dialogue* and the

Discourses, talks about heavy bodies [*corps graves*] but avoids talking about *gravity*.

At the beginning Galileo says that heaviness is nothing other than the natural tendency of bodies to move towards the centre of the earth, or towards the centre of heavy things whether this be the earth's centre or that of the whole Universe.[201] For is it not necessary, precisely in order to be able to extend the domain of gravity to all bodies, to start by speaking a language which is acceptable and comprehensible to everybody, even to the Aristotelian? Thus Galileo says that it is certainly necessary for a body to have a specific tendency towards some particular place in order for it thereby to be set in motion; if it were not for this it would remain untroubled where it is.[202] This tendency is the explanation of both the acceleration of the motion and of the fact that fall is accomplished in a straight line. However, we should not take these explanations too seriously; we are still at the very beginning of the *Dialogue* and later in the book things will be totally different. The first thing we must do is to distinguish between the centre of the earth and that of the world (if there is in fact any such centre, Galileo remarks, then it is no doubt in the Sun[203]) and, following Copernicus, to explain the motion of fall by a natural tendency of parts to unite with the Whole.[204] But once again this is but a stage in the argument; for later Galileo's critique, which progressively dissolves the basic, traditional ideas of physics in order to rebuild and reorder them, will arrive at a point at which he will deny any positive, explanatory value to the idea of gravity.

Bodies fall, which is to say that the parts of the earth are thrust in a 'downwards' direction. This is common knowledge. But that is all that it is. Because the 'cause' of this motion, whether it be internal or external, is totally unknown. To point to 'heaviness', 'gravity', 'downwards tendency' or 'tendency towards the centre' is to refer to the phenomenon; it does not explain it. Therefore, to Simplicio's indignant objection that everybody knows the cause of this effect (downwards motion), that it is well known that it is gravity, Salviati replies:[205] 'You are wrong, Simplicio; what you ought to say is that everybody knows that it is called 'gravity'. What I am asking you for is not the name of the thing, but its essence, of which essence you know not a bit more than you know about the essence of whatever moves the stars around.[206] I except the name which has been attached to it and which has been made a familiar household word by the continual experience that we have of it daily. But we do not really understand what principle or what force it is that moves stones downward, any more than we understand what moves them upward after they leave the thrower's hand, or what moves the moon around. We have merely, as I said, assigned to the first the more specific and definite name 'gravity', whereas to the second we assign the more general term 'impressed force', and to the last-named we give 'spirits' (*intelligenza*), either 'assisting' (*assistente*) or 'abiding' (*informante*); and as the cause of infinite other motions we give 'Nature'.'

We can see how far Galileo has come since Pisa. Then it was 'lightness' which was said to be no more than a 'name', falsely substantialised, and used in referring to the effects (upwards motion) of an underlying cause. Henceforth 'heaviness' or 'gravity' will share its fate, for it is also merely a 'name', as is the celebrated *vis impressa*, the *impetus* of the Parisian School, alleged internal cause of the motion of projectiles. The ultimate conclusion towards which Galileo is surreptitiously leading us is clear: all of these 'internal causes' are merely 'names'.[207]

No doubt the term 'gravity' does refer to something in reality, something of great importance even. But this absolutely fundamental property does not constitute the 'nature' of bodies; it is not one of their *essential* properties. In fact in the well known and justly famous passage in *The Assayer* in which Galileo gives an account of his philosophy of nature (and which is moreover reproduced word for word in his *Letter to the Grand Duchess Christina of Tuscany*[208]) *there is no mention of gravity*. In this text, which is strangely and significantly reminiscent of similar passages in Descartes, Galileo explains that:[209] '. . . whenever I conceive any material or corporeal substance, I immediately feel the need to think of it as bounded, and as having this or that shape; as being large or small in relation to other things, and in some specific place at any given time; as being in motion or at rest; as touching or not touching some other body; and as being one in number, or few, or many. From these conditions I cannot separate such a substance by any stretch of my imagination. But that it must be white or red, bitter or sweet, noisy or silent, and of sweet or foul odour, my mind does not feel compelled to bring in as necessary accompaniments. Without the senses as our guides, reason or imagination unaided would probably never arrive at qualities like these. Hence I think that tastes, odours, colours, and so on are no more than mere names so far as the object in which we place them is concerned, and that they reside only in the consciousness. Hence if the living creature were removed, all these qualities would be wiped away and annihilated'.

It can be seen that in Galileo's view (as in Descartes', and for the same reasons) what constitutes the *essence* of bodies, or of matter, that which we cannot think of them as being without, and consequently that without which they could not exist, are their mathematical properties. Number, shape, motion: arithmetic, geometry, kinematics. Gravity is not included.

Nor is it among the purely sensible qualities, such as colour, odour, heat or sound, which are declared by Galileo to be purely subjective and dependent for their very existence on that of living creatures.

So where then is it to be found? Nowhere. Or rather, somewhere between non-being and being; gravity occupies an intermediate place between the non-being of sensible appearances and the being of the mathematically real. Or one might say, the place of an intermediary. Therefore it exists only as a phenomenon.

Obviously its existence cannot be denied. Bodies fall; *physical* bodies, that is: as for geometrical bodies, they do not 'fall' at all. It is precisely this, the fact that bodies 'fall' (which means 'are spontaneously set in motion'), which makes physics a specific science and distinguishes it from geometry.[210] Bodies are heavy (*graves*): so although *gravity* is certainly not a clear, mathematical idea, and does not designate an essential quality of bodies, yet *physics*, the science of motion and rest, cannot do without it. In what way can it use it? The bodies of mathematical physics, Galilean bodies, or to give them their real name, Archimedean bodies, are no more nor less than geometrical, Euclidean 'bodies' endowed with gravity. In other words, gravity is the only 'physical' property they have.

Thus 'physical', Archimedean bodies are, in a way, heavy (*graves*) by definition.[211] This is why they are 'movable bodies', which geometrical bodies are not at all.[212] Thus they fall and have a natural tendency to move in a downwards direction, something which geometrical bodies do not do at all.

So the gravity of bodies is presented as being connected with their motion. Or, to put it another way, motion, without which there would be no physics, is presented as connected with the phenomenon of gravity. This deep-seated Archimedeanism of Galileo's thought, which we have already emphasised above, together with its realism,[213] is the explanation, even more than is his having been unconsciously influenced by empirical phenomena, of why it was impossible for Galileo to formulate correctly the principle of inertia.

In Galilean physics, then, gravity remains, in spite of everything, a source of motion. It is even, as we have said, the only source of motion which it recognises. For impact only transfers a motion (a speed) which already exists from one movable body to another. Fall, on the contrary, creates it. So to produce motion, i.e., to confer a speed on a body, it is necessary according to Galilean physics to let it fall from 'up' to 'down'.[214]

It is very easy to accept the proposition that gravity is a source of motion. It is a reasonable proposition. It is even common sense. It is also a proposition in Aristotelian physics. Aristotelian physics, however, obviously cannot accept that it is the only one. This would be to accept the oneness of matter, to give up the division of the Cosmos into two regions, celestial and sublunary, and to admit that the same laws and the same physics hold equally on earth and in the heavens.

This is precisely Galileo's thesis. The odd mythical cosmogony with which the *Dialogue* opens (and which Galileo, in order to indicate once again his philosophical allegiance, attributes to Plato, although Plato in fact never taught anything like it), in which we see God letting the planets fall before conferring circular motion on them in their respective orbits,[215] is certainly only offered as a way of posing very clearly for us this opposition between the Aristotelian and the Galilean perspectives, between ancient and classical science, and to allow us to clearly grasp the philosophical implications of the fundamental principles of the latter, and in particular of the principle of the uniformity of law.

It could be said that Galileo's thought covers the same ground as that on which Copernicus set out, but in the opposite direction. The latter applied the laws of the heavens to the earth,[216] while the former applied terrestrial principles to the heavens.

It had been accepted that the motion of fall is the only natural motion on earth. Galileo declared that this is also true in the heavens, that the circular motion of the planets is in no way a 'natural', i.e., spontaneous, motion, and that in order to generate motion nobody, not even God, could employ means different from those, or more accurately different from that, which we use on earth.

Galileo, of course, is tactful. He is not calling in question God's omnipotence. God could, of course, create motion directly. But this would be, as it were, an extra miracle.[217] The miracle of the creation of bodies is difficult enough. It is not good science to require that God perform a second one, and one which is moreover completely unnecessary. Furthermore, if God were to create this motion directly it would not be a natural motion.

The reversal of the situation, compared with Aristotelianism, is complete. The circular motion of the planets, seen as spontaneous motion, demonstrated according to Aristotle that the earth and the heavens are of different natures. By contrast, when it is seen as secondary motion it demonstrates, according to Galileo, that they possess a common nature. For the special status of circular motion (motion around a centre) is explained precisely by the phenomenon of gravity.[218]

The phenomenon of gravity determines and explains the phenomenon of motion. The motion of fall is the natural motion of all bodies left to themselves. Moreover, the motion of fall, simply as a form of motion (i.e., kinematically), possesses specific features which make it exceptional. It is a kind of motion quite unlike any other. Not only is it a constantly and continuously accelerated motion, which means that a body impelled by this motion progressively acquires all degrees of speed or slowness without missing any out and without pausing in any one (so that it is capable of conferring on a given body a specific degree of motion, i.e., of speed[219]), but also it represents a quite specific form of motion which is always the same in all its instances, wherever a body is in free fall or moves down an inclined plane.[220] Yet still more remarkable is the fact that it is not just the general form of the motion which is realised in each instance, but the very motion itself, each time identical whatever body it is that is falling. For whatever body falls, i.e., whatever its weight and whatever it is physically made of, it always falls with the same speed.[221]

When one reflects on all this, when one reflects in particular on the fact that all bodies whatever fall with the same speed and according to the same law, one comes to appreciate why it was that Galileo tried to elaborate his dynamics as a dynamics of fall. One comes to understand also the pride with which Galileo declares, through his spokesman Salviati, that of course everybody has always observed that the motion of falling bodies starting from rest is not uniform but is continuously accelerated, but that this general knowledge is of no value unless one knows the ratio according to which the increase in speed takes place; in particular, unless one knows that it does so in the ratio of the odd numbers *ab unitate*, i.e., that the ratio of the distances covered is equal to that of the squares of the times.[222]

To discover the mathematical laws of motion, to discover that the motion of fall *follows a numerical law*, that really is something to be proud of.

The whole of Galileo's dynamics is based on the 'postulate' that 'the degrees of speed acquired by one and the same body moving down planes of different inclinations are equal when the heights of these planes are equal',[223] and Salviati's commentary adds:[224] 'By the height of an inclined plane we mean the perpendicular let fall from the upper end of the plane upon the horizontal line drawn through the lower end of the same plane. Thus, to illustrate, let the line AB be horizontal, and let the planes CA and CD be inclined to it; then the Author calls the perpendicular CB the 'height' of the planes CA and CD; he supposes that the speeds acquired by one and the same body, descending along the planes CA and CD to the terminal points A and D are equal since the heights of these planes are the same, CB; and also it must be understood that this speed is that which would be acquired by the same body falling from C to B'.

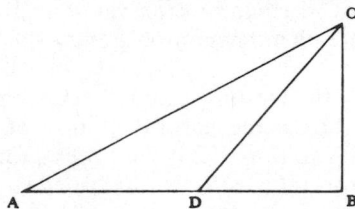

This 'postulate' of Galileo's does not explicitly employ dynamical ideas any more than does the well known definition of uniformly accelerated motion.[225] It shows us, even better than this definition (or more accurately, even better than the arguments which introduce the latter), the extent to which Galileo's thought is governed by the phenomenon of gravity, by the conception of the natural downwards motion of bodies. For Galileo's postulate makes no mention at all of any cause, of any force. It avoids not only the term 'gravity' but even 'heavy body' (*grave*); and it takes it for granted that any body placed on an inclined plane will move down it with increasing speed!

To us Galileo's postulate does not seem at all self-evident, and we would certainly not think of putting it at the beginning of a treatise on mechanics. But this is just what Galileo does, and Sagredo's judgement is that[226] 'Your assumption appears to me so reasonable that it ought to be conceded without question, provided of course that we set aside all external and chance resistance, and that the planes are hard and smooth, and that the shape of the moving body is perfectly round, so that neither plane nor moving body is rough. All resistance and opposition having been removed, my reason tells me at once that a heavy and perfectly round ball descending along the lines CA, CD, CB would reach the terminal points A, D, B, with equal *impetuses*'.

Sagredo is quite right to emphasise the necessity for setting aside all 'external resistance', because the laws of Galilean physics are 'abstract'laws which do not hold as they stand for real bodies. There is, no doubt, some reality to which they refer, but this is not the reality of everyday experience. It is an ideal and abstract reality. As for us we have no need to be reminded of this. We are perfectly used to such abstraction. In fact what we need is the opposite; we need to be reminded that the ideal and abstract world of mathematical physics is not, strictly speaking, the real world.[227] But Galileo's postulate does not seem to be something that we could take for granted, not even for the ideal and abstract world. It is not self-evident for us; it is not immediately obvious in the light of our 'natural reason'. This is because we were not brought up on Benedetti and Archimedes. We ceased to be Archimedeans a long time ago.

Let us return now to the study of motion. We have just seen that speed is acquired in and by descent. But how is it lost? In the Archimedean world of Galilean physics, this world in which all external resistance to motion is 'set aside', it is only lost by re-ascent. Galileo's postulate implies, in fact, that simple translation, i.e., horizontal translation, is accomplished without expenditure of energy. The distance covered by the heavy body is irrelevant, for the *impetus* or the acquired *momenta* remain the same.[228] Conversely it is clear that, whatever the distance covered, the energy used in raising a body to a given height will always be the same; and that this energy will be exactly equal to that acquired by the body in descending this same distance. In other words it is clear that in descending a body acquires an *impetus* or *momentum* which is exactly sufficient for it to re-ascend the slope.[229]

These are necessary consequences of the Galilean concept of motion. This is why Galileo makes no effort to prove them for us. In fact he says no more than that if one imagines a perfect ball descending the whole length of one inclined plane and re-ascending another, and if one discounts any obstacle which interferes with the experiment (in particular, any loss of *impetus* at the point of intersection of the two planes) then 'it is clear that the *impetus* acquired (which gains in strength with descent) will be able to carry the body up to the same height'.[230] In other words, for Galileo, the proposition is self-evident. It is true that he suggests that we should take it only as a 'postulate', for its absolute truth will not be established until later. But we know that this is mere rhetoric. It is also true that he has previously clarified the postulate for us by means of the extremely ingenious experiment of the pendulum, which, when falling from any given point, always re-ascends to the same height, i.e., to the same horizontal plane, regardless of the arc of swing.[231] This experiment is wonderfully ingenious. It is, however, and Galileo does not disguise the fact, a thought-experiment. We can add the conclusive point that Galileo's argument in fact presupposes the postulate which it is supposed to prove.

We should be clear, however, that we are not raising an objection against Galileo. The object of this investigation is not to uncover the formal errors in Galileo's arguments, but to disclose the real underlying structure of his thought, and in particular to enable us to grasp the role that was played in it by the phenomenon and the idea of gravity. We could, in fact, have made this job easier for ourselves; to grasp this role and to see the central importance of gravity in his dynamics it would be enough to cite Galileo's proof of his first 'postulate'.

This postulate, which is later transformed into a theorem,[232] says that the speed of a descending body depends on the height of its fall, regardless of the distance travelled. Now, the law of falling bodies which Galileo had established in the meanwhile, says that this speed depends on the time elapsed, i.e., on the duration of the descent, where this quite clearly is not the same for vertical descent (free fall) as for descent on an inclined plane. Galileo, therefore, will show that the law of fall, which he accepts as holding for the descent of a heavy body on an inclined plane,[233] leads precisely to the theorem in question. 'It has been shown', says Salviati on his behalf, 'that on a plane of any inclination whatever the increase in speed or in quantity of *impetus* of a body starting from rest is in proportion to the time, in agreement with the definition of naturally accelerated motion given by the Author. Hence, as he has shown in the preceding proposition, the distances traversed are proportional to the squares of the times and therefore to the squares of the speeds. The speed relations are hence the same as in the motion first studied [i.e., vertical motion], since in each case the gain of speed is proportional to the time'.[234] Now, since the speed of the body depends on the initial *impetus* or 'moment', and since these vary with the inclination of the plane, it follows that the body descending down an inclined plane, while moving less rapidly, will move for a longer time and will arrive at the bottom of its trajectory with the same final speed as if it had descended by free fall.

Thus the proof of Galileo's postulate, the relation between distance and duration, depends on dynamical concepts; the speed of the descending body is explicitly related to the magnitude of the initial *impetus*.

Have we, then, reverted to *impetus* physics? Or have we, as Duhem thinks,[235] never left it at all? This is a serious problem, and it requires very close examination. What, in fact, is this Galilean *impetus*?[236]

'Let us consider first of all', says Galileo,[237] 'the well known fact that the moments or speeds of a given moving body are different on planes at different inclinations. The speed reaches a maximum along a vertical direction, and for other directions diminishes as the plane diverges from the vertical. Therefore the *impetus*, ability, energy, [*l'impeto, il talento, l'energia*] or, one might say, the momentum of descent of the moving body is diminished by the plane upon which it is supported and along which it rolls.

'For the sake of greater clearness erect the line AB perpendicular to the horizontal AC; next draw AD, AE, AF, etc., at different inclinations to the horizontal. Then I say that all the momentum of the falling body is along the vertical and is a maximum when it falls in that direction; the momentum is less along DA and still less along EA, and even less yet along the more inclined plane FA. Finally, on the horizontal plane the momentum vanishes altogether; the body finds itself in a condition of indifference as to motion or rest; has no inherent tendency to move in any direction, and offers no resistance to being set in motion. For just as a heavy body or system of bodies cannot of itself move upwards, or recede from the common centre toward which all heavy things tend, so it is impossible for any body of its own accord to assume any motion other than one which carries it nearer to the aforesaid common centre. Hence, along the horizontal, by which we understand a surface, every point of which is equidistant from this same common centre, the body will have no momentum whatever.'

So the *impetus* of the moving body is nothing other than the dynamic impulse given to it by its gravity. It is no longer in any way the internal cause producing the motion, as it was in Parisian physics. It is the same thing as its 'moment', i.e., the product of its weight and its speed. In the moving body at the end of its descent it is the total energy, or total *impetus*; in the body at the beginning of its motion it is the product of its weight and its initial speed (i.e., the differential of the speed). Finally, for the body at rest the *impetus* is none other than the virtual speed.[238]

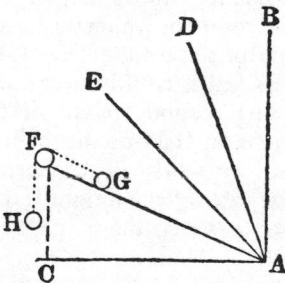

The *impetus* or initial motion, the impulse or differential of speed, varies with the inclination of the plane which the body is on. In order to measure it, and at the same time to measure its variation, one has only to recall the fact that it is obvious that the *impetus* of descent of a heavy body is of just the same magnitude as the resistance or the minimum force required to prevent its motion or to stop it. Now, 'in order to measure this force and resistance I propose to use the weight of another body. Let us place upon the plane FA a body G connected to the weight H by means of a cord passing over the point F; then the body H will ascend or descend, along the perpendicular, the same distance which the body G ascends or descends along the inclined plane FA; but this distance will not be equal to the rise or fall of G along the vertical in which direction alone G, as other bodies, exerts its resistance. This is clear. For if we consider the motion of the body G, from A to F, in the triangle AFC to be made up of a horizontal component AC and a vertical component CF, and remember that this body experiences no resistance to motion along the horizontal (because by such a motion the body neither gains nor loses distance from the common centre of heavy things) it follows that resistance is met only in consequence of the body rising through the vertical distance CF. Since then the body G in moving from A to F offers resistance only in so far as it rises through the vertical distance CF, while the other body H must fall vertically through the entire distance FA, and since this ratio is maintained whether the motion be large or small, the two bodies being inextensibly connected, we are able to assert positively that, in case of equilibrium (bodies at rest) the momenta, the speeds, or their tendency to motion, i.e., the spaces which would be traversed by them in equal times, must be in the inverse ratio to their weights. This is what has been demonstrated in every case of mechanical motion.[239] So that, in order to hold the weight G at rest, one must give H a weight smaller in the same ratio as the distance CF is smaller than FA. If we do this, FA:FC=weight G:weight H; then equilibrium will occur, that is, the weights H and G will have equal moments, and the two bodies will come to rest.

'And since we are agreed that the *impetus*, energy, momentum or tendency to motion of a moving body is as great as the force or least resistance sufficient to stop it, and since we have found that the weight H is capable of preventing motion in the weight G, it follows that the smaller weight H whose entire moment is along the perpendicular, FC, will be an exact measure of the partial moment which the larger weight G exerts along plane FA. But the measure of the total moment on the body G is its own weight, since to prevent its fall it is only necessary to balance it with an equal weight, provided this second weight be free to move vertically; therefore the partial moment on G along the inclined plane FA will bear to the maximum and total *impetus* on this same body G along the perpendicular FC the same ratio as the weight H to the weight G. This ratio is, by construction, the same which the height, FC, of the inclined plane bears to the length FA.'[240]

Galileo's reasoning here, which takes *impetus* to be a magnitude and which, in thus linking dynamics with statics,[241] measures *impetus* by resistance, i.e., in the last analysis, by the *weight* which counteracts the impulse to motion,[242] is a transposition of Archimedean reasoning. The *gravitas secundum situm* becomes an *impetus secundum situm*, and statics is transformed into dynamics, because Galileo gives to gravity a dynamic interpretation.

Now if this is so, if the very basis of Galileo's dynamics is profoundly Archimedean, and is wholly based on the idea of gravity, then it follows that Galileo would not be able to formulate the principle of inertia: and in fact he never did so.

For, in order to have been able to do this, i.e., to have been able to assert the eternal persistence not of motion in general but of *rectilinear* motion, to have been able to imagine a body, left to itself and *unsupported*, remaining at rest or continuing to move in *a straight line* and not *in a curve*,[243] it would have been necessary for him to have been able to conceive of the motion of fall not at all as natural but, on the contrary, as 'adventitious' and 'violent', i.e., as caused by an external force. It would have been necessary, in other words, for Galileo to have taken the mathematicism of his philosophy of nature to the limit, by excluding gravity not only from the essential constitution of bodies but even from their 'effective' constitution, i.e., it would have been necessary for him to have been able to reduce the manifest properties of bodies to those which are determined by their essence. This in turn means: it would have been necessary for him to have no longer been Archimedean, but to have become Cartesian.

It has sometimes been said, and we have said it ourselves, that for Galileo the path towards the principle of inertia was barred by the astronomical observation of the circular motion of the planets,[244] a motion which is inexplicable and therefore eminently 'natural'. This seems indisputable. This was not, however, the only obstacle that astronomy, or more accurately the study of the heavenly Universe, placed in the way of the invention of the principle of inertia. The belief in the finiteness of the Universe constituted an insurmountable barrier to Galileo's thought. This would have been enough in itself to have brought about its failure. Moreover, celestial physics was fully in agreement with terrestrial physics: for the latter, based entirely on a dynamical concept of gravity as a source of motion and as a constitutive, yet inadmissible, property of bodies, could not allow the special status of rectilinear motion.[245]

We have just seen that the impossibility of Galileo's arriving at the principle of inertia is to be explained on the one hand by his refusal to completely abandon the idea of the Cosmos, i.e., the idea of a well-ordered world,[246] and to unambiguously accept the infinity of space; and, on the other hand, by his inability to conceive of physical bodies (or the bodies of physics) as being without the constitutive property of gravity.

Why did Galileo refuse to accept the infinity of space? This is a question which we find it impossible to answer. We must be satisfied with merely taking note of the fact; the Galilean Universe is a finite Universe.[247] Perhaps—though this is no more than a hypothesis—he had been frightened by the example of Bruno, by the consequences which Bruno had been led to draw from the doctrine of infinity.[248]

Why was he unable to abstract from gravity? Quite simply because he did not know what it was. He could certainly abstract from all theories of heaviness or gravity, but not from the phenomenon of gravity itself, which is an immediate given of observation and of common sense. He could no more explain it than could his master, Archimedes. He had no conjectures about it to offer.

It might be objected, of course, that this explanation, while it is valid for Archimedes, is not so in the case of Galileo. In the absence of any physical theory of gravity Archimedes was indeed obliged to treat it simply as a fact. But what was true in his time was so no longer in that of Galileo. A physical theory of gravity did then exist, Gilbert's theory, which was adopted with modifications by Kepler. Why then did Galileo, who admired Gilbert almost as much as he admired Copernicus,[249] and who was convinced that

Gilbert was right, and that the earth is indeed a large magnet (Sagredo declares this on his behalf[250]), why did he not adopt this theory? The answer is immediately obvious: while Galileo could admire Gilbert and could perfectly well accept his doctrine of the magnetic nature of gravity, he could not use it because it was neither mathematical nor even mathematisable.[251] Attraction, for Gilbert, was an *animate* force.[252] No doubt this was no longer so in the case of Kepler's attraction, but from its animist inheritance it still retained the faculty of directing itself towards its object. It knew, in a manner of speaking, where it had to go, knew where the body that had to be attracted was located.[253] This mysterious faculty, which Galileo's own investigations on the magnet had not succeeded in clarifying, nor in mathematising, remained unusable for physics.

So, on three occasions Galileo came close to the principle of inertia, so close as to be, as it were, within touching distance. Each time, at the last moment, he drew back. Nothing is more instructive, we believe, than the analysis of these three denials.

On the first occasion the principle of circular motion becomes vulnerable during a discussion on centrifugal force. It will be recalled that Ptolemy had based on this force an argument against the earth's motion, asserting that the vast speed of this motion would cause the earth to fly to pieces. Salviati, following Galileo's usual method of strengthening his opponent's arguments, attempts to show[254] 'more sensibly how true it is that heavy bodies, whirled quickly around a fixed centre, acquire an *impetus* to move away from that centre even when they have a natural tendency to go toward it. Tie one end of a cord to a bottle containing water and, holding the other end firmly in your hand (making your arm and the cord the radius, and your shoulder knot the centre), cause the vessel to go around swiftly so that it describes the circumference of a circle. Whether this is parallel to the horizon, or vertical, or slanted in any other way, the water will not spill out of the bottle in any event; rather, he who swings it will always feel the cord pull forcibly to get farther away from his shoulder. And if a hole is made in the bottom of the bottle, the water will be seen to spurt forth no less toward the sky than laterally or toward the ground. And if in place of water you put pebbles in the bottle, upon your turning it in the same manner it will be felt to exert the same force against the cord. Finally, small boys may be seen throwing rocks a great distance by whirling a slotted stick with a stone in the end. All these are arguments of the truth of the conclusion that whirling confers an *impetus* upon the moving body toward the circumference, if the motion is swift. And since, if the earth revolved upon itself, the motion of its surface (especially near the equator) would be incomparably faster than the objects mentioned, it would necessarily throw everything into the sky.'

Taken literally this argument recounted by Salviati is invalid (and nobody before Galileo had pointed this out) because it confuses the linear speed of a point on the surface of the earth with the latter's angular speed of rotation. As Salviati will say later:[255] 'Up to this point we have made no issue about granting it to Ptolemy as an unquestionable fact that since the casting off of the stone is caused by the speed of the moving wheel about its centre, the cause of this casting off is augmented as the speed of whirling is increased. From this it was inferred that on account of the rapidity of the terrestrial whirling being very much greater than that of any machine which we can rotate artificially, the consequent extrusion of stones, animals, etc. should be very violent.

'I now take note that there is a very gross fallacy in this argument when we indiscriminately compare such speeds with each other absolutely. It is true that if I make a comparison between speeds of the same wheel, or of two equal wheels, then that which is turned the more rapidly will hurl stones with the greater *impetus*, and when the speed increases the cause of projection will increase also in the same ratio. But now suppose the speed to be made greater not by increasing the speed of a given wheel (which would be done by making it have a larger number of revolutions in the same time), but by increasing the diameter and enlarging the wheel, preserving the same time for each revolution of the large wheel as of the small one. The speed would now be greater in the large wheel merely by reason of its greater circumference. No one would suppose the cause for extrusion to increase in the ratio of the speed of its rim to that of the smaller wheel; that would be quite false, as may be shown at once by a ready experiment, roughly as follows. We can throw a stone better with a stick a yard long than with one six yards long, even if the motion of that end of the long stick[256] where the stone is stuck is more than twice as fast as the motion of the end of the shorter stick—as it would be if the speeds were such that during one complete revolution of the larger stick, the smaller one made three turns'. For it is only the speed of rotation (the angular speed) which is relevant here, and as Sagredo observes,[257] 'the whirling of the earth would no more suffice to throw off stones than would any other wheel, as small as you please, which rotated so slowly as to make but one revolution every twenty-four hours'.

The Ptolemaic argument is clearly fallacious. But this does not prevent its being of capital importance, and its enabling us to uncover something which is completely incompatible with Galileo's repeated assertions. For if, as Galileo has stated over and over again, motion as such is as nothing and non-existent for things which all participate in it together, and if, in particular, everything which happens on the rotating earth happens in exactly the same way as it would if the earth were stationary, in other words if the principle of the relativity of motion holds universally and absolutely, and holds, in particular, for circular motion 'around a centre', then the earth's rotation could no more produce a centrifugal force than could any other motion. The existence of this force goes without saying in Aristotle's and Ptolemy's physics. For them, circular motion (around a centre) is natural only for the bodies and spheres of the heavens, for they have no gravity; it is not at all natural for heavy bodies. Now, Galileo has shown us that this is not so at all, and that it is precisely for heavy bodies that circular motion has a special status. No doubt, given the very slow rotation of the earth, the centrifugal force produced by its motion is very weak; but however weak it may be it would, nevertheless, produce observable effects;

and what if the earth were to turn more rapidly?

Galileo, therefore, is obliged to demonstrate that, whatever the speed of the earth's rotation, the effects predicted by Ptolemy could never occur. But his demonstration, which is so ingenious that one wishes it were not invalid, will disclose for us a very significant fact: namely, that every impulse or tendency to motion is *rectilinear*,[258] and that the circular motion of heavy bodies is only the resultant of two rectilinear motions.[259] We are on the threshold of the principle of inertia, a threshold, however, which Galileo will refuse to cross!

Ptolemy's reasoning, though incorrect, is nevertheless plausible. Sagredo's assertion that the extremely rapid motion of the earth's surface is no more able to project a stone than is the extremely slow motion of the circumference of a wheel one metre in diameter, is correct. But it does seem rather paradoxical: are not the speeds impelling the stones vastly different in the two cases? Of course. But Galileo explains that this is of no importance, and in order to explain this more clearly he even provides a diagram.[260]

'Let there be two unequal wheels around this centre A, BG being on the circumference of the smaller, and CEH on that of the larger, the radius ABC being vertical to the horizon. Through the points B and C we shall draw the tangent lines BF and CD, and in the arcs BG and CE we take two arcs of equal length, BG and CE. The two wheels are to be understood as rotating about their centre with equal speed in such a way that two moving bodies will be carried along the circumferences BG and CE with equal speeds. Let the bodies be, for instance, two stones places at B and C, so that in the same time during which stone B travels over the arc BG, stone C will pass the arc CE.

'Now I say that the whirling of the smaller wheel is much more powerful at projecting the stone B than is the whirling of the larger wheel at projecting the stone C. And since, as already explained, the projection would be along the tangent, if the stones B and C should be separated from their wheels and commence motions of projection from the points B and C, they would be flung along the tangents BF and CD by the *impetus* received from whirling. The two stones therefore have equal *impetuses* for travelling along the tangents BF and CD, and if no other power were to deviate them, it is along these that they would travel. Isn't that so, Sagredo?

Sagredo: That is the way it seems to me the thing takes place.

Salviati: But what power do you think could deviate the stones from moving along the tangents, where the *impetus* of whirling actually casts them?

Sagredo: Either their own weight, or some glue which may hold them in place attached to the wheels.

Salviati: But to deviate a moving body from a motion for which it has the *impetus*, is not a greater or a lesser power needed, according as the deviation must be greater or less? That is, according as they must in this

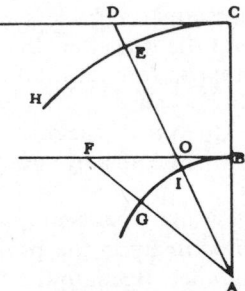

deviation pass through a greater or a lesser space in a given time?

Sagredo: Yes. For it was already concluded above that in order to make a body move, the faster it is to be moved the greater must be the moving force.

Salviati: Well, consider how in order to deviate the stone on the smaller wheel from the motion of projection that would be made along the tangent BF, and to keep it attached to the wheel, its weight would have to be pulled back as far as the secant FG, or rather the perpendicular drawn from the point G to the line BF, whereas on the larger wheel the withdrawal would need to be no more than the secant DE, or rather the perpendicular drawn from the point E to the tangent DC. This is much less than FG, and always less and less, the larger the wheel is made. And since these withdrawals have to be made in equal times (i.e., while the two equal arcs BG and CE are being traversed), that of stone B (viz., the retraction FG) will have to be much faster than the other, DE. Therefore much more force is needed to hold the stone B joined to its small wheel than the stone C to its large one, which is the same as to say that a smaller thing will hinder projection from the large wheel than will prevent it on the small one. And thus it is obvious that the larger the wheel becomes, the more the cause for projection is diminished.'

Salviati's reasoning is faultless, but in order to make it comprehensible he has had to develop a whole theory of centrifugal force, and to show that it is not directed radially, outwards towards the circumference, but on the contrary, tangentially, perpendicular to the radius of the wheel.[261]

However, it seems to follow from this (and actually does follow from it in fact) that a body placed on a large wheel, and therefore moving with a greater linear speed than the same body would be if it were placed on a smaller wheel (and given that the two wheels have the same angular speed) would have a very much greater *impetus*. Therefore a body would be thrown very much further by a long stick or sling than by a short one with the same angular speed. No doubt, Galileo replies, this would happen if the body managed to actually leave the wheel (or sling); but it cannot do this by itself, because even the smallest force would be enough to hold it there.

So, the *impetus* of the body in circular motion is directed along the tangent to the circle of its motion, and seeks to separate it from this circle. But how does this separation take place? Simplicio, to whom the question is addressed, does not understand it very clearly. He cannot give an answer, never having thought about the question. But Salviati reassures him; all that he is lacking is the appropriate terminology. As for the root of the problem, he tells him:[262] 'In the same way that you knew what went before, you will know—or rather, do know—the rest too. And by thinking it over for yourself you would likewise recall it by yourself. But to save time, I shall help you to remember it.

'Up to this point you knew all by yourself that the circular motion of the projector impresses an *impetus* upon the projectile to move, when they separate, along the straight line tangent to the circle of motion at the point of separation, and that continuing with this motion, it travels ever farther from the thrower. And you have said that the projectile would continue to move along that line if it were not inclined downward by its own weight, from which fact the line of motion derives its curvature. It seems to me that you also knew by yourself that this bending always tends toward the centre of the earth, for all heavy bodies tend that way.

'Now I shall pass on a little further and ask you whether the moving body in continuing its straight motion after the separation goes uniformly farther from the centre (or from the circumference, if you like) of that circle of which its previous motion was a part. That is to say, do you believe a body which leaves from the point of tangency and moves along the tangent goes uniformly away from the point of contact and from the circumference of the circle?' Simplicio has now understood. So he replies: 'No, indeed; because the tangent when close to the point of contact is very little distant from the circumference, with which it makes an extremely small angle. But as it goes farther and farther away, the distance increases in an increasing ratio . . .'

Galileo is not interested in the ultimate fate of the projected stone. What he is interested in is what happens at that very instant of separation, when the stone no longer moves in a circle and begins its rectilinear motion. So he brings the discussion back to this.[263]

'*Salviati*: Then the departure of the projectile from the circumference of its previous circular motion is extremely small at first?
Simplicio: Almost imperceptible.
Salviati: Now tell me something else. How far away after the separation would the projectile commence to sink downward, having received from the thrower's motion an *impetus* to move straight along the tangent, as indeed it would move if its own weight did not draw it down?
Simplicio: I think it would begin at once, for having nothing to sustain it, its own weight could not help acting.
Salviati: So that if the rock thrown from a rapidly moving wheel had any such natural tendency to move toward the centre of the wheel as it has to go toward the centre of the earth, it might very well return to the wheel, or rather never leave it. For the distance travelled being so extremely small at the beginning of its separation (because of the infinite acuteness of the

angle of contact), any tendency that would draw it back toward the centre of the wheel, however small, would suffice to hold it on the circumference.'

Even though Galileo's argument is not correct it is nevertheless plausible. For the angle between the circumference of the wheel and the direction of motion (*impetus*) impressed on the stone by the rotation is infinitely small; the radial component of this motion is thus infinitely small: therefore, Galileo concludes, an infinitely small force is enough to counteract it.

In order for projection to occur it would be necessary, and sufficient, for the speed produced by the rotation to be greater than the speed of fall. Of course the relevant speed here is not the tangential speed, but the speed of separation, the radial speed. But why is it that this, even though it is infinitesimal, would not be greater than the speed of fall?

In Galileo's view this is impossible; and it would still be impossible even if, as the Aristotelian believes, the speed of fall were to decrease with the weight of the body. This would even be so if the body were so light as to infinitely reduce its speed of fall, and if the projection were favoured by two causes, 'the lightness of the moving body, and its closeness to the point of rest . . . both . . . infinitely susceptible of increase':[264] this double infinity would not be enough. So, *a fortiori*, one of them by itself would be insufficient.

Galileo's proof is very odd:[265] 'So let us mark thus a perpendicular line toward the centre, AC, and let the horizontal line AB be at right angles to this, along which the motion of projection is made, and which the projectile would continue to follow with uniform motion if its weight did not bend it downward.

'Now suppose a straight line AE to be drawn from A, making any desired angle with AB, and let us mark off on AB some equal spaces, AF, FH, and HK, drawing from these the perpendiculars FG, HI, and KL, down as far as AE. And since as we have remarked on other occasions the falling body starting from rest acquires always a greater degree of speed as time goes on, according to the time elapsed, we can picture the spaces AF, FH, and HK, as representing equal times, and the perpendiculars FG, HI, and KL as representing the degrees of speed acquired in the said times. Thus the degree of speed acquired in the whole time AK will, by the line KL, be represented relatively to the degree HI acquired in the time AH, and to the degree FG acquired in the time AF; which degrees KL, HI, and FG obviously have the same ratios as the times KA, HA, and FA. And if other perpendiculars are drawn from arbitrary points marked on the line FA, smaller and smaller degrees will be found *ad infinitum,* always proceeding toward the point A, which represents the first instant of time and the original state of rest. This withdrawal toward A represents the infinite diminution of the original tendency toward downward motion with the

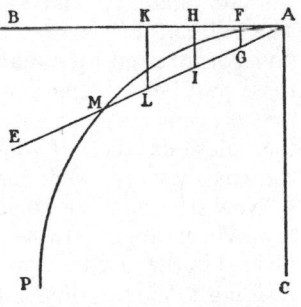

approach of the moving body to the original state of rest, which approach is infinitely augmentable. . .

'Therefore the downward speed of motion can indeed be diminished so much (admitting of a twofold diminution *ad infinitum*) that it no longer suffices to restore the moving body to the surface of the wheel, and consequently to impede its projection or prevent it.

'On the other hand, then, in order to prevent projection taking place it is necessary that those spaces through which the projectile has to descend in order to get back to the wheel must be made so short and close that however slow the descent of the moving body may be, even if infinitely diminished, it still suffices to take it back there. Hence it would be necessary to find a diminution of these spaces which was not merely infinite, but of an infinity such as to overcome the double infinity accomplished in decreasing the downward speed of the body. But how is a magnitude to be diminished still more than one which is doubly diminished *ad infinitum*? Take note, Simplicio, just how far one may go without geometry and philosophise well about nature!

'The degrees of speed, infinitely diminished . . . are always determinate. They correspond proportionately to the parallels included between the two straight lines meeting in an angle such as the angle BAE, . . . or some other angle infinitely acute but still rectilinear. But the diminution of the spaces through which the moving body must go to return to the surface of the wheel is proportional to another sort of diminution included between lines which contain an angle infinitely narrower and more acute than any rectilinear angle whatever, which is as follows: take some point C on the perpendicular AC, and with it as centre describe the arc AM or radius CA. This will cut the parallels which determine the degrees of speed, no matter how compressed they may be within the most acute rectilinear angle. Of those parallels, the parts which lie between the arc and the tangent AB are the amounts of the spaces of return to the wheel. They grow always less than these parallels of which they are parts, and diminish in an increasing ratio as they approach the point of contact.

'Now, the parallels included between the straight lines, as they retreat toward the angle, always diminish in the same ratio; that is, AH being divided in the middle by the point F, the parallel HI will be double FG, and dividing FA in the middle, the parallel drawn from the point of division will be one-half FG. Continuing this division *ad infinitum*, each subsequent parallel will be half of the next preceding one. But it is not thus with the line intercepted between the tangent and the circumference of the circle; for making the same division on FA and assuming, for example, that the parallel through H to the arc is double that through F, this latter will then be more than double the next one. And continually as we come closer to the contact A, the preceding line will contain the following line three, four, ten, a hundred, a thousand, a hundred thousand, a hundred million times, and more *ad infinitum*. Thus the shortness of such lines is reduced until it far surpasses what is needed to make the projectile, however light, return to (or rather be kept on) the circumference'.

Galileo's argument—and we have made a point of quoting it at length, for nothing is more instructive than error—is, as we have already said, extremely subtle and seductive. Unfortunately it is incorrect; and, what is worse, it is manifestly incorrect. Certainly, arguments concerning infinitesimals are difficult, and the temptation to thorough-going geometrisation is very strong. However, it is not insuperable and Galileo is more than anyone aware of its pitfalls.

Galileo's error is not merely a slip. He knows perfectly well that the rapid motion of the wheel (or of the sling) can break the tie which holds the stone there.[266] So he knows that any given force can be defeated and overcome by the centrifugal force if the rotation is sufficiently rapid. If he does not accept that this is possible in the case of the earth's rotation, and does not even notice the contradiction that he thereby gets into (to us this contradiction is blatant), this is because for him the natural force of gravity, which attracts, or thrusts, heavy bodies toward the centre of the earth, cannot be considered on the same level as the external, adventitious, violent action of a tie which holds a stone on the wheel. Gravity acts constantly and naturally. For the centrifugal force to be able to overcome it it would be necessary, he tells us, for the heavy body to be able to *defeat itself, to overcome itself.*[267] This means that for Galileo gravity is the basis and the explanation of the body's faculty to receive and to accumulate motion; it is one and the same body, and by virtue of one and the same gravity, which receives the linear impulse of the earth's rotation and which tends toward its centre. Therefore, as he explains to Sagredo, any decrease in the gravity is irrelevant, for accompanying this, and in the same degree, there is also a decrease in the capacity to receive the motion's *impetus.*[268]

Of course the *impetus* is rectilinear. But it is so only for an instant.[269] Now, no motion takes place in an instant; and no real motion can take place in a straight line; gravity prevents this. Rectilinear motion would only be possible for a body without gravity. But such a body, alas, would not be a real body; and would not be able to receive *impetus.*

A strange thing: it is the progress made by Galileo in his analysis of motion in general, and of that of projectiles in particular, which causes him to fail to grasp the role of rectilinear motion, because it leads him to believe that it does not exist in reality.

For violent motion, or at least the *impetus* of violent motion, is always rectilinear. A bullet from a gun sets off to go in a straight line, as does an arrow, a thrown stone, etc. But they never actually move in a straight line. In contrast to his predecessors, engineers and artillery-men, who split up a cannon-ball's trajectory into a rectilinear segment and a curved segment, Galileo does away with the rectilinear segment. The principle of the relativity of motion leads him to understand that, since horizontal and vertical motion do not impede each other, and since gravity is always at work, the trajectory will be curved right from the start.[270] The cannon-ball could only travel in a straight line if it had no weight. But then, obviously, it could not be propelled.

However, the non-existence, or more accurately the impossibility, of 'inertial' rectilinear motion on the earth, does not explain, or not adequately, that error of Galileo which we are in the process of investigating. No doubt tangential motion is impossible. But Galileo is a good enough geometer to know that between the tangent and the circumference (the earth's surface) there is an infinite number of curves (even an infinite number of circular curves) that could be followed by the motion of a stone when it is projected by the earth's rotation. Why is it that he refuses to accept, or even to consider, this possibility? Basically we have already answered this; to accept it would be to give up the general relativity of motion in favour of a partial relativity, limited to an unrealisable, and strictly speaking impossible, case, that of rectilinear motion. It would be to give up seeing the motion of a heavy body around a centre, this motion which neither raises nor lowers weight, as a motion with a physically special status. It would be to admit that things do not happen on a moving earth just the same as they do on a stationary earth:[271] and, in particular, that bodies falling from the top of a tower would not, strictly speaking, arrive at its foot, and nor would they ever arrive at the centre of the earth.

Now, Galileo is so convinced of this that he is led to commit yet another error; he postulates for the complex motion of a projectile (or, which comes to the same thing, for the real complex motion of fall on a rotating earth) a law which is observably incorrect; he states that the trajectory of the motion in question will be the circumference of a circle and not, as we know it to be, and as he will later show himself, a parabola.[272] This error is explained by the fact that in his argument Galileo takes it for granted (a) that a heavy body, tending naturally toward the centre of the earth, does in the end arrive at it, and (b) that, if gravity did not carry it to the centre of the earth, i.e., if something (for example, the earth's surface) prevented it from proceeding there, its motion would be, naturally, in a circle. Here is this odd passage which is, to tell the truth, usually misunderstood.[273]

'If the straight motion toward the centre of the earth were uniform, and the circular motion toward the east were also uniform, the two could be compounded into a spiral line; one of those defined by Archimedes in his book about the spirals. . . . But since the motion of the falling weight is continually accelerated, the line compounded of the two motions must have an ever-increasing ratio of successive distances from the circumference of that circle which would have been marked out by the centre of gravity of the stone had it always remained on the tower. It is also required that this departure be small at the beginning—or rather minimal, even the least possible. For leaving from rest (that is, from the privation of downward motion) and entering into motion straight down, the falling weight must pass through every degree of slowness that exists between rest and any speed of motion. These degrees are infinite, as was discussed at length and decided already.

'Supposing, then, that such is the progress of acceleration; it being further true that *the descending weight tends to end at the centre of the earth*,[274] then the line of its compound motion must be such as to travel away from the top of the tower at an ever-increasing rate. To put it better, this line leaves from the circle described by the top of the tower because of the revolution of the earth, its departure from that circle being less *ad infinitum* according as the moving body is found to be less and less removed from the point where it was first placed. Moreover, *this line of compound motion must tend to terminate at the centre of the earth.*[274] Now, making these two assumptions, I draw the circle BI with A as a centre and radius AB, which represents the terrestrial globe. Next, prolonging AB to C, the height of the tower BC is drawn; this, carried by the earth along the circumference BI, marks out with its top the arc CD.

'Now dividing line CA at its mid-point E, and taking E as a centre and EC as radius, the semicircle CIA is described, along which I think it very probable that a stone dropped from the top of the tower C will move, with a motion composed of the general circular motion and its own straight one.

'For if equal sections CF, FG, GH, HL are marked on the circumference CD, and straight lines are drawn to the centre A from the points F, G, H, and L, the parts of these intercepted between the two circles CD and BI represent always the same tower CB, carried by the earth's globe toward DI. And the points where these lines are cut by the arc of the semicircle CIA are the places at which the falling stone will be found at the various times. Now these points become more distant from the top of the tower in an ever-increasing 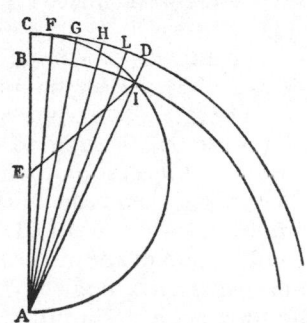 proportion, and that is what makes its straight motion along the side of the tower show itself to be always more and more rapid. You may also see how, thanks to the infinite acuteness of the angle of contact between the two circles DC and CI, the departure of the stone from the circumference CFD (that is, from the top of the tower) is very, very small at the beginning, which is the same as saying that the downward motion is extremely slow; in fact, slower and slower *ad infinitum* according to its closeness to the point C, the state of rest. Finally, one may understand how such motion tends eventually to terminate at the centre of the earth.'

It is clear that for real motions, motions of heavy bodies on the earth, the horizontal plane, as we have said above, is and always will be the circumference of a circle.

Perhaps it will be objected that in the *Discourses* Galileo manages to free himself from this obsession with the spherical and the circular. This is, of course, true. The *Discourses* represent not only a later stage in Galileo's thought but, even more importantly, a very much higher level of 'abstraction'.[275] Thus in the *Discourses* the straight line is not a circle and the horizontal plane is not a sphere. This is because the Archimedean world studied in the *Discourses* is not the world of terrestrial reality: it is a world in which heavy bodies do not fall toward the centre of the earth. However, they do fall. But the direction of gravity is not towards a 'centre';[276] the 'lines of force' of gravity are parallel, and this is why the horizontal plane of this world is Euclidean. But these 'lines of force' do exist; and this is why rectilinear inertial motion is impossible there.

If we look at the two passages in the *Discourses* in which Galileo comes closest to rectilinear inertial motion we will see that he strongly asserts there the natural character of downwards motion; and we will see yet again that he is unable to abstract from gravity.

First we will quote this admirable passage from the 'Third Day' of the *Discourses* in which we are given, in an exciting, abbreviated form, the fundamental principles of Galilean physics, the principles of the relativity and the conservation of motion:[277] '. . . note that the degree of speed of a moving body is, by its nature, ineradicably impressed on it so long as external causes of acceleration or retardation do not destroy it, a condition which is found only on horizontal planes; for in the case of planes which slope downwards there is present a cause of acceleration, while on planes sloping upwards there is a cause of retardation; from this it follows that motion along a horizontal plane is perpetual; for if it is uniform it will not be weakened or slowed down, much less destroyed. Furthermore, we must examine the case in which some degree of speed, by its nature ineradicable and perpetual, is acquired by the moving body in its natural descent, and where after its descent the body is deflected onto an upwards sloping plane: there will be then a cause of retardation. For on such a plane this same body would naturally descend. Accordingly there is formed here a certain combination of contrary tendencies, namely the degree of speed acquired during the preceding fall, which if acting alone would carry the body with uniform speed to infinity, and the natural tendency towards downwards motion with the rate of acceleration which always affects this motion. It seems altogether reasonable, therefore, if we wish to discover the accidents of the motion of a body which, having descended an inclined plane, is deflected onto an upwards sloping plane, for us to assume that the maximum degree of speed acquired during descent is conserved perpetually the same, and that during the ascent, however, there is superimposed upon it a natural downwards tendency, i.e., a naturally accelerated motion starting from rest and accelerating at the usual rate'.

It can be seen that in the Archimedean world of the *Discourses* the horizontal plane, on which uniform motion continues perpetually, is no longer a spherical surface; it is an infinite geometrical plane. The degree of speed acquired by the heavy body is conserved on it for ever, regardless of the direction of its motion, and this means that all heavy bodies, in other words all bodies, once set in motion on a horizontal plane, will move perpetually with a rectilinear and uniform motion. We are, as we have said, on the threshold of the principle of inertia; but we are not about to cross it. For Galileo immediately adds that the body in question will move naturally downwards, that it will naturally accelerate as it descends and that its motion will become slower as it re-ascends. Furthermore, this rectilinear motion does not last, or does not remain rectilinear, unless the body remains in motion *on this plane*. Now, what would happen if this *plane* were lacking, if it were not there to support the moving body? A famous passage from the 'Fourth Day', (this is another passage in which, it has been claimed, one can find a statement of the principle of inertia) will answer this for us:[278] 'Imagine a body projected along a horizontal plane, setting aside all obstacles. We know, from what has been said above, that its motion on this plane will remain uniform indefinitely, provided that the plane is infinitely extended. But if the plane is limited in extent and is elevated, then the moving body, *which we assume to be subject to gravity*,[279] when it passes over the edge of the plane will acquire in addition to its first uniform and indestructible motion a downward propensity *due to its own gravity*.[279] There thus results a motion, made up of the horizontal motion and the naturally accelerated motion of descent. I call this motion projection'; and this motion, as Galileo shows, using a proof which has been considered classic ever since, will be a semi-parabola.

What we see here is that once the plane is removed and no longer supports the heavy body, the body falls. Its motion only continues in a straight line so long as it remains on the horizontal plane; once it is no longer on it the motion carries on, by itself, but the body no longer moves in a straight line.

Certainly it could be objected here that Galileo argues *ex hypothesi* that bodies 'are subject to gravity' (which is, after all, a reasonable assumption) and that we ourselves would not argue otherwise.[280] This is true of course. This is why Galileo's argument seems so 'modern' to us. But we forget that in our case we *explain* 'gravity' or 'heaviness'—if only by replacing it with Newtonian mutual attraction between bodies—and that if we are able to imagine bodies as subject to gravity, we are equally able to imagine them as not being so. This is what we do, or at least used to do, when, in stating the fundamental principles of our physics, we distinguish gravity from mass. Now this, precisely, is what Galileo does not do; and cannot do, since for him, to use modern terminology, gravity and mass are one and the same thing. This is why, for him, gravity is not a *force* which *acts* on a body; it is something to which the body is 'subject', something which belongs to the body itself. Thus it is not subject to variation, either in time or space. A body weighs what is weighs, everywhere and always, and it falls with the same speed wherever it is located, whether it is close to the centre of the earth or, on the contrary, among the stars.[281] Of course Galileo, following Archimedes, can abstract from reality and can ignore the real direction which gravity takes on earth (which in fact both Simplicio and Sagredo take as cause for complaint[282]); in justification for this he can offer us his Archimedean world as a first approximation (and he would be right in this, even doubly right: the Archimedean law of fall is an approximation to the real, more complex law; and the Archimedean world, being at one remove from the purely geometrical world, is a first approximation to the physical world); but he cannot carry the 'abstraction' beyond this, because for him gravity, as we have seen over and over again, is a constitutive and inseparable property of physical bodies.

Galileo's physics explains that which is by that which is not. Descartes and Newton go further: their physics explain that which is by that which cannot be; they explain the real by the impossible. Galileo, we have seen, does not do this. But we should not blame him for this. For this impossible thing, i.e., inertial motion in a straight line, is in a manner of speaking less impossible for Newton or Descartes than it is for Galileo. Or one could put this by saying that the impossibility of such a motion was not the same, did not have the same structure.

For Newton, the rectilinear motion of a body projected into space is impossible because the action of other bodies alters it, deflects it, interferes with it. A body could only move in a straight line if it were alone in space; an impossible condition, of course. But it is only contingently impossible. For, strictly speaking, God would certainly be able to realise this condition.

In the case of Descartes, the impossibility of inertial motion is much deeper. It is certainly for him, as for Newton, a matter of an impossibility which is in some sense external. A body cannot move in a straight line because other bodies, those which surround it, prevent it from doing so. But for Descartes an isolated body is inconceivable. God himself could not remove the obstacles which of necessity stand in its path. Finally, with Galileo the impossibility is no longer external. If no body can move in a straight line this is not because bodies inevitably run up against obstacles, or are subject to attractions, which prevent them from doing so. It is the body itself which rejects rectilinear motion. Its weight draws it down: and if, which is impossible, its weight were eliminated, then its motion would not thereby be straightened out; it would disappear together with the *physical* existence of the body.

So, as we have seen, *Galileo did not formulate the principle of inertia*. He did not travel the whole distance along the road from the well-ordered Cosmos of ancient and mediaeval science to the infinite Universe of classical science. It was Descartes who was destined to achieve this.

CONCLUSION

However, the traditional historical view, which sees Galileo as the father of classical science, is not wrong. For it is in his work—and not in that of Descartes[283]—that the idea of mathematical physics, or rather the idea of the mathematisation of the physical, was realised for the first time in the history of human thought.

Thus, the question of greatest consequence under discussion throughout the *Dialogue* and underlying every step in its argument, a question of far greater importance than that of the merits of the two astronomical systems under consideration, which after all was really of limited significance, was that of the respective merits of two philosophies. For the resolution of the astronomical problem depends on the constitution of a physical science; and this in its turn presupposes the prerequisite resolution of the *philosophical* problem of the nature and structure of this science. In practice this amounts to understanding the role to be played by mathematics in the constitution of scientific knowledge of the real world.

The role of mathematics in physical science was not a new problem. On the contrary, it had been the subject of philosophical meditation and debate for two thousand years: and Galileo was perfectly well aware of this. He had known this since the time when, as a young student attending the philosophy lectures of Francesco Bonamico at Pisa, he had learned that the question of the role and nature of mathematics constituted the main subject of disagreement between Plato and Aristotle.[284]

Some years later, when he returned to Pisa as a teacher, Galileo was able to hear this confirmed by his friend and colleague Jacopo Mazzoni, author of a book on the relations between Plato and Aristotle:[285] 'No question or debate between Plato and Aristotle has given rise to so many of the finest and noblest of speculations . . . as the question of whether the use of mathematics in physical science, as a method of reasoning and as a middle term in demonstration, is appropriate or inappropriate; i.e., whether it leads to truth or is detrimental and harmful. Plato believed that mathematics is particularly well suited to physical speculations. This is why on several occasions he made use of it in resolving questions in physics. But Aristotle's view was diametrically opposed to this, and he attributed Plato's errors to his love of mathematics'.

It is clear that in the eyes of the philosophy and science of the time (for Bonamico and Mazzoni were merely expressing a generally held view[286]) there was a sharp dividing line between Aristotelians and Platonists. Those who attributed to mathematics a *higher value*, and who in addition saw it as having *real* value and a dominant position in and for physics, were

Platonist: those who, on the contrary, saw mathematics as an 'abstract' science, and consequently as of *lower value* than the sciences (physics and metaphysics) which were concerned with the real, and in particular those who claimed to base physics *directly on experience* and attributed to mathematics only a subordinate role, were Aristotelian.

It is worth pointing out in passing that the problem of certainty was in no way involved here. No Aristotelian had ever doubted the certainty of geometrical proof. What was involved was the problem of reality. Furthermore the use of mathematics in physical science was not in question; no Aristotelian had ever refused to measure whatever is measurable or to count whatever is countable. It was a matter of the role of mathematics in and for the very structure of scientific knowledge, i.e., necessarily, of reality itself.

At the same time we should also say that in our opinion this conception of epistemology and history held by Galileo's contemporaries is not without merit. To tell the truth we are fully in agreement with them: the mathematisation of physics *is* Platonism, even if it is not aware of it. Therefore, from our vantage point we can see that the advent of classical science was a return to Plato.

These debates to which we have just referred are alluded to in the *Dialogue* right from the beginning, where Simplicio points out that 'in physical [*naturali*] matters one need not always require a mathematical demonstration'.[287] To which Sagredo, who pretends not to have understood, replies, 'Granted, where none is to be had. But when there is one at hand,[288] why do you not wish to use it?' Obviously if, in physical matters, it is possible to achieve a proof having the force of mathematical necessity it would be wrong not to press on and do so. But is this possible? As Galileo well knows this is the key question, and in a marginal note he sums up the Aristotelian view quite differently from Sagredo: 'Geometrical exactitude should not be sought in physical proofs'.[289]

Should not be sought. Because it is impossible. Because physical reality itself, which is qualitative and indefinite, does not conform to the rigour of mathematical concepts. Thus Simplicio will say later that philosophy, i.e., physics, should not concern itself with numerical detail, and should not seek numerical exactitude, in the laws of motion: it should be limited to establishing its general categories (natural, violent, rectilinear, circular) and its general laws (relation between force and speed, force and resistance).[290] Why? Simplicio does not tell us, and the astonished modern reader might ask why one should remain at the level of vague and abstract generality and not advance to that of precise and concrete universality.

The modern reader does not know the answer, but a reader in Galileo's time would have filled it in for himself: because it is not possible, because quality and form cannot be geometrised. Terrestrial matter is never the embodiment of precisely defined forms, and the forms never inform it perfectly; there is always some 'play'. It is no doubt quite different in the heavens; consequently astronomy is possible.[291] But, therefore, astronomy is not physics. Not having realised this was precisely the reason for Plato's failure. Seeking to mathematise nature one will get nowhere.

The Aristotelian attitude is by no means absurd. As far as we are concerned it seems perfectly reasonable, and the objections that Aristotle had addressed to Plato cannot be refuted, except in practice. For one cannot logically prove something to be possible. The only proof that something is possible is that it exists. So in order to show that it is possible to formulate exact mathematical laws which apply to reality it is necessary to actually formulate them. Galileo understands this perfectly well, and it is by giving a mathematical treatment to a concrete problem in physics (the problem of fall, or that of projectiles) that he will bring Sagredo to announce that 'it must be admitted that trying to deal with physical problems without geometry is attempting the impossible'.[292]

Sagredo, the *bona mens*, is easy to convince. Too easy. For the Aristotelian is not without further ammunition. So the argument continues:[292] '*Salviati:* "Simplicio will not say so, though I do not believe he is one of those Peripatetics who discourage their disciples from the study of mathematics as a thing that disturbs contemplation." *Simplicio:* "I would not do Plato such an injustice, although I should agree with Aristotle that he plunged into geometry too deeply and became too fascinated by it. After all, Salviati, these mathematical subtleties do very well in the abstract, but they do not work out when applied to sensible and physical matters. For instance, mathematicians may prove well enough in theory that *sphaera tangit planum in puncto*, a proposition similar to the one at hand;[293] but when it comes to matter, things happen otherwise. What I mean about these angles of contact and ratios is that they all go by the board for material and sensible things."'

Simplicio's argument deserves our attention. From the point of view of Aristotelianism, and even from that of ancient Platonism, it is decisive and irrefutable. For in the real world, the physical world, there are no straight lines, nor planes, nor triangles, nor spheres. The bodies of the material world do not have regular geometrical forms. Therefore, geometrical laws cannot be applied to them. Of course the Platonist would reply, as we have seen that Galileo does, that mathematical laws, in their application to physical reality, are *approximations*. This is a defensible position if one accepts, and to the extent that one accepts, that physical entities 'imitate' and 'approximate' geometrical entities, i.e., if one is already a Platonist and accepts that the real is ultimately mathematical in essence. But this is not enough. For we have no way whatsoever of determining the degree of approximation, or one could say, of divergence between geometrical figures and real shapes, while at the same time we are obliged to affirm the reality, and even the necessity, of this divergence, which is a consequence of the very existence of matter. For the real is not only not regular, it is also not exact. It is for this reason that there can only be science of the general and that the particular cannot be the object of scientific knowledge. Between the essence and its realisation there is always some 'play'. The individual always differs from the norm, and this difference, which explains the existence of *monstra*, is never predictable or calculable.

If this is so then the views of those Aristotelians to whom Galileo-Salviati alludes, and with whom Simplicio is well acquainted,[294] are not so absurd as may seem at first sight. On the contrary, in this light they seem perfectly reasonable. For would not a mind used to the precision and rigour of geometrical reasoning be to that extent less likely to grasp the subtle and indeterminate multiplicity of reality?[295] It is well known that this was the view of Pascal, and even of Leibniz.[296]

Let us look now at Galileo's reply. It is of the greatest importance and interest, for while it is deeply Platonist it is not restricted to the classical counter-arguments; on the contrary, it presents a decisive innovation. For Galileo denies the premise, accepted by both Platonists and Aristotelians, on which the argument was founded. He denies the 'abstract' character of mathematical ideas, and he denies the special ontological status of regular figures.

A sphere is no less a sphere for being real: its radii are not thereby unequal; if they were it would not be a sphere. A real plane, if it is a plane, is as much a plane as is a geometrical plane; if it were not, it would not be a plane.[297] This seems obvious. How could Simplicio have denied it? It is because for him a real sphere is impossible, as impossible as a real plane; whereas Galileo's argument implies, on the contrary, that the real and the geometrical are in no way heterogeneous and that a geometrical figure can exist in material form. In fact it implies much more than this, it implies that they always do. For even though it may be impossible for us to make a perfect plane or a true sphere, those material objects, while not 'sphere' or 'plane', are not for all that without geometrical form. They are irregular but not at all indeterminate. The most irregular stone has as definite a geometrical form as a perfect sphere. It is merely infinitely more complicated.[298]

Geometrical form is homogeneous with matter.[299] This is why geometrical laws have real significance and are of the first importance in physics. This is why, as Galileo tells us in a justly famous passage in *The Assayer*, that nature speaks a mathematical language, a language of which the letters and syllables are triangles, circles and lines. This is also why we must ask questions of nature in this language:[300] the theories of mathematics precede observation.

This conception, it goes without saying, implies a completely new concept of matter. It is no longer the bearer of Becoming and of quality but, on the contrary, the bearer of unchangable and eternal Being.[301] One could say that terrestrial matter is henceforth elevated to the rank of the heavenly. Thus, we have seen the new science—geometrical physics, physical geometry—born in the heavens, descend to earth, and rise up again to the skies.

So in Galileo's time the belief in mathematisation was identified with Platonism. Therefore when Torricelli said that 'of the liberal disciplines *only* geometry exercises and sharpens the mind and makes it fit to ornament the city in time of peace and to defend it in time of war', and that 'other things being equal, the mind trained in geometrical gymnastics is quite specially powerful and *virile*', not only did he thereby show himself to be an authentic disciple of Plato, but also that he considered himself and declared himself to be such.[302] In doing so he remained a faithful disciple of his master, Galileo, who in his reply to Antonio Rocco's *Esercitationi filosofiche* had asked this author to judge for himself the merits of the two methods, namely the purely physical method and the mathematical method, and added, 'and at the same time see whose argument you consider the more correct; Plato's, when he said that without mathematics one cannot learn philosophy, or Aristotle's when he criticised this same Plato for having studied geometry too much'.[303]

Right from the beginning the *Dialogue* makes a point of alerting us to the fact that Galileo is a Platonist. In its opening pages, in fact, Simplicio remarks that Galileo, being a mathematician, is probably inclined to regard the numerological speculations of the Pythagorians sympathetically, which allows Galileo-Salviati to announce that he considers them absolutely worthless but to point out at the same time:[304] 'That the Pythagorians held the science of numbers in high esteem, and that Plato himself admired the human understanding and believed it to partake of divinity simply because it understood the nature of numbers, I know very well; nor am I far from being of the same opinion'.

For how could he not be of the same opinion, since he himself thought that the human mind, through knowledge of mathematics, could achieve the perfection of divine understanding? Does he not tell us, through Salviati:[305] *Extensively,* that is, with regard to the multitude of intelligibles, which are infinite, the human understanding is as nothing even if it understands a thousand propositions; for a thousand in relation to infinity is zero. But taking man's understanding *intensively,* in so far as this term denotes understanding some proposition perfectly, I say that the human intellect does understand some of them perfectly, and thus in these it has as much absolute certainty as Nature itself has. Of such are the mathematical sciences alone; that is, geometry and arithmetic, in which the Divine intellect indeed knows infinitely more propositions, since it knows all. But with regard to those few which the human intellect does understand, I believe that its knowledge equals the Divine in objective certainty, for here it succeeds in understanding necessity, beyond which there can be no greater sureness.[306]

Simplicio: This speech strikes me as very bold and daring.

Salviati: These are very ordinary propositions and far from any shade of temerity or boldness. They do not detract in the least from the majesty of Divine wisdom, just as saying that God cannot undo what is done does not in the least diminish His omnipotence. But I question, Simplicio, whether your suspicion does not arise from your having taken my words equivocally. So in order to explain myself better, I say that as to the truth of the knowledge which is given by mathematical proofs, this is the same that Divine wisdom recognises; but I shall concede to you indeed that the way in which God knows the infinite propositions of which we know some few is exceedingly more excellent than ours. Our method proceeds with reasoning by steps from one conclusion to another, while His is one of simple intuition. We, for example, in order to win a knowledge of some properties of the circle (which has an infinity of them), begin with one of the simplest, and, taking this for the definition of circle, proceed by reasoning to another property, and from this to a third, and then a fourth, and so on; but the Divine intellect, by a simple apprehension of the circle's essence, knows without time-consuming reasoning all the infinity of its properties. Next, all these properties are in effect virtually included in the definitions of all things; and ultimately, through being infinite, are perhaps but one in their essence and in the Divine mind. Nor is all the above entirely unknown to the human mind either, but it is clouded with deep and thick mists, which become partly dispersed and clarified when we master some conclusions and get them so firmly established and so readily in our possession that we can run over them very rapidly.[307] For, after all, what more is there to the square on the hypotenuse being equal to the squares on the other two sides, than the equality of two parallelograms on equal bases and between parallel lines? And is this not ultimately the same as the equality of two surfaces which when superimposed are not increased, but are enclosed within the same boundaries? Now these advances, which our intellect makes laboriously and step by step, run through the Divine mind like light in an instant; which is the same as saying that everything is always present to it.

'I conclude from this that our understanding, as well in the manner as in the number of things understood, is infinitely surpassed by the Divine; but I do not thereby abase it so much as to consider it absolutely null. No, when I consider what marvellous things and how many of them men have understood, inquired into, and contrived, I recognise and understand only too clearly that the human mind is a work of God's, and one of the most excellent'.

Galileo might well have added that the human understanding is a work of God of such excellence that, *ab initio*, it possesses those 'clear and distinct' ideas of which the clarity is a guarantee of *truth,* and that he has only to turn inwards in order to discover in his 'memory' the foundation of his knowledge of reality, i.e., the alphabet, the elements of the language (the language of mathematics) spoken by nature, God's creation. For we should not be misled: this is not a matter of a kind of truth which is purely immanent in Reason, inherent in the proofs and theorems of mathematics, unaffected by the non-existence in nature of the objects in question. Neither Galileo nor Descartes would ever have agreed to be satisfied with *ersatz* truth or science like this. It is a matter of truth about nature and knowledge of the real world. This is the knowledge, this truly 'philosophical' knowledge, knowledge concerning the very essence of reality, about which the true Platonist, consciously Platonist, Galileo speaks when he says, through Sagredo:[308] 'I say to you that if one does not know the truth by himself, it is impossible for anyone to make him know it. I can indeed point out things to you, things being neither true nor false; but as for the true—that is, the necessary; that which cannot possibly be otherwise— every man of ordinary intelligence either knows this by himself or it is impossible for him ever to know it. And I am sure that Salviati holds this opinion too'.

He certainly does. For the allusions to Plato, the references to the Socratic method (of maieutics) of assisting at the birth of men's thoughts, the reminders of the doctrine of anamnesis, are not at all simply appetisers, or stylistic ornaments, arising from a superficial infatuation with the works of Plato, an infatuation which merely reflects the influence of the 'Platonism' of the Florentine Renaissance. Equally they do not simply arise from the desire in a facile way to win over the 'cultivated' audience which had been for so long exhausted by the aridity of Aristotelian scholasticism, or the desire to be armed, against the authority of Aristotle, with that of his master and main opponent, the divine Plato. Quite the contrary: these allusions and reminders must be taken absolutely seriously: and so that no doubts about this will linger in the reader's mind Galileo addresses to him the following remarks:[309] 'The solution [of this problem] depends upon some data well known and believed by you just as much as me, but because they do not strike you, you do not see the solution. Without teaching them to you then, since you already know them, I shall cause you to resolve the objection by merely recalling them.

Simplicio: I have frequently studied your manner of arguing, which gives me the impression that you lean towards Plato's opinion that *nostrum scire sit quoddam reminisci*. So please remove all question for me by telling me your idea of this.

Salviati: How I feel about Plato's opinion I can indicate to you by means of words and also by deeds. In my previous arguments I have more than once explained myself with deeds. I shall pursue the same method in the matter at hand, which may then serve as an example, making it easier for you to comprehend my ideas about the acquisition of knowledge'.

The matter at hand is the deduction of the laws of mechanics, and we have given an account of this above. We can see that Galileo believes that he has done much more than simply announce his support for Platonist epistemology. He believes that he has demonstrated the truth of Platonism 'by deeds', in applying his method, in discovering the real laws of physics, and in bringing Sagredo and Simplicio, which is to say also *the reader*, to rediscover them for themselves. The *Dialogue,* and also the *Discourses,* tell us the story of a spiritual experience; it is a decisive experience, for it concludes with Simplicio's regretful admission; his admission of the necessity for the study of mathematics, and his regret at not having done this himself.[310] The *Dialogue* and the *Discourses* tell us the story of the discovery, or rather of the rediscovery, of the language spoken by nature, and explain how questions should be put to nature, in a theory of experiment in which the statement of 'postulates' and the deduction of their consequences precedes recourse to observation.[311] This also is a proof 'by deeds', an experimental proof of Platonism.

In the light of this we can understand the profound meaning of this fine passage by Cavalieri:[312] 'Now as for knowledge of the mathematical sciences, which was believed by the celebrated Schools of the Pythagorians and the Platonists to be a paramount necessity for the understanding of physical things, I hope that, by the publication of the new doctrine of motion promised by the most admirable Assayer of Nature, I mean by Signor Galileo Galilei, in his *Dialogues,* [the role of mathematics] will shortly be made manifest'.

In fact Platonist mathematisation had always come up against two main obstacles; the first was qualities, the second was motion. Aristotle had opposed the attempt to mathematise nature on the grounds that it is impossible either to mathematise qualities or to deduce motion.[313] There is no motion of numbers. Mathematical entities do not move. How could they since they are eternal and non-temporal?[314] Furthermore, the Aristotelian of Galileo's time could add that the greatest of Platonists, the divine Archimedes himself, had only been able to produce a statics, to mathematise rest, but not motion. As is well known, *ignoto motu ignoratur natura.* Therefore, mathematical physics, Platonist physics, had remained a pious hope that nobody had even attempted to realise.

In the light of this we can understand the pride of the Platonist Galileo when he says:[315] 'My purpose is to set forth a very new science dealing with a very ancient subject. There is, in nature, perhaps nothing older than *motion,* concerning which the books written by philosophers are neither few nor small; nevertheless I have discovered by experiment some properties of it which are worth knowing and which have not hitherto been either observed or demonstrated. Some superficial observations have been made, as, for instance, that the natural motion of a heavy falling body is continuously accelerated; but the proportion with which this acceleration occurs has not yet been announced; for so far as I know, no one has yet pointed out that *the distances traversed, during equal intervals of time, by a body falling from rest, stand to one another in the same ratio as the odd numbers beginning with unity'.*

Motion obeys a mathematical law. Times and distances are related by a numerical law. Galileo's discovery changed the failure of Platonism into a victory. His science is Plato's revenge.

It was, of course, only a partial, incomplete revenge. For as we have explained elsewhere[316] it was Descartes, and not Galileo, who assured the definitive victory of Platonism, and who swept Aristotelianism from the territory it had occupied for so long.

Descartes would probably not have accepted our way of viewing the matter, and would have refused to see himself as a disciple of Plato:[317] and he would not have been altogether wrong in this. However, are Descartes' 'innate ideas' not a distant relation to Plato's anamnesis? Does not the Cartesian idea of extension refer us back to Plato's χωρα ?[318] Was not their conception of science the same? In the well known and justly famous passage in the *Discourse on Method* in which Descartes rejects the

Scholastic conception of mathematics (which viewed it as a science useful only for the mechanical arts[319]) is he not thereby in continuity with the Platonist tradition which had come down to him via Clavius.[320] Finally, from the point of view which concerns us here, in declaring the priority of mathematics in physics, and even the possibility of reducing the latter to the former,[321] is he not thereby locating himself in the Platonist camp?

And yet how remote this new Platonism is from the ancient! For, in fact, while thanks to Descartes we can henceforth understand space by an act of pure thought, and no longer via a hybrid knowledge, and can thereby replace myth by science, and while thanks to Galileo motion is henceforth governed by an arithmetical law, on the other hand this space and these numbers have lost that cosmic significance that they once had, once were able to have, for Plato.

Galilean science and Cartesian science are victorious. But never was a victory so dearly paid for.

NOTES
GALILEO AND THE LAW OF
INERTIA

Introduction

1 Cf. E. Wohlwill, 'Die Entdeckung der Beharrungsgesetzes', *Zeitschrift für Völkerpsychologie,* etc., vol. XV, p. 379 ff.

2 See above, p. 79 ff.

3 We had thought that we had expressed our admiration for the genius of Galileo sufficiently clearly to make any misapprehension impossible, at least as far as the unbiased reader is concerned. Unfortunately there are others . . . Thus we see ourselves recruited to the ranks of the 'detractors' and 'enemies' of Galileo by M. A. Mieli: cf. his 'Il tricentenario dei *Discorsi e dimostrazioni matematiche* di Galileo Galilei', *Archeion*, vol. XXI, fasc. 3, Rome, 1938.

4 See Pascal, *Pensées et opuscules*, ed. Brunschvicg, Paris, 1907, p. 193.

5 This means that a body left to itself remains at rest or moves indefinitely with a *rectilinear* and *uniform* motion; in other words its motion retains its speed and direction. See V. Laplace, *Exposition du système du monde, Oeuvres*, vol. VI, p. 155 ff.; Lagrange, *Mécanique analytique*, Paris, 1853, p. 308 ff.

6 It is precisely because rest and motion have the same ontological status in classical science, that of a state, that motion can be conceived as persisting for ever, like rest, without modification and without cause or motor. To put it in mediaeval terminology, with Galileo and Descartes motion ceases to be a *forma fluens* and becomes a *forma stans*. See above, p. 91 and below, p. 255 ff.

7 This is true, of course, only if motion is considered in itself, without the intervention of forces: in other words, only if one is restricted to kinematics or pure phoronomy, and is not yet concerned with dynamics.

8 In fact it is *speed* and *direction* which are conserved.

9 For Aristotelian physics, as for *impetus* physics, two motions *always* interfere with each other.

10 We apologise for emphasising something which goes, or should go, without saying. Unfortunately the failure to grasp the radical difference between the assertion of the persistence of circular motion and that of rectilinear motion (the two propositions are, in fact, incompatible) has vitiated the majority of works, even the best, devoted to the study of the origins of the principle of inertia.

11 Once again we apologise for recalling that for classical physics circular motion is not uniform, but is accelerated motion.

12 In classical science the action of a force produces not motion but acceleration.

13 The term *inertia*, which for Kepler, its inventor, means *natural resistance to motion* (cf. above, p.144 ff.) for classical physics means being unaffected by the states of motion and rest, the persistence of these states, and resistance to any change from one state to another.

14 The history of the introduction of the principle of inertia has been studied often before. See the fine works by E. Wohlwill, 'Die Entdeckung des Beharrungsgesetzes', *Zeitschrift für Völkerpsychologie und Sprachwissenschaft*, vols. XIV, XV, and by E. Mach, *Die Mechanik in ihrer Entwicklung*, 8th edition, Leipzig, 1921, and the well known works by

P. Duhem, 'De l'accélération produite par une force constante', *Còngrès International de Philosophie*, 2nd Session, Geneva, 1905, and *Etudes sur Léonard de Vinci*, 3 vols., Paris, 1909-13; finally, the admirable chapter devoted to the principle of inertia by E. Meyerson in his *Identité et Réalité*, 3rd edition, Paris, 1926.

15 Cf. E. Bréhier, *Histoire de la Philosophie*, vol. II, section I, Paris, 1929, p. 95: 'Descartes rescued physics from the obsession with the Greek Cosmos, i.e., from the picture of a certain privileged state of things which served our aesthetic purposes . . . There is no privileged state because all states are equivalent. Therefore there is no room in physics for the study of final causes or for the contemplation of the Good'.

The Physical Problem of Copernicanism

16 Cf. N. Copernicus, *De Revolutionibus orbium coelestium*, Paris, 1934, Book 1, chapter 5, p. 76 ff.; chapter 6, p. 81.

17 *Ibid.*, Book 1, chapter 7, p. 85 ff.

18 Motion which is the consequence of the body's nature, and which expresses this nature, when it is a *simple* body, cannot give rise to any other than an equally simple motion. See Aristotle, *De Caelo*, 296 b 31, and *Physics*, II, 1 and V, 2.

19 Ptolemy, *Almagest*, I, 7.

20 Emphasis added.

21 Copernicus, *De Revolutionibus*, Book 1, chapter 9, p. 101.

22 The heavens spin round by virtue of their nature, and moreover they are weightless; therefore they are not subject to the effects of centrifugal force.

23 Copernicus, *De Revolutionibus*, Book 1, chapter 8, p. 89 ff.

24 Introduction to Copernicus, *De Revolutionibus*, Paris, 1934, p. 19 ff.

25 No doubt this idea seems rather strange to us. It no longer does, however, if we think of the motion on the model of the propagation of a wave.

26 Copernicus, *De Revolutionibus*, Book 1, chapter 8, p. 93 ff.

27 Thus the motion of bodies would generally be a mixed motion, and Copernicus says that the circular is combined with the rectilinear 'as the illness to the animal'.

28 Because of their 'terrestrial' nature, terrestrial bodies perform the same circular motion as the earth itself. This is the reason why this motion remains invisible to us who also participate in it.

29 It seems to us that Bruno's influence was very much greater than is usually admitted, and than it appears to have been judging from the texts themselves. For example, it seems certain that Galileo was well acquainted with him; if he never mentions him this is not through ignorance, but prudence. For instance, he carefully avoided taking up (even to oppose it) the Brunian interpretation given by Mattheus Washer, and elsewhere by Kepler, of the discoveries announced in the *Nuntius Sidereus*: cf. Kepler, *Dissertatio cum Nuntio Sidereo, Opere*, voi. III, p. 105 ff.

30 Bruno's name was not mentioned in connection with the condemnation of Copernicus (1616), nor with that of Galileo. But there is no doubt at all that it was only Bruno's example which warned the Church that the new astronomy was a danger to religion.

31 Giordano Bruno, *La Cena de la Ceneri*, III, 5 (*Opere Italiane*, ed. Wagner, Lipsiae, 1830, vol. I, p. 169 ff.): « Da quel, que rispondete a l'argomento tolto da venti et nuvole, si prende ancora la risposta de l'altro che nel secondo libro del cielo e mondo apporto Aristotele, dove dice, che sarebbe impossibile, che una pietra gittata a l'alto potesse per medesma rettitudine perpendicolare tornare al basso ; ma surebbe necessario, che il velocissimo moto della terra se la lasciasse molto a dietro verso l'occidente. »

32 Aristotle, *De Caelo*, 296 b 24.

33 G. Bruno, *op. cit.*, p. 170: « Per che essendo questa projezione dentro la terra, è necessario, che col moto di quella, si venga a mutar ogni relazione di rettitudine et obbliquità. »

34 Emphasis added.

35 *Ibid.*: « per che é differenza tra il moto del nave, e moto di quelle cose, che sono ne la nave, il che se non fusse vero, seguitarebbe, che, quando la nave core per il mare, giammai alcuno potrebbe trare per dritto qualche cosa da un canto di quella a l'altro, e non

sarebbe possibile che un potesse far un salto, o ritornare co' piè, onde li tolse. »

[36]   *Ibid.* : « Con la terra dunque si muovano tutte le cose, che si trovano in terra. »

[37]   *Ibid.*: « Se dunque dal loco estra la terra qualche cosa fusse gittata in terra, per il moto di quella perderebbe la rettitudine. Come appare ne la nave, la qual, passando per il fiume, se alcuno, che si ritrova ne la sponda di quello, venga a gittar per dritto un sasso, verrà fallito il suo tratto, per quanto comporta la velocità del corso. Ma posto alcuno sopra l'arbore di detta nave, che corra quanto si voglia veloce, non fallirà punto il suo tratto : di sorte che per dritto dal punto, che'è ne la cima de l'arbore, o ne la gabbia al punto, che'e ne la radice de l'arbore o altra parte del ventro e corpo di detta nave, la pietra o altra cosa grave gittata non vegna. Cosi se dal punto de la radice al punto de la cima de l'arbore, o de la gabbia, alcuno che'è dentro la nave, gitta per dritta una pietra, quella per la medesima linea ritornarà a basso, muovasi quanto si voglia la nave, pur che non faccia de gl'inchini. »

[38]   G. Bruno, *La Cena de la Ceneri*, III, 5, p. 171: « *Teo.* Or per tornare al proposito, se dunque saranno dui, de quali l'uno si trova·dentro lo nave, che corre, e l'altro fuori di quella, de' quali tanto l'uno, quanto l'altro abbia la mano circa il medesmo punto de l'aria, e da quel medesmo loco nel medesmo tempo ancora l'uno lasci scorrere una pietra, e l'altro un' altra, senza che le donino spinta alcuna, quella del primo, senza perdere punto, nè deviar da la sua linea, verrà al prefisso loco ; e quella del secondo si trovarà tralasciata a dietro. Il che non procede da altro, eccetto che la pietra, ch'esce da la mano de l'uno, ch'è sustentato da la nave, e per conseguenza si muove secondo il moto di quella, ha tal virtù impressa, quale non ha l'altra, che procede da la mano di quello, che n'è di fuora, ben che le pietre abbino medesmo gravità, medesmo aria tramezzante, si partano—possibil fia—dal medesmo punto, e patiscano la medesma spinta. De la qual diversità non possiamo apportar altra ragione, eccetto che le cose, che hanno fissione, o simile appartenenze ne la nave, si muovono con quella ; e l'una pietro porta seco la virtù del motore, il quale si muove con la nave, l'altro di quello, che non ha detta participazione. Da questo manifestamente si vede, che non dal termine del moto, onde si parte, nè dal termine dove va, nè dal mezzo, per cui si muove, prende la virtù d'andar rettamente, ma da l'efficacia de la virtù primieramente impressa, da la quale dipende la differenza tutta. E questo mi par che basti aver considerato, quanto a le proposto di Nundiano. »

[39]   G. Bruno, *Acrotismus Camoerracensis*, art. XXXV (*Opere latina*, Naples, 1879, vol. I, 1, p. 138): « virtus impressa quandiu durat, tandiu pellat : ut ubi quis pilam sursum jaciat, illi levitati proportionale impressit ; ad cuius certe lationis differentiam nihil facit medium, quamvis ad lationem simpliciter sit necessarium, quia, nisi sit spatium per quod feratur, nulla latio esse potest. »

[40]   This same doctrine can be found in the young Galileo. See above, p. 29.

[41]   'Space' here, for Bruno, as for Benedetti (see above, p.27, note 93) is *intervallum* and not *locus;* which demonstrates a Platonist inspiration.

[42]   See above, p. 4 ff.

[43]   G. Bruno, *Acrotismus*, art. LXXIV, p. 185: « Disciplina de gravi et levi, quae est apud Aristotelem, prorsus perversa est, pro quo hasce verissimas ponimus propositiones. Grave et leve non dicuntur de corporibus naturalibus, naturaliter constitutis, nec de ipsis integris sphaeris, nec partibus earum : si terreno globo et cuicunque astro constantes in una sede conveniat habere partes. »

[44]   G. Bruno, *Acrotismus*, art. LXXX, p. 189: « Gravitas et levitas nihil aliud est praeter appulsum partium ad locum suum, in quo vel moveantur, vel quiescant, et per quod ferridebeant, pro quo quaelibet pars tum gravis tum levis esse intelligitur, quae, ubi nata est, esse degens, neque gravis est neque levis ; relinquitur ergo gravis levisque ratio respectiva tantum, per absolutas enim differentias mundi locales nullum est. Quocirca bene Plato in Timaeo dicit : in coelo non esse aliud quidem sursum, aliud vero deorsum, si ex omni parte simile est et undique oppositis pedibus ambulabat unusquisque ipse sibi. Hunc frustra refricat Aristoteles, sicut etiam, cum gravius bene dicebatur in Timaeo, esse quod ex pluribus est, levius autem quod ex paucioribus est. »

[45]   G. Bruno, *Acrotismus*, art. LXIII, p. 175: « Mundus, quem antiqui philosophorum parentes genitum esse dicunt, postmodumque sempiternum, inter quos est Empedocles

non est UNIVERSUM, sed haec machina huicque machinae similes. »

46 G. Bruno, *Acrotismus*, art. l LXXII, p. 183: « Certis ergo legibus infinita astra in immenso spatio feruntur, universo uno infinito, immobileque manente; cujus sicut nulla est circumferentia, ita nec ulla forma, et in quo aeteris est finire atque terminare singula; quae non minus apta sunt ad motum (sive per se moveantur per aetereum campum, sive magis secundum deferentis lationem), si angularis, quam si sphaericae sint figurae. Nullum interea astrorum, quodcunque et qualecunque sit illud, sive sol, inquam, fuerit, sive tellus, in medio vel in universi circumferentia dicere possis, ubi omnium singula circumquaque infinitum spatium habere convincentur. Hinc habes, quomodo omnia dicere possis in medio, vel nulla. Apparebit autem omnibus astrorum incolis se universi medium obtinere. »

47 G. Bruno, *Acrotismus*, art. LIII, p. 169: « Quam levi persuasione motus, ipse movetur et nos movere contendit Aristoteles ! ubi trium suarum lationum differentias concludit ex trium magnitudinum seu dimensionum differentia. Nos enim nullum sursum vel deorsum nisi respective intelligimus, neque diceremus unquam principium, unde motus, esse rationem dextri : ad unum quippe situm quod est dextrum ad alium secundum alias loci differentias invenietur, puta sinistrum, ante, supra. Mitto quod, cum infinita sint mundana corpora et infinita mundi dimensio, nec deorsum esse poterit, neque medium, neque sursum. »

48 G. Bruno, *Acrotismus*, art. XXXV, p. 138: « Spacium. . . nullam ad motum differentiam habet. »

49 G. Bruno, *Acrotismus*, art. XXXII, p. 130: « Minime verum est, quod recta movetur magis mutare locum, quam quod circulo torqueatur »

50 G. Bruno. *Acrotismus*, art. XXXIV, p. 133 : « Vacuum est spatium, in quo tot corpora continentur. Ipsum est unum infinitum, cujus partes ibi tantum sine corpore esse intelligimus, ubi corpora corporibus continguntur et alia moventur intra alia. » *Ibid.*, art. XXXV, p. 140 : « Vacuum vero spacium, in quo corpora continentur, est unum infinitum cujus partes alicubi sine corpore esse intelligantur.» *Ibid.*, art. XXXVII, p. 142 : « Vacuum tum separatum quid a corporibus, tum ipsis imbibitum, tum unum continuum dicere non formidamus : id enim necesse est. »

51 G. Bruno, *Acrotismus*, art. XXVII, p. 123 : « Infinitum dicimus non solum ut materiam, sed et ut actum.—*Ratio* : Non est materia infinita sine aliqua potentia et actu, sed ubique actus, alicujusque formae participes : non est enim vacuum sine aere vel alio corpore ; sive vacuum capias ut spacium, sive ut disterminans ; non est locus sine locato. »

52 G. Bruno, *Acrotismus,* art. XXIV, p. 121 : « Nobis non impossibile est simul infinitum dicere corpus, et locum quemdam corporibus esse.—*Ratio* : Si non superficies, sed spacium quoddam locus est, nullum corpus, neque ulla pars corporis illocata erit, sive maximum, sive minimum, sive finitum sit ipsum, sive infinitum. » *Ibid.,* art. XXIV, p. 122 : « Finitum Aristotelis est ignotum, falsum et impossibile : notum, verum atque necessarium est infinitum plurium philosophorum : . . . Finitum ipsum et terminus universalis est inconveniens, falsus et impossibilis . . . » Cf. the passage by Benedetti quoted above, p. 26.

53 G. Bruno, *Acrotismus*, art. XXX, p. 126 : « *Ratio*: Potuit sane Plato dixisse, materiam esse receptaculum quoddam et locum quoddam receptaculum esse. » Cf. also p. 130.

54 G. Bruno, *Acrotismus*, art. XXIII, p. 120: « *Ratio* : Infinitum, quia infinitum, maxime non nutat, non trepidat ; infinitas enim est maxima immobilitatis ratio, ideo infinitum seipsum, firmare dicitur : quia ex sua ratione habet, atque natura firmitatem. » *Ibid.*, art. XXXIV, p. 134 : « *Ratio* : Vacuum est, a quo corpora recipiuntur, et in quo corpora continentur ; recipiuntur autem ab eo, dum eodem spatio semper immobili permanente (quo nihil fixius esse potest) aër vel aliud alii in ipso cedit. Interim igitur nihil per vacuum feri intelligitur, quasi ante ibi nihil extiterit, quia aër est ubi nullum aliud corpus sensibile apparet. »

55 G. Bruno, *Acrotismus*, art. XXXV, p. 135 : « Non igitur ullus erit motus, si non si vacuum, omne enim movetur aut e vacuo, aut ad vacuum, aut in vacuo. » *ibid.,* art. XXVIII. p. 123 : « Translatio corporum indicat magis locum esse spacium, quam quidcunque aliud. Est igitur receptaculum corporum magnitudinem habentium, ad nullam quattuor causarum reducibile, sed per se quintum causae genus referens.—*Ratio:*

Hoc (spacium) neque elementum est, neque ex elementis, non enim elementa corporea habet, nec incorporea ; haud quidem corporea, quia non sensibile : haud incorporea, quia magnitudinem habet. Porro vacuum est, seu spacium, in quo sunt corpora magnitudinem habentia. »

56   G. Bruno, *Acrotismus*, art. XXXV, p. 136 : « Non necesarium est moveri in instanti quod movetur per vacuum' ; *ibid*, p. 137 : 'In his omnibus quod ad motum spectat, vacuum 'nihil conducere videtur, cui non motum vel quietem sed locum et continentiam tantum est administrare. »

57   Thus he explains (*Acrotismus*, art. LXV) the possibility of the circular motion of the planets by the fact that heavenly bodies have no weight: « Tellures superiores igitur non sunt graves neque leves, sicut neque terra ista, ubi mole sua in regione infinita consistit. »

58   *Ibid.*, art. LXXIV, p. 176 ff.

59   It was Tycho Brahe who invented and inserted into the debate the celebrated arguments about the cannon shot which were to become so popular later.

60   Cf. Tychonis Brahe, *Astronomicarum Epistolarum liber*, Uranienburgi, MDXCII, p. 188 ff.; ed. Dreyer, Hafniae, 1919, p. 218 ff.

61   *Astronomicarum Epistolarum liber*, p. 188; ed. Dreyer, p. 218.

62   *Ibid.*

63   Imagine a ship moored beneath a bridge. Clearly a cannon-ball falling from the bridge (from point A) will hit the ship at a point (point B) directly underneath its starting point, and in the same way a cannon-ball which is dropped from the top of the mast of a stationary ship will hit the ship's deck at the foot of the mast. But now let us imagine that the ship is moving. It is quite clear that a cannon-ball dropped from point A could never hit the deck at point B, the point which was directly beneath A at the instant when the cannon-ball started its fall: for during the time when the cannon-ball was falling the ship, and hence point B, were moving away. Can we affirm then that it would be any different when the cannon-ball fell from the top of the mast? The Aristotelian could not admit this. Assume that the ship's mast is the same height as the bridge which the ship is passing underneath; and assume, as does Bruno, that at the same instant, just as the top of the mast is touching point A, we drop *two* cannon-balls, one from the bridge, the other from the mast. The Aristotelian would never agree that of these two cannon-balls *in free fall* at *the same instant* and from *the same place*, one would fall straight down and into the water, while the other, falling through some strange curve, would end up at the foot of the mast. What evidence is there that this would happen? What reason could there be for accepting that there might be this difference? Would this not be to affirm that the cannon-ball 'knows' where to go and 'remembers' its past association with the ship and its mast? Such a conception would, quite reasonably, seem to the Aristotelian to be anthropomorphic and mythical in the extreme.

64   Tychonis Brahe, *Astronomicarum Epistolarum liber*, p. 189; ed. Dreyer, p. 219: « Et quid, quaeso, fiet, si Tormento Bombardico majori versus Ortum directo, explodatur globus ferreus, sive plumbeus, sive etiam lapideus, atq : ex eo ipso versus Occasum in eodem loco disposito, idque utrinque ad pariles cum horizonte angulos respectu prioris inclinationis elevato? Au fieri posse putandum, ut globus utrinque eadem pulveris quantitate et vi emissus tantudinem in terra permeet spatii, ob naturalem motus scientiam qua globus quilibet e terrestribus formatus totam terram concomitaretur? »

65   Tycho's point is that the circular motion of the cannon-ball (the motion which it performs in following the earth), while natural, is of the same order of rapidity as the violent motion produced by the explosion of the gun-powder, and would therefore have an effect analogous to this; namely it would *prevent* the cannon-ball from falling to the ground.

66   Tycho Brahe, *op. cit.*, p. 189 (ed. Dreyer, p. 219) : « Ubi igitur manebit violentissimus ille motus e puluere Bombardico praeter Naturam concitatus, qui sane alteri illi naturali, quo Terra in gyrum verti deberet, ut admodum pernici, quodammodo aemulus est ? Sunt igitur iam in globo sic emisso tres motus : Vnus quo is ratione gravitatis per lineam rectam centrum Terrae peteret : Alter quo per consensum, totius Terrae convolutionem ad amussim imitaretur : Tertius vero ille, qui fit per violentiam, quam vis Nitri sulphurosata, et carbonibus inflammata, instar Tonitrui et Fulminis, cogit globum rapidissimo impetu eo pergere, quo minime suapte Natura vellet. Cumque is

violentissimus motus alterum, quo gravia necessario, et naturaliter recta descendunt, adeo impediat, ut nisi post longe emensum spatium, imo vix quidem antequam violentia illa se remiserit, atq ; in quietem paulatim desierit, Terram contingere possit, quidnam quaeso, obtinebit secundus ille motus, si et is naturalis esset (in circuitum videlicet convolutio) privilegii, ut in Aëre etiam tam tenui per violentissimam illam concitationem, contra Naturam factam, nihil prorsus impediatur ! Experientia enim testatur, quod globus eiusdem magnitudinis et ponderis, eo, quo diximus modo, vice versa vi pulveris bombardici ejusdem quantitatis, et validitatis emissus, idem proxime spatium de superficie Terrae post se relinquat, tam versus Ortum, pari, ut dixi, ejusdem Tormenti inclinatione, quam versus Occasum eiaculatus, Aëre presertim satis tranquillo existente, et hanc, vel illam impulsionem nihil per accidens promovente, vel retardante : cum tamen ob Terrae motum diurnum (si quis esset) concitatissimum, globus versus Ortum emissus nequaquam tantum spatii de superficie Terrae emetiri posset, praeveniente nonnihil suo motu Terra, atque is, qui versus Occasum pariformiter explosus est, Terra tunc aliquid de superficie, motu proprio subtrahente, et ob id spatium interceptum augente. Nam ut dilucidius haec intelligantur ; e maxima Bombarda quam duplicem Cartoam vocant, globus ferreus, ad obliquum emissus, intra duo minuta temporis vix motu fessus Terram pertingit quibus viginti millia passuum majorum motu diurno in parallelo Germaniae convolvi deberet, si motioni diurnae obnoxia esset Tellus. »

67 *Ibid* : « Donecis, vel quispiam alius invictis rationibus liquido ostenderit, qui fieri possit, ut supra modum violentus ille, de quo dixi, motus, a duobus istis quos ille assumit, naturalibus, omnino nihil impediatur, vel etiam hos nullo vestigio interturbet . . : »

68 It is well known that Kepler's *inertia* was something quite different from the inertia of classical physics. Kepler's *inertia* is the expression of the resistance of heavy bodies to *motion* (and not to their being set in motion or accelerated), their natural tendency to *rest*. Therefore, precisely because of *inertia*, all motion implies a motor and is used up and disappears when this is removed. The eternal persistence of motion, of any motion whatever, is inconceivable for Kepler. Inertia, internal resistance to motion, plays in Kepler's physics a role analogous to that of the external resistance of the medium in that of Aristotle. Thus it was Kepler's opinion that if bodies did not have inertia motion would be instantaneous. Cf. note 72 below.

69 [The full title of Kepler's *The New Astronomy* was *Astronomia Nova* ΑΙΤΙΟΛΟΓ− ΗΤΟΣ *seu Physica Caelestis, tradita commentariis de motibus stellae Martis. Trans.*]

70 Kepler, *De Fundamentis Astrologiae certioris*, Thesis XX (*Opera*, ed. Frisch, vol. I, p. 423) : « Ubi materia, ibi geometria »; *Mysterium Cosmographicum*, note, 1621 (*Opera*, vol. I, p. 134) : « Omnis numerorum nobilitas (quam praecipue admiratur Theologia Pythagorica rebusque divinis comparat) est primitus a geometria. » Cf. *Apologia adversus Robertum de Fluctibus* (*Opera*, vol. V, p. 421 ff.).

71 See above, p.146 ff.

72 Coming to rest is explained precisely by a body's natural *inertia*. *Inertia* (cf. *Opera*, ed. Frisch, vol. II, p. 674; vol. III, pp. 305, 374, 459; vol. VI, pp. 167, 174, 181) is, for Kepler, an absolutely general property of matter, a consequence of its 'impotence'. Since matter is the same throughout the Universe he attributes this inertia to the heavenly bodies, which like any others must be moved by an active force (which Kepler saw as emanating from the sun) and which would stop moving if they were no longer moved in this way. Cf. *Epitome Astronomiae Copernicanae*, Book 4, p. 2 (*Opera*, vol. VI, p. 342): « Si nulla esset inertia in materia globi coelestis, quae sit ei velut quoddam pondus, nulla etiam opus esset virtute ad globum movendum : et posita vel minima virtute ad movendum, jam causa nulla esset, quin globus in momento verteretur. Jam vero cum globorum conversiones fiant in certo tempore, quod in alio planeta est longius, in alio brevius, hinc apparet, inertiam materiae non esse ad virtutem motricem ut nihil ad aliquid. »

73 In Kepler attraction replaces the unity or commonness of nature which had been invoked by Copernicus.

74 In fact Tycho saw gravity as a *tendency of heavy bodies* to go towards a specific place; for Kepler, gravity was an interaction between the heavy body and the earth, and it was even a passion much more than an action; cf. Kepler, *Astronomia Nova*, Introduction (*Opera*,

ed. Frisch, vol. III, p. 151) : « Vera igitur doctrina de gravitate his innititur axiomatibus : omnis substantia corporea, quatenus corporea, apta nata est quiescere omni loco, in quo solitaria ponitur extra orbem virtutis cognati corporis.

Gravitas est affectio corporea mutua intercognata corpora ad unitionem seu conjunctionem (quo rerum ordine est et facultas magnetica), ut multo magis Terra trahat lapidem, quam lapis petit Terram.

Gravia (si maxime Terram in centro mundi collocemus) non feruntur ad centrum mundi, ut ad centrum mundi, sed ut ad centrum rotundi cognati corporis, Telluris scilicet. Itaque ubicunque collocetur seu quocunque transportetur Tellus facultate sua animali, semper ad illam feruntur gravia. Si Terra non esset rotunda, gravia non undiquaque ferrentur recta ad medium Terrae punctum, sed ferrentur ad puncta diversa a lateribus diversis.

Si duo lapides in aliquo loco mundi collocarentur propinqui invicem extra orbem virtutis tertii cognati corporis, illi lapides ad similitudinem duorum magneticorum corporum coirent loco intermedio, quilibet accedens ad alterum tanto intervallo, quanta est alterius moles in comparatione. » Cf. Letter to Fabricius, 11 October 1605 (*Opera*. vol. III, p. 459) and vol. III, p. 511. It is obvious that there is no room for *lightness* in this conception: light bodies are only *minus gravia*. Cf. vol. III, p. 152.

[75] Kepler, *Astronomia Nova*, Introduction (*Opera*, vol. III, p. 152): « Etsi virtus tractoria Terrae, ut dictum, porrigitur longissime sursum, tamen si lapis aliquis tanto intervallo abesset, quod fieret ad diametrum Telluris sensibile, verum est, Terra mota lapidem talem non plane secuturum, sed suas resistendi vires permixturum cum viribus Terrae tractoriis, atque ita se explicaturum nonnihil a raptu illo Telluris : non secus atque motus violentus projectilia nonnihil a raptu Telluris explicat, ut vel praecurrant, projecta versus orientem, vel destituantur, si in occidentem projiciantur : atque ita locum suum, a quo projecta sunt, vi compulsa deserant : neque raptus Terrae hanc violentiam in solidum impedire possit, quam diu violentus motus in suo vigore est.

Sed quia nullum projectile centies milesimam diametri Terrae partem a superficie Terrae separatur, ipsaeque adeo nubes atque fumi, quae minimum terrestris materiae obtinent, non millesima semidiametri parte evolant in altum : nihil igitur potest nubium, fumorum et eorum, quae perpendiculariter in altum projiciuntur, resistentia et naturalis ad quietem inclinatio, nihil inquam potest ad impediendum hunc sui raptum ; utpote ad quem haec resistentia in nulla proportione est. Itaque quod perpendiculariter sursum est projectum, recidet in locum suum, nihil impeditum motu Telluris, ut quae subduci non potest, sed una rapit in aere volantia, vi magnetica sibi non minus concatenata, quam si corpora illa contingeret.

Hisce propositionibus mente comprehensis et diligenter trutinatis, non tantum evanescit absurditas et falso imaginata impossibilitas physica motus Terrae, sed etiam patebit, quid ad objecta physica quomodocunque informata sit respondendum. »

[76] It is the *strength* of the *raptus* which explains the absence, or rather the insignificance, of the displacement. It follows that if there were no attraction, or rather traction, of the heavy body by the earth, then Aristotle and Tycho would be right. It also follows that Bruno was wrong; the case of the ship is very different from that of the earth. See above, p. 150.

[77] Kepler, *op. cit.*, p. 152 : « Etsi Copernico magis placet, Terram et terrena omnia, licet avulsa a Terra, una et eadem anima motrici informari, quae Terram, corpus suum, rotans rotet una particulas istas a corpore suo avulsas : ut sic per motus violentos vis fiat huic animae per omnes particulas diffusae, quemadmodum ego dico, vim fieri facultati corporeae (quam gravitatem dicimus seu magneticam) itidem per motus violentos. Sufficit tamen pro solutis a Terra facultas ista corporea ; abundat illa animalis. »

[78] Kepler, *In commentaria de Motibus Martis*, note 21 (*Opera*, vol. III, p. 458) : « D. Fabricius in epistola (d. d. 26 jan. 1605) hanc movit quaestionem, spectans locum Tychonis in Epistolarum collectione (p. 189), ubi Tycho refert, quibus rationibus innixus ipse Rothmannum refutaverit Copernicum defendentem. Fabricii verba haec sunt : qua ratione tu Copernico addictus argumentum Tychonis de explosione tormenti solvere vis? Certe si versus ortum cartrana explodatur, fiet ut ob celeriorem motus Terrae emissus globus versus occasum potius locum quietis inveniat, tantum abest ut versus ortum proferatur. Herculeum certe est argumentum adversus motum Terrae diurnum, quo

destructo cetera facile cadunt. »

79 Quoted above, note 66.

80 Kepler, *op. cit.*, p. 458 ff. : « De objectione Tychonis, qui tormento impugnat motum Terrae, rogas eadem quae Cancellarius Bavariae nuperrime. Respondeo eadem, misceri motus, non impugnari aut aboleri alterum ab altero. Terra movetur ab occasu in ortum, cum ea omnis copia aeris circumfusi, omne grave, sive jacens sive pendens. Nam cur non et pendens quid impedit ? Num gravitas ? At ea tendit ad centrum Terrae, ad centrum faciei Telluris, quae lapidi est exposita, quod vi magnetica lapidem attrahit fortius quam si centum catenarum nervorum tensissimorum vinculis quaquaversum esset annexus Telluri. Nam igitur impediet ipsum aer, qui est trajiciendus ? At Terram et ipse sequitur, saltem in hac propinquitate. Quid igitur impedit? Nihil tu potes ostendere. Ergo quid impediat ostendem, sed simul et respondebo. Quodcunque materiatum corpus se ipso aptum natum est quiescens, quocunque loco reponitur. Nam quies ut tenebrae privatio quaedam est, non indigens creatione, sed creatis adhaerens, ut nullitas aliqua : motus vicissim est positivum quippiam ut lux. Itaque si lapis loco movetur, id non facit ut materiatum quippiam, sed ut vel extrinsecus impulsus vel attractus vel intrinsecus facultate quadam praeditus ad aliquid respiciente. Hanc dicunt Aristotelici appetentem centri mundi. Nego, sic enim vere impediretur sequi Terrae notum. Probent, scio, futiles ipsorum probationes ab ignis natura contrarii, quae est *petitio principii*. Nam ignis non petit coelum, sed fugit Terram . . . Ergo aliter ego definio gravitatem, seu illam vim, quae intrinsece movet lapidem, vim magneticam coagmentantem similia, quae eadem numero est in magno et parvo corpore, et dividitur per moles corporum accipitque dimensiones easdem cum corpore. Itaque si lapis aliquis esset pone Terram positus in notabili aliqua proportione magnitudinis ad molem Telluris, et casus daretur, utrumque liberum esse ab omni alio motu : tunc ego dico futurum, ut non tantum lapis ad Terram eat, sed etiam Terra ad lapidem, dividantque spatium interjectum in eversa proportione ponderum, sitque ut A ad B causa molis, sic BC ad CA et C locus ubi jungentur, plane ea proportione qua statera utitur. »

81 One can see just how Aristotelian Kepler is; rest is a privation, motion is something positive! Fire flees the earth; the same doctrine as in Copernicus or the young Galileo (cf. above, p.15 ff. and 33 ff.) But there is some progress as regards the idea of 'gravity'. Kepler's gravity is coextensive with matter; it is a universal force common to all bodies and *proportional to their masses*, and no longer a vague tendency of similar things to come together. For Kepler, as for Galileo, all bodies are 'similar'.

82 Kepler, *op. cit.*, p. 459 : « Sed contrahe vela. Dixi, si a lapide removeas animo facultatem illam jungendi similia, remansuram in lapide meram impotentiam ad mutandum locum. Ut igitur illa expugnetur, vi et contentione extranea opus est. Dum ergo fingimus lapidem in aere pendentem, negamus ei vim conjungendi similia, hoc est gravitatem, et tamen eam vim Terrae in lapidem relinquimus. Esto hoc ita, quamvis re vera absurdum sit, tantummodo ut nobis casus constet. Habebit igitur pendulus iste lapis adhuc vim quiescendi in suo loco, ea repugnabit virtuti Telluris circumacturae. Ex pugna materialium et corporearum proprietatum fiet permixtio, ut quaelibet vincatur et vincat vim suam corporum proportione. Itaque hinc evincitur, quod dixi me indicaturum, impedimentum nempe, quo minus pendulus hujusmodi lapis perfectissime sequi possit circularitatem Telluris. Atque hoc impedimentum est verissimum. Quare jam destruamus casum nostrum fictitium et sint illae lineae a superficie Terrae in lapidem tendentes non tantum ut fulcra, sed vere id quod per naturam nobis indicatur, nempe instar nervorum tensissimorum, sic ut lapis iste sit in actu descensus ad superficiem et centrum Terrae : dico, propter hanc impotentiam ad motum omnino futurum, ut lapis hic in descensu nonnihil aberret a perpendiculo ex centro Terrae per superficiem in centrum lapidis ducto et sic Terra ab occasu in ortum eunte, lapidis perpendiculum paulatim in occidentales superficiei partes deveniet : nec Terram omnimode sequetur, sed ab ea relinquetur. Habes causam cur lapis non debeat sequi Terram, qualem tu ad tuae sententiae confirmationem non potuisti dicere. Audi nunc solutionem. Verum est, si lapis notabili intervallo a Terra distaret, fore ut hoc accidat. At nunc sunt 860 milliaria a centro ad superficiem, et vero nulla avis tam alte volat ut dimidium unius milliaris absit a solo ; sane quia in aetere non magis apta est volare, quam nos in aere, quam lapis in aqua aptus est natare. »

83 It can be seen that the resistance to motion, 'impotence' of matter, is something very positive; it is even proportional to the body's mass, as is attraction. Cf. E. Meyerson, *Identité et Réalité*, 3rd edition, Appendix III, p. 534 ff.

84 There is a mathematical equivalence between these two concepts, but Kepler, like Newton, sees the heavy body not as *tending towards* but as *attracted by* the other body.

85 Kepler, *op. cit.*, p. 461 : « Nunc tandem ad tormentum Tychonicum. Cum demonstratum sit, lapidem in perpendiculo cadentem non debere illam lineam egredi in casu, jam facile expeditur et globus tormenti (lapis in obliquam jactus ; nubes vento impulsa ; avis in aëre volans). Nempe illud verum est, quod statim initio coepi dicere, misceri motum utrumque, et eum qui a Tellure est in globo, et eum qui a tormento. Itaque et miscentur spatia. Nam respectu totius universi plane plus spatii conficitur eodem tempore, cum globum in ortum ejaculamur, quam cum in occasum ; quia illic et Terra in ortum tendit, hic Tellus derogat motui in occasum, volvens globum in ortum. Imo vero plane nunquam ullus globus respectu totius universi in partem tendit contrariam viae Telluris, quia Tellus multo est celerior quam ullius globi jactus. Quod vero spatium in ipsa Telluris superficie attinet, cum quiescens lapis, quamvis in aere pendens, demonstratus sit plane sequi debere Terram, omnino etiam eadem vis per idem Ţelluris spatium tam in ortum quam in occasum abripiet globum. Nam quacunque globum impellat, invenit eandem vim lapidis attractricem, eundem etiam effectum promotionis lapidis. Si autem supra casus lapidis in perpendiculo aberasset sensibiliter a suo perpendiculo, sane etiam hoc fieret, ut brevius esset spatium jactus in occasum quam in ortum ; non quidem ob causam a Tychone allegatam, sed ob hac ipsam quam ego diligenter hic explicui. »

86 Kepler, Letter to Fabricius of 10 November 1608 (*Opera*, vol. III. p. 462) : « Cupis tibi declarari solutionem argumenti Tychonici contra motum Terrae. Non est ita horribile, ut illius machinae ictus. Plane coïncidit cum illa objectione, cur globus sursum missus ad perpendiculum recidat ad locum eundem, si Terra interim abit. Respondendum enim, non tantum Terram interim abire, sed unam cum terra etiam catenas illas magneticas infinitas et invisibiles, quibus lapis alligatus est ad partes Terrae subjectas et circumstantes undique, quibusque retrahitur proxima id est perpendiculari via ad Terram. Quemadmodum igitur hic vis infertur catenis illis a motu violento sursum, quo fit ut omnes illae aequaliter quasi extendantur, ita quoque vis infertur catenis occidentalibus, cum globus vi tormenti in orientem truditur, et vis infertur orientalibus, cum vapor globum protrudit in occidentem. Nihil nec impedit hic nec illic promovit motus universalis Telluris et catenarum omnium. Nam haec motus violentia, quae globum projicit, versatur intus in complexu catenarum omnium, quae tam sunt fortes, ut parum contra illas possit etiam ventus validissimus contrarius, nedum aura quieta et cum. Tellure circumiens. »

87 In fact Kepler is wrong about this: as we have shown above Tycho's argument is not completely identical with the old Aristotelian argument.

88 *Ibid.*, p. 462 : « Si vero nullae tales essent catenae, remaneret sane lapis in aethere pendulus abeunte Terra, nec recideret ulla ratione. Facit ad hanc considerationem et hoc, quod nullus jactus, neque quoad lineae longitudinem sensibilis est ad Telluris diametrum, neque quoad motus pernicitatem Telluris catenarumque seu virtutis magneticae. Sic igitur cum habeat hoc negotium et animi mei sententia, noli a me petere, ut veritatem prodam ad comparandum vulgi favorem. Si consuli arti non potest nisi per fraudes, pereat sane : reviviscet nempe. »

89 *Ibid.*, p. 462 ff. : « Objectio tua a ventis plane ventorum naturam imitatur, nihil efficit nisi strepitum. Quidquid enim de ventis tute ipse judicas, et ego judico : si Tellus per vapidum aërem moveretur, jure objiceres ventorum experientiam. At nunc vapor, materia ventrorum, consistit intra complexum virtutis magneticae Telluris ; cumque sit substantiae tenuis uti non valde attrahitur ad Terram, sic facile transfertur et abripitur a qualicunque virtute magnetica Telluris. Nam vis magnetica fortissima quidem est ratione suae propriae sedis, nempe Telluris, corporis densissimi : illa tamen langescit in objectu materiae rarioris. Exemplo sit vis illa motus violenti auctor. Puer manu projiciens lapidum propellit illum quam longissime. Idem totis viribus connixus, ut pumicem ejusdem molis eodem projiciat, scopum nunquam assequetur. Sed ad vapores redeo. Illi igitur asportantur cum locis Terrarum sibi subjectis a virtute magnetica Telluris, et sic

quiescunt incumbentes iisdem Terrarum locis, quantisper non a causis aliquibus impelluntur, quae causae ex eodem cum ipsis origine nascuntur. Impulsi vero ab iis causis, quae ventum faciunt, facillime e catenis illis magneticis avelluntur in plagam quamcunque, idque aequali spatio, si causa aequalis. Quippe in eorum motu non consideratur longitudo tractus per aetherem, sed multitudo catenarum seu longitudo tractus Terrarum. »

90   *Ibid.*, p. 463 : « Nam ad trajiciendum per aetherem non indigent sua opera, contentae virtute Telluris, seu navi. Adeoque genuinum est exemplum navis et vectorum in ea discursantium, nisi quod vectores navis non attrahit magnetica virtute, sed solo contactu rapit, eosdem vero Tellus adhuc attrahit per gravitatis virtutem, quam Tellus non communicat motu navis, vapores vero et projectilia non attrahit aether, itaque a sola sua navi (id est a Tellure) attrahuntur. Non itaque ut in navi ex motu navis contingunt corporum jactationes, dum abripiuntur corpora a locis iis Terrarum, ad quae tendunt, gravitatis momentis, non, inquam, sic etiam jactari necesse est corpora nostra, dum a Terra abripiuntur, neque enim tendunt ea ad ullam partem aetheris, sed ad solius Terrae subjectum planum per catenas magneticas attrahuntur : quo fit demonstratione geometrica, ut ad centrum tendant gravia ; etsi non tendunt ad centrum tanquam ad rem geometricam, sed tanquam ad medium corporis rotundi. Nisi enim Terra rotunda esset, ad idem ejus commune punctum omnia gravia non tenderent. »

91   Kepler is saying that when it leaves the top of the mast of the moving ship the heavy body is drawn along by the action of the 'chains' of the earth's gravity, which are not connected to the ship. Therefore Bruno was wrong: the ship's motion does necessarily make a difference to the motion of projectiles on it.

92   Kepler, *Epitome Astronomiae Copernicanae*, Book 1, chapter 5 (*Opera*, vol. VI, p. 181) : « Si Terra volveretur circa axem, tunc ea, quae recta sursum projiciuntur, non reciderent in locum pristinum, unde sunt projecta, quippe centro quidem persistente, loco vero superficiei, in quo stat projiciens, interim *se subducente ex linea ducta ex centro Telluris ad projectile*. Si gravia centrum per se peterent nihilque praeterea, sequeretur argumentum. At dictum in priori themate, motus gravium scopum non esse centrum per se primo, sed per accidens et secundario, quia scilicet centrum est medium et intimum corporis, quod gravia per se et primo petunt et a quo gravia attrahuntur.

Cum autem gravia petant Terrae corpus per se petanturque ab illo, fortius itaque movebuntur versus partes viciniores Terrae, quam versus remotiores. Quare transeuntibus illis partibus vicinis perpendiculariter subjectis, gravia inter decidendum versus superficiem transeuntem illam insuper etiam circulariter sequentur, perinde ac si essent alligata loco, cui imminent, per ipsam perpendicularem, adeoque per infinitas circum lineas, ceu nervos quosdam obliquos, minus illa fortes, qui omnes in sese paulatim contrahi soleant.

Atqui dixisti, corpora materiata naturali sua inertia reniti motui sibi ab extra illato; id si verum est, gravia igitur extricabunt sese nonnihil ex hoc raptu exque suo illo perpendiculo ceterisque vinculis. Extricarent sese nonnihil, si abscederent a Terra intervallo tanto, quod ad semidiametrum Terrae vel saltem horizontis visibilis proportionem haberet sensibilem. »

93   Kepler, *ibid.*, p. 182 ff. : « At saltem emissi globi bombardici, alter in ortum, alter in occasum, cadent inaequalibus intervallis a loco primo ; longius in occasum, quippe partes Terrae versus occasum sitae obviabunt globo, tendentes in ortum, brevius in ortum, quia partes orientales Terrae, in quas, si immobiles starent, globus fuerat casurus, fugiunt globum versus ortum. Non recte fit, quod comparantur spatia mundi, quasi Terra longissime absente ab emisso globo, cum de hoc solo agatur, pomum quod alter tenet manibus, quorsum ei facilius excutiatur a socio ejusdem navis vectore, non quam longe a navi aut per quantum spatium inter navem et litora. Nam si litora consideres, quantum fugit navis a loco superiore, in quem excutitur pomum, tanto fere langidior, respectu litorum quiescentium, est excussio, cedente quippe deorsum, quod excutienti substernitur, enervata resistentia ; ita quod erat defluxus navis adjecturus saltui pomi, detrahit iterum cessio ejus, quo nitebatur flictus. Et vicissim, quod erat pernicitas navis praereptura saltui pomi deorsum, hoc addit resistentia fortior violentiae flictus ; fortius enim deorsum excutit vis eadem, cum a navi deorsum et rapitur, quam cum in litore stat

immobilis. At cum, ut par est, vires nudae considerantur manus pomum prehendentis ipsiusque pomi pondus, vis equidem infertur utrinque eadem, nihil ad hanc magnitudo effectus, qui foris extra navem, compositis causis, est secuturus, etsi respectu navis solius (non etiam litorum) idem proxime futurum est ab ipsa intervallum.

Idem igitur judicium mutatis mutandis et de bombardis esto. Equidem globus magnus, duobus minutis horae unius perdurans in volatu per aerem, trajicit in occidentem per unum milliare Germanicum in Terra, interimque Terra, subjecta aequatori, obviat per octo milliaria ; quare respectu spatii mundani rapitur globus adhuc in contrariam motus violenti plagam, scilicet in orientem, septem milliaribus nihilque prodest ei aliud explosio in contrariam plagam, nisi quod octavum milliare absumit facitque, ut globus tardius in orientem sequatur ; excutere non potest pulvis globum penitus veluti de manibus Telluris, semper ille in virtute trahente haeret irretitus; si rupit prehensionem indicis, haeret in prehensione succedentis minimi digiti. E contra globus, in orientem emissus ejusdem temporis intervallo, promovetur raptu ipsius Terrae pro octo milliaria additque nonum ipse, violenter quippe explosus itidem in ortum. Ita sive in orientem sive occidentem explodatur, semper in orientem fertur, tantum paulo plus hic quam ille. At hoc compositum spatium mundanum nihil attinet ad spatium in Terra, quod homines metiri possunt ; hoc utrinque fere idem est, quia vis eadem, quia vincula magnetica utrinque eadem, ex quibus globus velut eripitur inque ulteriora transponitur.

Concurrunt tamen in occasum promotionis duae causae. Nam globus, se ipso iners ad motum, si non raperetur versus ortum, permaneret se ipso in occidente, loco in ortum abeunte, facilius igitur de loco in occasum promovebitur a violento motu ; at in ortum vincenda est illi motui non tantum prehensio magnetica Telluris, sed etiam inertia materialis globi, restitantis in occasu. Esto hoc ut supra de oceano concessum ; at quidquid sit, in globo certe bombardico inaestimabile quippiam est, nec ulla proportio sensibilis alterius pugnae ad alteram. Nam si globus bombardicus exploderetur eadem vi pulveris, positus extra virtutem Telluris attractoriam, transvolaret is non tantum per unum aut per octo milliaria spatii mundani, sed plane per incredibilem eorum numerum.

Posito etiam, quod differentia sit perceptibilis se ipsa, tandem deerit occasio experimentandi. Quis enim certum me reddet de eadem vi pulveris in utraque explosione ceterisque circumstantiis utrinque iisdem ? »

The *Dialogue Concerning the Two Chief World Systems* and the Anti-Aristotelian Polemic

94　See T. H. Martin, *Galilée . . .*, Paris, 1868

95　E. Wohlwill, *Galileo Galilei und sein Kampf für die Copernikanische Lehre*, 2 vols., Hamburg-Leipzig, 1909-1926.

96　See E. Wohlwill, *op. cit.*, vol. I, p. 105 ff., and above, p. 35 ff.

197　Cf. Galileo, *De Motu, Opere*, vol. I, p. 304 (quoted above, Part I, note 142).

98　Cf. above, Part I, p. 35 ff.

99　See the excellent comment by P. Tannery, *Galilée et les principes de la dynamique* (vol. 6 of his *Mémoires scientifiques*), Paris, 1926, p. 399: 'If, in forming an opinion of Aristotle's system of dynamics, we discount the prejudices which derive from our modern education, if we try to recapture the intellectual situation of an independent thinker at the beginning of the seventeenth century, it would be difficult not to recognise that this system is in far greater conformity with direct observation of the facts than is ours'. Cf. above, Part I, note 19.

100　Cf. Galileo, *De Motu*, p. 305 (quoted above, Part I, note 142).

101　Similarly, heavenly bodies have no weight. In general a body located at its natural place 'has no tendency to move downwards', and is therefore without weight. Cf. above, Part I, p. 35 ff.

102　*De Motu*, pp. 300, 304 (quoted above, Part I, pp. 59, 60). Cf. *Le Mecaniche, Opere*, vol. II, p. 180 : « Nella superficie esatemente equilibrata detta palla resti come indifferente e dubbia tra il moto e la quiete, si che ogni minima forza sia bastante a muoverla, siccome all'incontro, ogni pochissima resistenza, e quale è quella sola dell'aria che la circonda, potente a tenerla ferma. Dal che possiamo prendere, come per assioma indubitato, questa conclusione : che i corpi gravi, rimossi tutti l'impedimenti esternied adventizii, possono esser mossi nel piano dell'orizonte da qualunque minima forza »

103 See *De Motu*, p. 300 (quoted above, Part I, p. 60.) The same argument can be found in the *Dialogo dei due massimi sistemi del mondo, Opere*, vol. VII, pp. 46f, 53f, 172, and in the *Discorsi e dimostrazioni matematiche intorno à due nuove scienze, Opere*, vol. VIII, p. 268.

104 As we shall see below, Galileo's dynamics could be called 'a dynamics of fall'.

105 *Dialogo*, p. 53 (*Dialogue*, p. 28). [All page references to the *Dialogo* are to *Opere*, vol. VII, and the equivalent references to the English translation, the *Dialogue*, are to the translation by Stillman Drake, Berkeley-Los Angeles, 1962. *Trans.*]

'06 *Dialogo*, p. 43 (*Dialogue*, p. 19): in the margin we can read: 'Moto retto di sua natura infinito. Moto retto impossibile per natura. Moto retto impossibile esser nel mondo ben ordinato ». In the body of the text we read: 'This principle being established then, it may be immediately concluded that if all integral bodies in the world are by nature movable, it is impossible that their motions should be straight, or anything else but circular; and the reason is very plain and obvious. For whatever moves straight changes place and, continuing to move, goes ever farther from its starting point and from every place through which it successively passes. If that were the motion which naturally suited it, then at the beginning it was not in its proper place. So then the parts of the world were not disposed in perfect-order. But we are assuming them to be perfectly in order; and in that case, it is impossible that it should be their nature to change place, and consequently to move in a straight line.'

107 Cf. *Dialogo*, p. 56 (*Dialogue*, pp. 31, 32): « Moti circulari finiti e terminati non disordinano le parti del mondo. Nel moto circolare ogni punto della circonferenza è principio e fine. Moto circolare solo uniforme. Moto circolare può continuarsi perpetuamente. Moto retto non può naturalmente esser perpetuo. Moto retto assegnato a i corpi naturali per ridursi al ordine perfetto, quando ne siano rimossi. La quiete sola e il moto circolare atti alla conservazione dell'ordine. » Cf. *ibid.*, p. 166 (*Dialogue*, p. 135): '*Salviati*: [it is] impossible that any movable body is eternally moved in a straight line'.

108 Cf. W. Wohlwill, 'Die Entdeckung des Beharrungsgesetzes', *Zeitschrift für Völkerpsychologie*, etc., vol. XV, p. 387. Cf. also A. Höfler, *Studien zur gegenwärtigen Philosophie der mathematischen Mechanik*, Leipzig, 1900, p. 111 ff.

109 Cf. E. Mach, *Die Mechanik in ihrer Entwicklung, historisch-kritisch dargestellt*, 8th edition, Leipzig, 1921, p. 133 ff., and especially p. 265 ff.

110 Cf. E. Cassirer, *Das Erkentnisproblem in der Philosophie und Wissenschaft der neueren Zeit*, Berlin, 1911, p. 397: « Die Entdeckung des Beharrungsgesetzes hängt . . . mit den Grundgedanken von Galileis Forschung innig und unverkennbar zusammen. Schon aus der Betrachtung dieses Zusammenhanges heraus sollte jeder Zweifel daran schwinden, ob Galilei die volle Einsicht von der Allgemeinheit und Tragweite seines neuen Grundsatzes gewonnen hat. » Yet it was Wohlwill who was right, even more than he himself realised.

111 Therefore we shall quote Galileo's own words abundantly, since it is precisely the movement of Galileo's thought, and not the results, in which we are interested here.

112 Cf. *Notae per il Morino* (Galileo's notes on the book by J. B. Morin, *Famosi et Antiqui Problematis de Telluris Motu vel Quiete hactenus optata solutio*, Paris, 1631), *Opere*, vol. VII, p. 565: « Noi non cerchiamo quello che Iddio poteva fare, ma quello che Egli ha fatto. Impero che io vi domando, se Iddio poteva fare il mondo infinito o no: se Egli poteva e non l'ha fatto, facendola finito e quale egli e de facto, non ha esercitato della Sua potenza, in farlo cosi, piu che se l'avesse fatto grande quanto una veccia. » Cf. *Dialogo*, p. 43 (*Dialogue*, p. 19), where Galileo states that among the things said by Aristotle 'I agree with him, and I admit that the world is a body endowed with all the dimensions, and therefore most perfect. And I add that as such it is of necessity most orderly, having its parts disposed in the highest and most perfect order among themselves. Which assumption I do not believe to be denied either by you or by anyone else'.

113 Cf. P. Tannery, *Galilée et les principes de la dynamique*, Mémoires scientifiques, vol. VI, Paris, 1926, p. 404 ff., and P. Painlevé, *Les axiomes de la mécanique*, Paris, 1922, p. 31 ff.

114 Cf. U. Forti, *Introduzione Storica alla lettura del* Dialogo Sui Massimi Sistemi *di Galileo*

Galilei, Bologna, 1931.

115   The astronomical part of the *Dialogue* is particularly weak. Galileo completely ignores not only Kepler's discoveries but also even the concrete content of the works of Copernicus. The heliocentrism offered us here by Galileo is of the very simplest form (the sun in the centre with the planets in circular motion around it), a form which Galileo knew to be false. This was a conscious simplification, analogous to that presented by Descartes in his *Principes*, a simplification which would be incomprehensible in an astronomical work but which we can easily understand in a philosophical work.

116   On the literary form of the *Dialogue*, and on its overall structure, see L. Strauss, in the introduction to his translation of Galileo's work, *Dialog über die beiden hauptsächlichsten Weltsysteme*, Leipzig, 1891, and more recently L. Olschki, *Galilei und seine Zeit*, Halle, 1927. The literary, or rather the dialogue, form of Galileo's work is as important for him as it was for Plato, and this for similar reasons, for very profound reasons having to do with their very conception of scientific knowledge. Therefore we must give samples of it here even though this will make our exposition slower and more repetitious. This cannot be helped. Galileo's work occupies a unique place in modern thought, and the latter cannot be understood without understanding the former.

117   Every work is written to be read by specific readers. The *Dialogue* was written for Italians seventeenth century, just as Plato's dialogues were for fourth century Athenians.

118   Cf. the whole of the discussion of the 'Second Day'.

119   This is why Galileo's arguments are not all on the same level.

120   On the philosophical character of Galileo's work, see E. Cassirer, 'Wahrheitsbegriff und Wahrheitsproblem bei Galilei', *Scientia*, September-October, 1937.

121   In the history of philosophy there are several Platos and several Platonisms. There are, in particular, these two different versions: the Platonism, or more accurately the Neoplatonism, of the Florentine Academy, a mixture of mysticism, numerology and magic; and the Platonism of the mathematicians, that of Tartaglia and Galileo, a Platonism which is a commitment to the role of mathematics in science but without all these additional doctrines. The failure to distinguish between these two Platonisms (for the former the *Timaeus* is a treatise of magical cosmology, for the latter it is an attempt at physical mathematics) is the venial sin of E. A. Burtt's fine book, *The Metaphysical Foundations of Modern Physical Science*, London, 1925, and also the mortal sin of E. Strong's *Procedures and Metaphysics*, University of California Press, Berkeley, 1936. On the two Platonisms see L. Brunschvicg, *Les étapes de la philosophie mathématique*, Paris, 1922, p. 69 ff. and *Le Progrès de la conscience dans la philosophie occidentale*, Paris, 1927, p. 39 ff.

122   Thus right from the beginning of the *Dialogue* Galileo shows that Aristotelian physics and cosmology are based on the belief in the *perfection* of the circle, and the belief that (*Dialogo*, p. 42; Dialogue, p. 18) 'circular motion is more perfect than straight. Just how much more perfect the former is than the latter, he determines from the perfection of the circular line over the straight line. He calls the former perfect and the latter imperfect; imperfect, because if it is infinite, it lacks an end and termination, while if finite, there is something outside of it in which it might be prolonged. This is the cornerstone, basis, and foundation of the entire structure of the Aristotelian universe, upon which are superimposed all other celestial properties—freedom from gravity and levity, ingenerability, incorruptibility, exemption from all mutations except local ones, etc. All these properties he attributes to a simple body with circular motion. The contrary qualities of gravity or levity, corruptibility, etc., he assigns to bodies naturally movable in a straight line. Now whenever defects are seen in the foundations, it is reasonable to doubt everything else that is built upon them'. Now for Galileo it is absurd to make a distinction between the *perfection* of mathematical lines (cf. *Il Saggiatore, Opere*, vol. VI, p. 293 = *The Assayer*, in ed. Stillman Drake, *Discoveries and Opinions of Galileo*, New York, 1957, p. 263, where Galileo says that he has never read the pedigrees and patents of nobility of mathematical figures); similarly there is no justification in seeing the immutability of the heavenly world (which is debatable anyway) as a sign of perfection. For why should not the life and changability of the sublunary world be, on the contrary, a greater perfection than the frozen immobility of the heavens? (Cf. *Dialogo*, p.

83; *Dialogue*, p. 58.) Does not Aristotelianism itself see motion as a reality and rest as simply a privation? Finally, we should note that in Galileo's opinion, in order to criticise Aristotle he needs not to confront him with the facts of experience but with *another system*, and that in the construction of this new system the arrogance of anthropocentrism should be rejected (*Dialogo*, p. 399; *Dialogue*, p. 367-8).

123 Galileo's Platonism, a fact which is in our opinion of the greatest importance and to which we shall return later, has been noted by some of the recent historians of the great Florentine. For instance, E. Strauss, who has produced an excellent, although occasionally 'modernised', German translation of the *Dialogo (Dialog über die beiden hauptsächlichsten Weltsysteme*, Leipzig, 1891, p. xlix) correctly points out the Platonist influence on the very form of the *Dialogue*, and adds: 'Die platonische Lehre von dem unbewussten Wissen und der Wiedererinnerung, die Galilei mit besonderer Vorliebe erwähnt, beinflusst seine Darstellung; er will nicht nur die erkannte Wahrheit überliefern, auch den psychologischen Vorgang bei dem Acte der Erkenntnis veranschaulicht er, er gibt uns ein litterarisches Gegenstück zu der berühmten Mathematikergruppe der Raphaelischen Schule von Athen, welche malerisch die Stufen der Erkenntnis darstellt. Die ganze Inscenierung, die an die platonischen Dialoge erinnert und erinnern will, legt ein rühmliches Zeichen für die künstlerische Befähigung Galileis ab'. E. Cassirer, in his *Erkenntnisproblem*, vol. I, expresses the opinion that Galileo resurrected the Platonist ideal of scientific knowledge; from which follows, for Galileo (and Kepler), the necessity for mathematising nature, for (p. 389) 'Das platonische Ideal des Begreifens ist nur von dem möglich, was in dauernder Einheit sich erhält'. Unfortunately (at least in our opinion) Cassirer turns Platon into Kant. Thus, for him, Galileo's 'Platonism' is expressed in his giving priority to function (p. 402) and law (p. 397) over being and substance. According to this interpretation Galileo inverted the Scholastic proposition *operatio sequitur esse*. L. Olschki (*Galilei und seine Zeit*, Leipzig, 1927) correctly speaks of the *Platonische Naturansicht* in Galileo (p. 350) and sees the essence of his work in a *Uebertragung mathematischer Denkmethoden auf die Erfassung der Naturvorgänge* (p. 360). He even notes (following Mach) that Galileo sometimes *vertraute der Theorie mehr als der Beobachtung* (p. 268), although this strangely does not prevent him from telling us that Galilean dynamics derived from the study of ballistics (p. 206), that technology was the *Vorbedingung seiner Forschung* (p. 207), that Galileo was in the line of the tradition of the Renaissance engineers, and that 'in Galilei's Methode dem Experimente das Uebergewicht zukommt und die geometrische Fassung seiner Ergebnisse lediglich deren Uebertragung in eine strenge Begriffsprache ist, die nur auf diesem konkreten Erfahrungsboden sinnvoll und zweckhaft erscheint' (p. 212). It is E. A. Burtt (see his *The Metaphysical Foundations of Modern Physical Science*) who seems to us to have best understood the underlying metaphysical structure—Platonist mathematism—of classical science.

124 This is true above all in the critique of Aristotelianism. L. Olschki, *Galilei und seine Zeit*, pp. 198-204, is of the opinion that Galileo is giving his own history, or even (p. 355) that the *Dialogue* brings together parts which in fact belong to different stages in the development of the author's thought.

125 Thus, for example, the meaning of the term *impetus* changes from *force* (i.e., the cause of the motion impelling the body) to *moment* (i.e., motion multiplied by mass, i.e., quantity of motion).

126 For example he never refers to Bruno; and very rarely to Kepler.

127 Moreover these interlocutors are by no means merely cardboard figures or straw men. Not only the historical characters Salviati and Sagredo, but even Simplicio, have well defined personalities: they are alive, just as alive as the characters in Plato's dialogues. Cf. E. Wohlwill, *op. cit.*, vol. II, p. 85 ff.; A. Favaro, 'Amici e correspondenti di Galileo', *Nuovo Archivo Veneto*, 1903, and G. Gabrieli, 'Degli interlocutori dei Dialoghi Galileani . . .', *Rendiconti dell'Accademia dei Lincei*, 1932.

128 Cf. the present author's note in *Annuaire de l'Ecole Pratique des Hautes Etudes*, 1936-37.

129 *Dialogo*, p. 150 ff. (*Dialogue*, p. 124 ff.)

130 Aristotle, *De Caelo*, II, 14, 296 a, 27-296 b, 12.

131 All heavenly bodies (all the planets) possess a double motion and retrogress in relation to that of the fixed stars.

132   If the earth rotates it must also move around in an orbit, and this would result in certain changes in the appearance of the heavens. Copernicus had already replied to this objection, postulating the immensity of the sphere of the fixed stars in relation to which the earth's orbit can be thought of as a point. Cf. Copernicus, *De Revolutionibus*, Book I, chapter 6.

133   Since the other arguments are specifically astronomical we will not investigate them. The discussion of them constitutes the content of the 'Third Day' of the *Dialogue*.

134   In particular the argument about the centrifugal force.

135   These proofs, as we know, contribute very little that is new, and are in fact only different versions of one and the same argument. Some have found it surprising that Galileo recounts and discusses them in such great detail, and have wondered what the reason might be for these tedious repetitions. The answer, however, is quite simple: these 'modern' arguments, about the cannon shot, which had been put into circulation by the great astronomer Tycho Brahe, were those which could be heard put forward by all the 'up to date' Aristotelians; they were also those which were most effective.

136   *Dialogo*, p. 151 ff. (*Dialogue*, p. 126 ff.)

137   See above, p. 133.

138   Cf. *Dialogo*, p. 153 (*Dialogue*, p. 127).

139   Cf. *Dialogo*, pp. 57, 101, 139, 141 (*Dialogue*, pp. 33, 96, 114, 116).

140   Copernicus, *De Revolutionibus*, Book I, chapter 5; Galileo, *Dialogo*, pp. 139, 141 (*Dialogue*, pp. 114, 116).

141   *Dialogo*, p. 139 ff. (*Dialogue*, p. 114 ff.).

142   *Dialogo*, p. 148 (*Dialogue*, p. 122).

143   *Dialogo*, p. 141 (*Dialogue*, p. 116).

144   In relation to the moving body itself the motion is 'nothing', and 'nothing' has no need of a 'cause'. In other words motion in itself no more acts than does rest, and this makes it possible to locate them both on the same ontological level. Cf. above, Part II, p. 91 ff., and below, p. 254 ff.

145   It seems to us certain that the Galilean and Cartesian conception of motion, taken literally, is self-contradictory, and that the law of inertia in the last analysis implies a Newtonian conception of absolute motion and rest. However, this is not the place to discuss this question on which a great deal of ink has already been spent. See the summary of these discussions in E. Mach, *Die Mechanik*, 8th edition, p. 231 ff. Cf. also P. Duhem, *Le mouvement absolu et le mouvement relatif*, Montligeon, 1907, and A. Sesmat, *Systèmes de références et mouvements*, fasc. II, 'Mécanique newtonienne et gravitation' and fasc. IV, 'Le système absolu de la mécanique', Paris, 1937.

146   *Dialogo*, p. 42 (*Dialogue*, p. 18).

147   The discussion of the classical arguments against the earth's motion has a double objective: first, and above all, to break down the traditional conception of the two worlds and the two physics, and to assert the fundamental unity of nature and nature's laws; second, to develop a new theory, or more accurately a new concept of motion, and to establish this theory in the minds of the interlocutors/readers.

   Now the argument for the unity of nature results, of course, in a kind of 'levelling down' in which celestial nature loses its special status and is demoted to the level of terrestrial nature. But it starts with a movement in exactly the opposite direction, a 'levelling up', with the attribution to the earth and to terrestrial nature of the properties and special features of the heavens. Thus it is not in the first instance the planets which are assimilated to the earth but, on the contrary, it is the earth which is transformed into a planet and thereby graced with a *natural* circular motion. It was only later that the thrust was reversed, that the assimilation changed direction, and that man came to be aware of the real meaning of the message of the stars. For if the earth is a planet then the planets are also, in their turn, merely of terrestrial nature.

148   *Dialogo*, p. 169 (*Dialogue*, p. 144).

149   *Dialogo*, pp. 171, 208 (*Dialogue*, pp. 144, 180). Galileo is right, nobody had ever performed the experiment, although this did not stop one Antonio Rocco from writing (and this was after the publication of the *Dialogue*) in his *Esercitazioni filosofiche* (published in Galileo, *Opere*, vol. VII; see p. 677): 'Che un sasso cadente dall' albero della

nave corrente venga direttamente al piede dell'albero, io non lo credo; e quando lo vedessi, m'ingegnerei trovarli altra cagione che la rivoluzione della terra'. In fact this experiment with a ship was only performed, by Gassendi, in 1641. It aroused great excitement: cf. *Recueil de Lettres des sieurs Morin, De la Roche, De Nevre et Gassend, et suite de l'apologie du sieur Gassend, touchant la question DE MOTU IMPRESSO A MOTORE TRANSLATO, Paris, MDCL*, Preface: 'Monsieur Gassendi having always been very interested to try to confirm the truth of philosophical speculations by experiment, and happening to be in Marseille with the Comte d'Allais in the year 1641, observed, on a galley which put to sea on the special orders of this prince, more illustrious by his love and knowledge of good things than by the elevation of his birth, that a stone dropped from the very top of the mast while the galley was proceeding with the greatest strength and speed possible, fell at a point no different from where it would have fallen had the same galley been stopped and immobile; so that whether the galley was moving or not the stone always fell alongside the mast to its foot and on the same side. This experiment, performed in the presence of the Comte d'Allais and of a large number of other people who were there, seems to many people who have never seen it to be somewhat paradoxical; and for this reason Monsieur Gassendi wrote a treatise *De Motu impresso a motore translato*, which we saw that same year in the form of a letter written to Monsieur du Puy'.

150 *Dialogo*, p. 169 (*Dialogue*, p. 145).
151 Emphasis added. [The Italian original has '*ma io son tanto buon cozzon di cervelli . . .*'; Koyré's French has '*mais je suis un si bon accoucheur des cerveaux . . .*'; both allude metaphorically to Socratic method and this is missed in Professor Drake's rendering, 'I am so handy at picking people's brains . . .'; I have therefore modified it. *Trans.*]
152 *Dialogo*, p. 171 ff. (*Dialogue*, p. 145 ff.); cf. Letter to Ingoli, *Opere*, vol. VI, pp. 542, 546.
153 Cf. *Il Saggiatore* (*Opere*, vol. VI, p . 328 ff.); Letter to Ingoli (*Opere*, vol. VI, p. 545): 'Io sono stato doppiamento miglior filosofo di loro, perchè loro al dir quello ch'è il contrario, in effeto hanno anco ajunto la buggia, dicendo d'aver ció veduto dall'esperienza, ed io ne ho fatto l'esperienza, avanti la quale il natural discorso mi aveva molto fermamente persuaso che l'effetto doveva succedere come appunto succede'.
154 Cf. E. Jouguet, *Lectures de Mécanique*, Paris, 1924, vol. I, p. 111.
155 Cf. E. Mach, *Die Mechanik*, 8th edition, p. 127 ff.
156 A considerable part of the 'First Day' of the *Dialogue* is devoted to optical experiments concerning the reflection of light on smooth and uneven surfaces and the experimental proof of the paradox that a mirror in sunlight usually appears darker than the wall which it is placed against, and of the fact that a polished sphere is invisible. Galileo draws the conclusion from this that if the moon were a smooth sphere one would probably not be able to see it at all. See *Dialogo*, p. 91 ff. (*Dialogue*, p. 70 ff.); *Il Saggiatore*, p. 281.
157 *Dialogo*, p. 134 (*Dialogue*, p. 108).
158 *Dialogo*, p. 138 (*Dialogue*, p. 111).
159 It is only after the deduction that one turns to experiment. Cf. the passage from the Letter to Ingoli quoted above, note 153.
160 *Dialogo*, p. 171 (*Dialogue*, p. 145), Cf. above, Part 1, p. 36 ff.
161 The reader again! We should never forget the role of the reader. He is, in fact, the most important character in the dialogue.
162 *Ibid.*, p. 172 (*Dialogue*, p. 146).
163 A perfectly smooth surface, a perfectly round ball, etc.—we are no longer in the world of perceptible reality. We are in the Archimedean world of realised geometry. Cf. above, Part 1, p. 37 ff.
164 Emphasis added. The text has *impulso*.
165 Emphasis added. The text has *movimento*.
166 Rest is conceived by Galileo as simply an infinite degree of slowness. Since Galileo does not accept that there is any discontinuity between the two states (of motion and rest) a cause which stops a body must cause it to progressively slow down. Conversely, if there is no cause slowing a body down then there is nothing to cause rest. The contrast with Aristotle, with *impetus* physics and with Kepler, could not be clearer.
167 Cf. E. Wohlwill, 'Die Entdeckung des Beharrungsgesetzes', *Zeitschrift für*

Völkerpsychologie, vol. XV, p. 14 ff., 132 ff., 134.
168 *Dialogo,* p. 173 (*Dialogue,* p. 147). Cf. *Ibid.,* p. 53 (*Dialogue,* pp. 31, 32).
169 This is, for the Aristotelian, an *essential* obstacle.
170 Emphasis added. The text has: *d'un moto indelebili in lei.*
171 Cf. above, p. 136 ff.
172 This is Duhem's interpretation; cf. *Etudes sur Léonard de Vinci,* vol. III, p. 560 ff.
173 *Dialogo,* p. 180 (*Dialogue,* p. 155). It is Sagredo, the *bona mens,* who is drawing out the consequences of Galileo's doctrine.
174 Emphasis added. The text has: *Moto participato dalla pietra.*
175 Emphasis added. [This phrase, emphasised by Koyré, (Italian: *indelebilimente*) is omitted from Drake's translation. *Trans.*]
176 *Dialogo,* p. 181 (*Dialogue,* p. 156).
177 Mersenne, in the Preface to his translation of Galileo's *Le Mecaniche* (Paris, 1634), remarks with astonishment on the fact adduced here by Simplicio! See also, Gassendi, *De motu impresso a motore translato,* Paris, 1642, p. 22 ff.
178 *Dialogo,* p. 182 (*Dialogue,* p. 156). The whole of the first part of Gassendi's *De motu* is devoted to establishing this equivalence.
179 Emphasis added.
180 *Dialogo,* p. 186 ff., p. 197 (*Dialogue,* p. 157 ff., p. 171).
181 *Dialogo,* p. 194 (*Dialogue,* p. 168). [Koyré has here *'par la bouche de Simplicio'*—which is a mistake. *Trans.*]
182 *Dialogo,* p. 194 (*Dialogue,* p. 168).
183 *Dialogo,* p. 195 (*Dialogue,* p. 169). We should note in passing that the experiment never went beyond conception. In fact in Galilean science the most important experiments are 'thought experiments'.
184 *Dialogo,* p. 195 (*Dialogue,* p. 169).
185 Emphases throughout this quotation are added.
186 For those who can 'be convinced by reason' this experiment, the *esperienza sensata* that Simplicio is looking for, is obviously unnecessary.
187 Cf. *Dialogo,* p. 53 (*Dialogue,* pp. 31-32).
188 Galileo's belief in the innateness of knowledge, like that of Descartes, reflects his Platonism.
189 More accurately it is not Simplicio but the reader who needs to be educated. But the reader can only be educated via Simplicio.
190 *Dialogo,* p. 193 (*Dialogue,* p. 167): '*Sagredo*: Well, Salviati, there is another remarkable thing which I have just been reflecting about. It is that, according to these considerations, straight motion goes entirely out the window and nature never makes any use of it at all. Even that use which you granted to it at the beginning, of restoring to their places such integral, natural bodies as were separated from the whole and badly disorganised, is now taken away and assigned to circular motion'. And in the margin: 'Straight motion seems entirely excluded from nature'.

Galileo's Physics

191 Cf. above, pp. 34 ff., 144 ff.
192 Cf. above, pp. 130 ff., 162.
193 Cf. above, p. 35 ff.
194 Cf. *Dialogo,* pp. 46-47, 253 (*Dialogue,* pp. 22, 227): '. . . if the terrestrial globe were perforated through the centre, a cannon ball descending through the hole would have acquired at the centre such an impetus from its speed that it would pass beyond the centre and be driven upward through as much space as it had fallen'. Cf. *Ibidem,* p. 262 (*Dialogue,* p. 236): 'Natural motion converts itself into motion that is called preternatural and violent'.
195 Cf. P. Duhem, *Etudes sur Léonard de Vinci,* vol. III, p. 185 ff. We hasten to point out that these examples were not accepted at all by the Aristotelians. For instance Antonio Rocco replied to Galileo in his *Esercitatione filosofiche* (in Galileo, *Opere,* vol. VII, p. 689): 'All'essempio della Terra forata, io negherei liberamente e senza scrupulo alcuno che, giunta la palle al centro, seguisse il suo mote dalla parte dell'altro emisfero verso il cielo'.

196 *Dialogo*, p. 53 (*Dialogue*, p. 32).

197 Cf. above, Part I, p. 33 ff.

198 See Benedetti, *Diversarum speculationum mathematicarum liber,* Taurini, 1585. Cf. above, Part I, pp. 25 ff., 15 ff.

199 Cf. *Dialogo*, p. 44 ff. (*Dialogue*, p. 20 ff.), and *De Motu*, p. 300.

200 Cf. *Dialogo*, pp. 48 ff., 171 ff. (*Dialogue*, pp. 23 ff., 145 ff.), and *Discorsi*, p. 205.

201 Cf. *Dialogo*, p. 58 (*Dialogue*, p. 33). Gravity or heaviness is the natural tendency of the parts of all world globes to go towards their centres.

202 Cf. *Dialogo, pp.* 44, 56 (*Dialogue*, pp. 20, 31). Like Aristotle Galileo believed that it is impossible for there to be a tendency to remain at rest in a particular place (with the exception of the sun). Cf. *ibid.*, p. 44 (*Dialogue*, p. 20): '*Salviati*: Every body constituted in a state of rest but naturally capable of motion will move when set at liberty only if it has a natural tendency toward some particular place; for it it were indifferent to all places it would remain at rest, having no more cause to move one way than another. Having such a tendency, it naturally follows that in its motion it will be continually accelerating. Beginning with the slowest motion, it will never acquire any degree of speed without first having passed through all the gradations of lesser speed—or should I say of greater slowness? For, leaving a state of rest, which is the infinite degree of slowness, there is no way whatever for it to enter into a definite degree of speed before having entered into a lesser, and another still less before that. It seems much more reasonble for it to pass first through those degrees nearest to that from which it set out, and from this to those farther on. But the degree from which the movable body began to move was that of most extreme slowness; that is to say, from rest. Now this acceleration of motion occurs only when the body in motion keeps going, and is attained only by its approaching its goal. So wherever its natural inclination draws it, it is conducted there by the shortest line; namely, the straight'.

203 Cf. *Dialogo*, p. 58 (*Dialogue*, p. 33): 'if any centre may be assigned to the universe, we shall rather find the sun to be placed there'. Cf. *ibid.*, p. 349 (*Dialogue*, p. 321).

204 *Dialogo*. p. 58 (*Dialogue*, p. 33). Cf. Copernicus, *De Revolutionibus*, Book I, chapter 5.

205 *Dialogo*, p. 260 (*Dialogue*, p. 234).

206 *Ibid.*, p. 260 (*Dialogue*, p. 234): '*Simplicio*: Very well, but as heavy and light bodies can have neither an internal nor an external principle of moving circularly, then neither does the earth move circularly . . . *Salviati*: I did not say that the earth has neither an external nor an internal principle of moving circularly; I say that I do not know which of the two it has. My not knowing this does not have the power to remove it'. But whatever it is, he adds, it will be the same as that which moves the other planets.

207 Galileo tells us that gravity is only a 'name'. What this means is that he adopts a position of positivist nominalism on the grounds that he does not know the *nature* of gravity, just as he does not know that of light. But in both of these cases Galileo only allows himself to accept this position because he realises that he cannot do otherwise. For he knows perfectly well that gravity is a force of the same nature as magnetic attraction. Thus he openly declares his allegiance to Gilbert's magnetic philosophy (*Dialogo*, pp. 431 ff., 429 ff.; *Dialogue*, pp. 400 ff.); he shares Gilbert's belief that the earth is a large magnet. But he does not know what magnetic force is, and his own investigations, recorded in the *Discourses*, did not enable him to construct a proper theory, i.e., a mathematical account of magnetism. As for Gilbert's theory, it was animist; and so was that of Kepler (see above, p. 145 ff.).

208 [An English translation of the Letter to Christina, Grand Duchess of Tuscany, can be found in ed. Stillman Drake, *Discoveries and Opinions of Galileo*, New York, 1957, but the passage in question is not in it.]

209 *Il Saggiatore, Opere*, vol. VI, p. 341 ff. (*The Assayer*, in ed. Stillman Drake, *Discoveries and Opinions of Galileo*, p. 274) : « Per tanto io dico che ben sento tirarmi dalla necessittà, subito che concepisco una materia o sostanza corporea, a concepire insieme ch' ella é terminata e figurata di questa o di quella figura, ch'ella in relazione ad altre é grande o piccola, ch'ella è in questo o quel luogo ch'ella si muove o sta ferma, ch'ella tocca o non tocca un altro corpo, ch'ella è una, poca o molta, nè per veruna imaginazione posso separarla da queste condizioni ; ma ch'ella debba essere bianca o rossa, amara o

dolce, sonora o muta, di grato o ingrato odore, non sento farmi forza alla mente di doverla apprendere da cotali condizioni necessariamente accompagnata : anzi, se i sensi non ci fussero scorta, forse il discorso o l'immaginazione per sè stessa non v'arriverebbe giammai. Per lo che vo io pensando che questi sapori, odori, colori, etc. per la parte del suggetto nel quale ci par che riseggano, non sieno altri che puri nomi, ma tengono solamente lor residenza nel corpo sensitivo, siche rimosso l'animale, sieno levate e annichilate tutte queste qualità. » Cf. *ibid.*, p. 350 : « molte affezzioni che sono reputate qualità risedenti ne'soggetti esterni, non ànno veramente altra essistenza che in noi, et fuor di noi non sono altro che nomi. » (*The Assayer*, p. 277: 'Many sensations which are supposed to be qualities residing in external objects have no real existence save in us, and outside ourselves are mere names'.)

210 When we come to deal with Descartes' dynamics we will see what follows from the fact that geometrical bodies do not have in themselves any propensity *either to motion or to rest*. Descartes thus finds himself obliged not only to have motion created by God but to have rest created by him also.

211 This was explicitly proposed by Torricelli. Cf. *Opera Geometrica*, Florence, 1642, p. 8 ff., quoted below, Appendix to Part III, note 10.

212 A purely mathematical 'body', without gravity, would be unable to move by itself. Cf. above, p.199.

213 Galileo does not attempt to construct an abstract world, but to grasp the mathematical essence of the real, moving, and hence *temporal,* world. Cf. above, Part II, pp. 108-9.

214 *Dialogo*, p. 53 (*Dialogue*, p. 28): 'But motion in a horizontal line which is tilted neither up nor down is circular motion about the centre; therefore circular motion is never acquired naturally without straight motion to precede it; but, being once acquired, it will continue perpetually with uniform speed'.

215 *Dialogo*, p. 44 (*Dialogue*, p. 21): 'We may therefore reasonably say that nature, in conferring a definite speed upon a body constituted at first in rest, gives it a straight motion through a certain time and space. This assumed, let us suppose God to have created the planet Jupiter, for example, upon which He had determined to confer such-and-such a velocity, to be kept perpetually uniform forever after. We may say with Plato that at the beginning He gave it a straight and accelerated motion; and later, when it had arrived at that degree of speed, converted its straight motion into circular motion whose speed thereafter was naturally uniform'. Note that this myth is repeated by Sagredo in the *Discorsi,* 'Fourth Day', p. 283.

216 Cf. above, pp. 131, 134, 165.

217 A miracle which Descartes requires of his God, and which Newton allows his to forgo.

218 For Bruno it was still the case that the planets move round because they are *weightless*. Cf. above, p. 139 ff.

219 *Dialogo*, p. 45 (*Dialogue*, p. 21), in the margin: 'Between rest and any assigned speed lie infinite degrees of lesser speed'.

220 *Dialogo*, pp. 46, 248 (*Dialogue*, pp. 22, 221); *Discorsi,* 'Third Day', p. 198 ff. Cf. above, Part II, p. 68.

221 Cf. *Dialogo*, p. 249 (*Dialogue*, p. 223): '. . . a ball of one, ten, a hundred, or a thousand pounds will all cover the same hundred yards in the same time'. Cf. *Discorsi*, p. 128 ff. Historians of Galileo, and of physics, usually mix up two entirely different propositions: (i) that which Galileo is alleged to have established at Pisa by experiments which in fact he never performed, and had no need to perform, (cf. L. Cooper, *Aristotle, Galileo and the Tower of Pisa*, Ithaca, 1935, and the present author's article, 'Galilée et l'expérience de Pise', *Annales de l'Université de Paris*, 1937); this proposition, to the effect that bodies *of the same nature* fall with the same speed, was actually already established by Benedetti (cf. above, Part I, pp. 26, 31); (ii) the second proposition, of which a proof is provided for the first time in the *Discorsi*, says that all bodies *regardless of their natures* fall with the same speed.

222 Cf. *Dialogo*, p. 248 (*Dialogue*, p. 221); and the demonstration of this, adds Salviati, as of many other things concerning motion, is purely mathematical. Cf. *Discorsi,* 'Third Day', p. 190.

223 Cf. *Discorsi*, p. 205. Note that it is the same for Torricelli. But what is for Galileo a

'postulate' is for Torricelli an axiom. Cf. Torricelli, *Opera geometrica*, p. 98.
224 *Discorsi*, p. 205.
225 Cf. above, Part II, p. 100.
226 *Discorsi*, p. 205. In the *Dialogue* Sagredo has some difficulty in understanding the
meaning of Galileo's postulate. However, once having understood it, he accepts it
immediately. See *Dialogo*, p. 47 (*Dialogue*, p. 23): '*Salviati:* You argue very well. And I
know that you will make no question of granting that the acquisition of *impetus* is
measured by the departure of the movable body from the point of origin and its approach
toward the centre to which its motion tends. So will you not put an end to your difficulty
by conceding that two equal movable bodies, descending by different lines and without
any impediment, will have acquired equal *impetus* whenever their approaches to the
centre are equal?

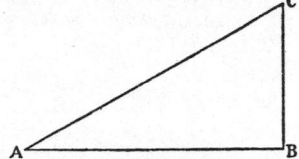

Sagredo: I do not quite understand the question.
Salviati: I shall express it better by drawing
a little sketch. Thus I take the line AB as
parallel to the horizon, and at the point
B I erect the perpendicular BC, and then I
add this slanted line CA. Now the line CA
is meant to be an inclined plane, exquisitely polished and hard, upon which descends a
perfectly round ball of some very hard substance. Suppose another ball, quite similar, to
fall freely along the perpendicular CB. I ask you to concede that the *impetus* of that
which descends by the plane CA, upon arriving at the point A, would be equal to the
impetus acquired by the other at point B after falling along the perpendicular CB.
Sagredo: I surely believe it would. In fact, they have both advanced equally toward the
centre; and by what I have already granted, the *impetus* of each should be equally
sufficient to carry it back to the same height.'
227 We are so used to hypostatising the results, or the premises, of our mathematical
calculations that we either simply accept that the limitations on the possibility of
determinacy represent aspects of reality (for example, we simply accept that we have
identified the ultimate constituents of matter, molecules, atoms, electrons) or we put this
forward as a 'principle'. The recent history of physics presents such striking examples of
this tendency of ours that there is no need to dwell on it.
228 Cf. *Dialogo*, pp. 46, 47 (*Dialogue*, 22, 23); *Discorsi*, p. 205.
229 Cf. *Dialogo*, p. 47 (*Dialogue*, p. 23), quoted above, note 226; *Discorsi*, p. 202.
230 Cf. *Discorsi*, pp. 218, 244.
231 *Discorsi*, p. 206: 'Imagine this page to represent a vertical wall, with a nail driven into it;
and from the nail let there be suspended a lead bullet of one or two ounces by means of a
fine vertical thread, AB, say from four to six feet long, on this wall draw a horizontal line
DC, at right angles to the vertical thread AB, which hangs about two finger-breadths in
front of the wall. Now bring the thread AB with the attached ball into the position AC
and set it free; first it will be observed to descend along the arc CBD, to pass the point B,
and to travel along the arc BD, till it almost reaches the horizontal CD, a slight shortage
being caused by the resistance of the air and the string; from this we may rightly infer that
the ball in its descent through the arc CB acquired an *impetus* on reaching B, which was
just sufficient to carry it through a similar arc BD to the same height. Having repeated
this experiment many times, let us now drive a nail into the wall close to the
perpendicular AB, say at E or F, so that it projects out some five or six finger-breadths in
order that the thread, again carrying the bullet through the arc CB, may strike upon the
nail E when the bullet reaches B, and thus compel it to traverse the arc BG, described
about E as centre. From this we can see what can be done by the same *impetus* which
previously starting at the same point B carried the same body through the arc BD to the
horizontal CD. Now, gentlemen, you will observe with pleasure that the ball swings to
the point G in the horizontal, and you would see the same thing happen if the obstacle
were placed at some lower point, say at F, about which the ball would describe the arc BI,
the rise of the ball always terminating exactly on the line CD. But when the nail is placed
so low that the remainder of the thread below it will not reach to the height CD (which

would happen if the nail were placed nearer B than to the intersection of AB with the horizontal CD) then the thread leaps over the nail and twists itself about it. This experiment leaves no room for doubt as to the truth of our supposition; for since the two arcs CD and DB are equal and similarly placed, the moment acquired by the fall through the arc CB is the same as that gained by fall through the arc DB; but the moment acquired at B, owing to fall through CB, is able to lift the same body through the arc BD; therefore, the moment acquired in the fall BD is equal to that which lifts the same body through the same arc from B to D; so, in general, every moment acquired by fall through an arc is equal to that which can lift the same body through the same arc. But all these moments which cause a rise through the arcs BD, BG, and BI are equal, since they are produced by the same moment, gained by fall through CB, as experiment shows. Therefore all the moments gained by fall through the arcs DB, GB, IB, are equal.' Cf. E. Jouguet, *Lectures de Mécanique*, vol. I, p. 98.

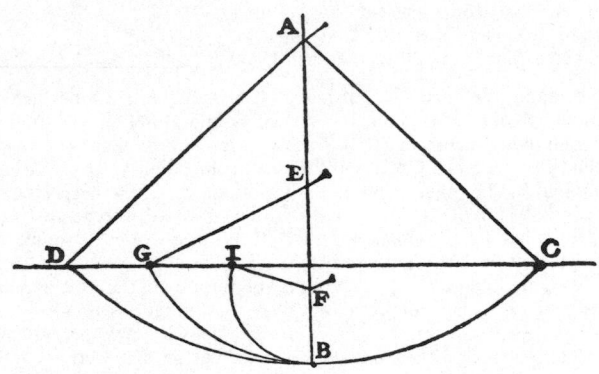

232   *Discorsi,* p. 215.
233   *Discorsi,* p. 218; Cf. *Dialogo,* p. 48 (*Dialogue,* p. 25).
234   *Discorsi,* p. 216.
235   Cf. P. Duhem, *Etudes sur Léonard de Vinci,* vol. III, p. 567.
236   Cf. above, Part II, p. 75 ff., and Part III, p. 174 ff.
237   *Discorsi,* p. 215.
238   Cf. *Discorsi,* p. 216.
239   Cf. *Le Mecaniche, Opere,* vol. II, pp. 156, 164, 168, 170, 185.
240   *Discorsi,* p. 216 ff.
241   Cf. E. Jouguet, *Lectures de Mécanique,* vol. I, p. 106, note 119, p. 111 ff.
242   Therefore he believes Kepler's *inertia* to be entirely superfluous.
243   Galileo's predecessors took the *rectilinear* motion of projectiles for granted: for them the trajectory became curved only towards its end. For Galileo it is curved from the beginning of the motion: therefore, for him, rectilinear projectile motion became *strictly impossible.*
244   Cf. E. Wohlwill, 'Die Entdeckung des Beharrungsgesetzes', *Zeitschrift für Völkerpsychologie,* vol. XV, p. 129 ff., 346 ff.; cf. above, p. 131.
245   Cf. *Dialogo,* p. 62 (*Dialogue,* p. 38), where circular motion is stated to be 'natural' for all bodies and not only for celestial bodies; and p. 193 (*Dialogue,* p. 167), where Galileo says that rectilinear motion does not exist in the world.
246   Cf. *Dialogo,* p. 42 ff. (*Dialogue,* p. 19 ff.), quoted above, note 106.
247   See the passage quoted above, note 112, and *Dialogo,* pp. 324, 375, 388 (*Dialogue,* pp. 290, 347, 361). Cf. Letter to Ingoli, *Opere,* vol. VI, pp. 518 ff., 524 ff.
248   We have pointed out above (p.159) that Galileo, who was certainly familiar with

Bruno's work, *never* mentioned him: when Kepler sent him a Brunoesque interpretation of the discoveries of the *Nuntius Sidereus* Galileo *did not reply*. One has a definite impression that Bruno's name was taboo. Cf. *Dissertatio cum Nuntio Sidereo, Opere*, vol. III, p. 105 ff.

249 Cf. *Dialogo*, p. 432 ff. (*Dialogue*, p. 406 ff.): '*Salviati*: I have the highest praise, admiration, and envy for this author, who framed such a stupendous concept regarding an object which innumerable men of splendid intellect had handled without paying any attention to it. He seems to me worthy of great acclaim also for the many new and sound observations which he made, to the shame of the many foolish and mendacious authors who write not just what they know, but also all the vulgar foolishness they hear, without trying to verify it by experiment . . .' And a little later, after having made certain criticisms of Gilbert, Galileo goes on: 'But this need not diminish the glory of the first observer. I do not have a lesser regard for the original inventor of the harp because of the certainty that his instrument was very crudely constructed and more crudely played; rather, I admire him much more than a hundred artists who in ensuing centuries have brought this profession to the highest perfection.' Galileo's feeling for Gilbert was certainly encouraged by his resolute Copernicanism. Cf. G. Gilbert, *De Magnete*, London, 1600, Book VI, chapter 3, p. 220 [an English translation by P. Fleury Mottelay, with the title *De Magnete*, was published in 1893, and republished by Dover Publications, New York, 1958]: « Jam vero cum coelum totum, et vastam mundi amplitudinem, in gyrum rotari, absurdius quam dici potest vulgares philosophi imaginentur : relinquitur ut terra diurnam immutationem perficiat. Dies igitur hic qui dicitur naturalis est meridiani alicuius telluris a sole ad solem revolutio. Revolvitur vero integro cursu, a stella aliqua fixa ad illam rursus stellam. Quae natura moventui corpora motu circulari, aequali et constanti, illa in suis partibus varijs instruuntur terminis. Terra vero non Chaos est, nec moles indigesta ; sed astrea sua virtute, terminos habet motui circulari inservientes, polos non mathematicos, aequatorem non imaginatione conceptum, meridianos etiam et parallelos ; quos omnes permanentes, certos, naturales in terra invenimus : quos tota philosophia magnetica plurimis experimentis ostendit. » See also pp. 225, 228.

250 *Dialogo*, p. 431 (*Dialogue*, p. 405).

251 *Dialogo*, p. 432 (*Dialogue*, p. 406): 'What I might have wished for in Gilbert would be a little more of the mathematician, and especially a thorough grounding in geometry'. On the non-mathematical character of Gilbert's physics, see the book we have already mentioned by E. A. Burtt, *The Metaphysical Foundations of Modern Physical Science*, p. 68 ff.

252 Cf. Gilbert, *De Magnete*, Book V, chapter 12, p. 209: « *Vis magnetica animata est, aut animatam imitatur, quae humanam animam dum organico corpori alligatur, in multis superat*. Admirabilis in plurimis experimentis magnes, et veluti animatus. Atque haec est una ex illis egregia virtus, quam veteres in caelo, in globis et stellis, in sole et luna animam existimabant. Suspiciabantur namque non sine divina et animata natura posse motus tam varios fieri, corpora ingentia certis temporibus torqueri, admirabiles potentias in alia corpora infundi. » It can be seen that Gilbert believed that the heavenly bodies have souls.

253 Newtonian attraction is not directed towards an object; it is a function of distance.

254 *Dialogo*, p. 216 (*Dialogue*, p. 190); cf. above, p. 132.

255 *Dialogo*, p. 237 (*Dialogue*, p. 211).

256 Notice that *impetus* is throughout represented as a function of speed.

257 *Dialogo*, p. 244 (*Dialogue*, p. 217).

258 *Dialogo*, p. 218 (*Dialogue*, p. 191).

259 *Dialogo*, p. 222 (*Dialogue*, p. 196).

260 *Dialogo*, p. 242 (*Dialogue*, p. 215).

261 Cf. *Dialogo*, p. 217 ff. (*Dialogue*, p. 191 ff.).

262 *Dialogo*, p. 219 (*Dialogue*, p. 193). Note Salviati's Socratic technique.

263 *Dialogo*, p. 220 ff. (*Dialogue*, p. 194 ff.).

264 *Dialogo*, p. 225 (*Dialogue*, p. 198).

265 *Dialogo*, p. 225 (*Dialogue*, p. 199).

266 Cf. *Dialogo*, pp. 216, 221 (*Dialogue*, pp. 190, 195).
267 Cf. *Dialogo*, pp. 220, 222 (*Dialogue*, pp. 194, 196).
268 Cf. *Dialogo*, p. 229 (*Dialogue*, p. 202). Cf. Kepler, quoted above, p. 150 ff.
269 Cf. *Dialogo*, p. 221 (*Dialogue*, p. 194).
270 Cf. *Dialogo*, p. 225 (*Dialogue*, p. 199).
271 There is, however, one phenomenon which, according to Galileo, does not take place on a moving earth the same as it would on a stationary earth: unfortunately this is that of the tides, which the 'Fourth Day' of the *Dialogue* represents as being the result of the *double motion* of the earth.
272 *Dialogo*, p. 191, marginal note (*Dialogue*, p. 165): 'The line described by a natural falling body, assuming the earth's motion about its own centre, would probably be the circumference of a circle'; *ibid.*, p. 192, marginal notes (*Dialogue*, p. 166): 'Body falling from the top of a tower moves along the circumference of a circle. It moves neither more nor less than if it had stayed on top. It moves equably and is not accelerated'. Cf. *Discorsi*, 'Third Day', p. 190, 'Fourth Day', p. 268 ff.
273 *Dialogo*, p. 190 ff. (*Dialogue*, p. 164 ff.). It was Cavalieri who was the first to prove that a projectile's trajectory is a parabola. See *Speccio Ustorio*, p. 151 ff. But there is evidence (cf. E. Wohlwill, 'Die Entdeckung des Beharrungsgesetzes', *Zeitschrift für Völkerpsychologie*, vol. XV, p. 107 and p. 109, note 2, and 'Entdeckung der Parabelform der Wurflinie', *Abhandlungen zur Geschichte der Mathematik*, Leipzig, 1899) that when *Speccio Ustorio* was published (1632) Galileo already knew this law, and had done for a long time (since 1610). So he bitterly reproached Cavalieri for having denied him the glory of being the first to publish this important discovery. But why then did he not publish it in the *Dialogue*? Why did he publish there an incorrect law? Neither Wohlwill, nor to our knowledge anyone else, has ever answered this question. However, the answer seems very simple: in the *Dialogue* Galileo investigates phenomena as they happen *on the earth*. Therefore he only gives a *probable* law.
274 Emphasis added.
275 The *Discourses* were published six years after the *Dialogue*. But the preparation was certainly at least in part much earlier than, or at the latest contemporaneous with, the drafting of the cosmological work.
276 At least not always; cf., however, the passage quoted above, p. 184.
277 *Discorsi*, 'Third Day', p. 243.
278 *Discorsi*, 'Fourth Day', p. 268.
279 Emphasis added.
280 Cf. E. Mach, *Die Mechanik*, 8th edition, p. 132 ff., p. 265 ff.
281 Thus when he calculates how long it would take for a stone to fall from the moon's sphere to the earth (*Dialogo*, p. 248 ff.; *Dialogue*, p. 221 ff.), he states that its acceleration *does not vary* with its distance from the earth. It is worth noting that Kepler already knew that this is incorrect.
282 Cf. *Discorsi*, 'Fourth Day', pp. 273-274. Cf. below, p. 242 ff., for Torricelli's reply to this objection.

Conclusion

283 We know that Descartes' physics ended in total failure (see above, Part II, p. 93 ff.). It was 'mathematical physics without mathematics' as P. Mouy has rightly said (*Le développement de la physique cartésienne*, Paris, 1934, p. 144).
284 F. Bonamico, *De Motu*, Florence, 1695, Book I, p. 54 ff. : « Mathematicae cum ex notis nobis, et natura simul efficiant id quod cupiunt, sic caeteris demonstrationis perspicuitate praeponentur, nam vis rerum quas ipsae tractant, non est admodum nobilis; quippe quod sint accidentia, id est habeant rationem substantiae quatenus subiicitur, et determinantur quanto; eaque considerentur longe secus atque in natura existant ; usque adeo ut nonnullis non naturae, sed mentis opera esse credantur. Attamen nonnullarum rerum ingenium tale esse comperimus, ut ad certam materiam sese non applicent, neque motum consequantur, quia tamen in natura quicquid est, cum motu existit ; opus est abstractione cuius beneficio quantum motu non comprehenso in eo munere contemplamur ; et cum talis sit earum natura nihil absurdi exoritur. Quod item

confirmatur, quod mens in omni habitu verum dicit ; atqui verum est ex eo, quod res ita est. Huc accedit quod Aristoteles distinguit scientias non ex ratione notionum, sed entium. Caeterum et mathematicae gradus habent : quando ea quae considerat quantum discretum certior est quam ea quae tractat continuum, cum superet perspicuitate demonstrationis, et simplicitate subjecti. nam quantum continuum se habet ad discretum ut includens positionem, punctus enim est unitas cum positione. Et multo praestantior est Astrologia, quippe quod sola ex mathematicis de substantia atque illa quidem perpetua et caussas invariabiles habentes disserat. ideoque sit omnium maxime affinis primae philosophiae. »

285 Jacobi Mazzonii, Caesenatis, in Almo Gymnasio Pisano Aristotelem ordinarie, Platonem vero extra ordinem profitentis, *In Universam Platonis et Aristotelis Philosophiam Praeludia, sive de Comparatione Platonis et Aristotelis*, liber primus . . . Venetiis, MDCXCVII, Apud Joannem Guerilium, p. 187 ff. « Disputatur utrum usus mathematicarum in Physica utilitatem, vel detrimentum afferat, et in hoc Platonis, et Aristotelis comparatio. Quartae sectionis. Caput sextum. Libri Decimumoctavum. Non est enim inter Platonem, et Aristotelem quaestio, seu differentia, quae tot pulcris, et nobilissimis speculationibus scateat, ut cum ista, ne in minima quidem parte comparari possit. Est autem differentia, utrum usus mathematicarum in scientia Physica tanquam ratio probandi, et medius terminus demonstrationum sit opportunus, vel importunus, id est, an utilitatem aliquam afferat, vel pitius detrimentum et damnum. Credidit Plato Mathematicas, ad speculationes physicas apprime esse accomodatas. Quapropter passim eas adhibet in reserandis mysteriis physicis. Ac Aristoteles omnino secus sentire videtur, erroresque Platonis adscribet amori Mathematicarum . . . Sed si quis voluerit hanc rem diligentius considerare, forsan, et Platonis defensionem inveniet, videbitque Aristotelem in nonnullos errorum scopulos impegisse, quod quibusdam in locis Mathematicas demonstrationes proprio consilio valde consentaneas, aut non intellexerit, aut certe non adhibuerit. Utramque conclusionem, quarum prima ad Platonis tutelam attinet, secunda errores Aristotelis ob Mathematicas male rejectas profitetur, brevissimis demonstrabo ». Cf. *ibid.*, p. 190 : « Nunc. . . videamus, quomodo Aristoteles ob non adhibitas opportunis locis mathematicas demonstrationes, maxime recesserit a vera philosophandi ratione. Ille itaque in quarto libro Physicorum multis rationibus probans vacuum non posse dari, illud inter cetera dicit, nempe quod si daretur vacuum, in eo motus fieret in instanti. Existimat enim successionem in motu ex medij, quando a mobili dividitur, resistentia provenire. Ita ubi medium majorem habet resistentiam : ibi mobile diutius moretur, ubi minorem, minus. Et ideo ubi nullam inveniet resistentiam, momento fiet motus. Hanc Aristotelis opinionem omnino falsam, et absurdam esse demonstrant Mathematici, quorum rationes ego compendio colligam. Illud itaque ; in primis supponunt ex libro Archimedis *de insidentibus* motum prodire a virtute motrice. Virtus autem deorsum impellens corpora est gravitas, quemadmodum et illa, quae rursus attolit corpora gravia, est vis corporis gravioris extrudens minus grave ex demonstratis ab Archimede in principio eiusdem libri *de insidentibus* ». Marginal note: '*Johannes Baptista Benedictus in disputationibus contra Aristotelem*'. Cf. *Dialogo*, p. 423 (*Dialogue*, p. 397).

286 Cf. E. Strong, *Procedures and Metaphysics*, chapter 4, p. 91 ff.

287 *Dialogo*, p. 38 (*Dialogue*, p. 14). Cf. p. 256 (*Dialogue*, p. 230).

288 The question under discussion is that of the number of the dimensions of space.

289 *Dialogo*, p. 38 (*Dialogue*, p. 14).

290 *Dialogo*, p. 249 (*Dialogue*, p. 222).

291 Similarly the science of music is possible. A mathematical science of music is possible because music obeys the laws of number. The error of Pythagoras and Plato was, then, that they extrapolated, and asserted the universal rule of mathematics, and failed to understand that mathematics comes to an end where matter begins.

292 *Dialogo*, p. 229 (*Dialogue*, p. 203). Cf. p. 423 (*Dialogue*, p. 397).

293 The matter under discussion is centrifugal force.

294 Cf. *Dialogo*, p. 423 (*Dialogue*, p. 397).

295 There are indeterminate realities; there are statistical concepts. It is just as absurd to seek to *precisely* determine the shape of a cloud as it is to calculate *precisely* a town's

population or the average temperature, to the nearest tenth of a degree, of a geographical region. Cf. the profound remarks by Gaston Bachelard, *La formation de l'esprit scientifique*, Paris, 1937, p. 216 ff.

296 Leibniz, Letter to Foucher, about 1668, *Philosophische Schriften*, ed. Gerhardt, vol. I, p. 392: 'I hold it to be demonstrable that bodies do not have exact shapes'.

297 *Dialogo*, p. 233 (*Dialogue*, p. 207), marginal note: 'Things in the abstract have precisely the same requirements as in the concrete'.

298 Cf. *Dialogo*, p. 234 ff. (*Dialogue*, p. 208 ff.).

299 Kepler similarly asserts this; see above, note 70.

300 *The Assayer, op. cit.*, p. 237: 'Philosophy is written in this grand book, the universe, which stands continually open to our gaze. But the book cannot be understood unless one first learns to comprehend the language and read the letters in which it is composed. It is written in the language of mathematics, and its characters are triangles, circles, and other geometric figures without which it is humanly impossible to understand a single word of it; without these, one wanders about in a dark labyrinth'. Cf. Letter to Liceti, Jan. 11 1641, *Opere*, vol. XVIII, p. 293.

301 *Discorsi*, 'First Day', p. 51: 'Since I assume matter to be unchangeable and always the same, it is clear that we are no less able to produce demonstrable proofs in relation to this constant and invariable property than if it belonged to simple and pure mathematics'.

302 Evangelista Torricelli, *Opera Geometrica*, Florentiae, Typis Amatoris Massae et Laurentii de Landis, 1644, II, p. 7 : « Sola enim Geometria inter liberales disciplinas acriter exacuit ingenium, idoneumque reddit ad civitates exornandas in pace et in bello defendendas: caeteris enim paribus, ingenium quod exercitatum sit in Geometrica palestra, peculiare quoddam, et virile robur habere solet : praestabitque semper, et antecellet, circa studia Architecturae, rei bellicae, nauticaeque, etc. »

303 Galileo's notes to Antonio Rocco, *Esercitationi filosofiche, Opere*, vol. VII, p. 744: '« ridottovi a memoria il detto del Filosoto, che ignorato motu ignoratur natura, guidicate con giusta lanze sig. Rocco, qual de' dua modi di filosofare cammini più a segno, o il vostro, fisico puro e semplice bene, o il mio, condito con qualche spruzzo di matematica ; e nell' istesso tempo considerate chi più giustamente discorreva, o Platone, nel dire che senza la matematica non si poteva apprender la filosofia, o Aristotele, nel tassare il medesimo Platone per troppo studio della geometria. »

304 *Dialogo*, p. 35 (*Dialogue*, p. 11), marginal note: 'Legendary character of Pythagorean number mysteries'.

305 *Dialogo*, p. 129 ff. (*Dialogue*, p. 103 ff.).

306 It is not really necessary to emphasise the kinship of inspiration of these texts with those of Descartes.

307 Cf. Descartes, *Regulae ad directionem ingenii*, VII, *Oeuvres de Decartes*, eds. C. Adam and P. Tannery, vol. X, p. 388.

308 *Dialogo*, p. 183 (*Dialogue*, p. 157). Cf. above, pp. 159, 167, 192 ff.

309 *Dialogo*, p. 217 (*Dialogue*, p. 190).

310 *Dialogo*, p. 333; *Discorsi*, p. 269 ff. [These references seem to be incorrect, but I am unable to identify which passages the author has in mind. *Trans.*]

311 Cf. *Discorsi*, 'Third Day', p. 212.

312 Bonaventura Cavalieri, *Lo Speccio Ustorio overo Trattato Delle Settioni Coniche e alcuni loro mirabili effetti intorno al Lume, Caldo, Freddo, Suono e Moto ancora*, Bologna, presso Clemente Ferroni, 1632, p. 152 ff. : « Ma quanto vi aggiunga la cognitione delle scienze Mathematiche, giudicate da quelle famosissime scuole de'Pithagorici, e de'Platonici, sommamente necessarie per intender le cose Fisiche, spero in breve sarà manifesto, per la nuova dottrina del moto promessaci dall'esquisitissimo Saggiatore della Natura, dico dal Sig. Galileo Galilei, ne'suoi Dialogi, protestando io haver'hauuto e motivo e lume ancora in parte intorno à quel poco, ch'io dico del moto in questo mio Trattato, per quanto alle settioni coniche si aspetta, da i sottilissimi discorsi di quello, e del Reverendiss. P. Abbate D. Benedetto Casteli Monaco Cassinenze, Matem. di N. S. e molto intendente di queste materie, ambidue miei Maestri. Rimetto dunque il Lettore in ciô, ch'io supporô al dottiss, libro, che da si grand'ingegno in breve dourà porsi in luce, e si contenterà di questo poco, ch'io dirô per manifestare, che cosa

habbino che fare le Settioni Coniche con cosi alto, e cosi nobile soggetto. »

[313] The second of the objections was by far the most important. For Galilean and Cartesian Platonisms could admit that they were unable to explain qualities; they could avoid the problem by consigning them to the realm of subjectivity. But they could not subjectivise motion.

[314] Cf. F. Bonamico, *De Motu*, Book I, chapter 11, '*Jurene mathematicae ex ordine scientiarum expurgantur*', p. 56: « . . .Itaque veluti ministrae sunt mathematicae, nec honore dignae, et habitae προπαιοεία id est, apparatus quidam ad alias disciplinas. Ob eamque potissime caussam, quod de bono mentionem facere non videntur. Etenim omne bonum est unis, is vero cuiusdam actus est. Omnis vero actus est cum motu. Mathematicae autem motum non respiciunt. Haec nostri addunt. Omnem scientiam ex propriis effici : propria vero sunt necessaria quae quatenus ipsum et per se insunt. Atqui talia principia mathematicae non habent. . . Nullum caussae genus accipit . . . propterea quod omnes caussae definiuntur per motum : efficiens enim est principium motus, finis cuius gratia motus est, forma et materia sunt naturae ; et motus igitur principia sint necesse est. At vero mathematica sunt immobilia. Et nullum igitur ibi caussae genus existit. »

[315] *Discorsi*, 'Third Day', p. 190: « *De Motu locali.* De subiecto vetustissimo novissimam promovemus scientiam. MOTU nil forte antiquius in natura et circa eum volumina nec pauca nec parva a philosophis conscripta reperiuntur ; symptomatum tamen, quae complura et scitu digna insunt in eo, adhuc inobservata, necdum indemonstrata, comperio. Leviora quaedam adnotantur, ut, gratia exempli, naturalem motum gravium descendentium continue accelerari ; verum, juxta quam proportionem eius fiat acceleratio, proditum hucusque non est : nullus enim, quod sciam, demonstravit, spatia a mobile descendente ex quiete peracta in temporibus aequalibus, eam inter se retinere rationem, quam habent numeri impares ab unitate consequentes. Observatum est, missilia, seu proiecta, lineam qualitercunque curvam designare ; verumtamen, eam esse parabolam, nemo prodidit. Haec ita esse, et alia non pauca nec minus scitu digna, a me demonstrabuntur, et, quod pluris faciendum censeo, aditus et accessus ad amplissimam praestantissimamque scientiam, cuius hi nostri labores erunt elementa, recludetur, in qua ingenia meo perspicaciora abditiores recessus penetrabunt. »

Cf. *Dialogo*, p. 248 (*Dialogue*, p. 221): '. . . the movement of descending bodies is not uniform, but starting from rest they are continually accelerated. This fact is known and observed by all, . . . But this general knowledge is of no value unless one knows the ratio according to which the increase in speed takes place, something which has been unknown to all philosophers down to our time. It was first discovered by our friend the Academician, who, in some of his yet unpublished writings, shown in confidence to me and to some other friends of his, proves the following. The acceleration of straight motion in heavy bodies proceeds according to the odd numbers beginning from one. That is, marking off whatever equal times you wish, and as many of them, then if the moving body leaving a state of rest shall have passed during the first time such a space as, say, an ell, then in the second time it will go three ells; in the third, five; in the fourth, seven, and it will continue thus according to the successive odd numbers. In sum, this is the same as to say that the spaces passed over by the body starting from rest have to each other the ratios of the squares of the times in which such spaces were traversed. Or we may say that the spaces passed over are to each other as the squares of the times.
Sagredo: This is a remarkable thing that I hear you saying. Is there a mathematical proof of this statement?
Salviati: Most purely mathematical, and not only of this, but of many other beautiful properties belonging to natural motions and to projectiles also. . . .'

[316] Cf. the present author's contribution to the IXe *Congrès International de Philosophie*, Paris, 1937, vol. II, p. 41.

[317] Descartes' idea of the history of philosophy does not always coincide with ours. Cf. *Principes, Oeuvres*, vol. IX, p. 5: 'The earliest and main authors of whom we have the writings are Plato and Aristotle, between whom there was no difference except that the former, following in the footsteps of his master, Socrates, ingenuously confessed that he had not been able to discover anything with certainty, and was content to write things

which seemed to him to be plausible, and in this way thought up several principles in terms of which he tried to explain other things: Aristotle, on the other hand, was less frank, and while he had been Plato's disciple for twenty years and had no principles other than his, he completely changed the manner of their presentation and put them forward as true and certain, even though there had never been any suggestion that Plato had believed them to be so'.

[318] Cf. G. Milhaud, *Les Philosophes-Géomètres de la Grèce,* Paris, 1900, p. 292, and L. Robin, *Platon,* Paris, 1935, p. 234.

[319] *Discours de la Méthode, Oeuvres,* vol. VI, p. 7.

[320] Cf. Ch. Clavius, S. J., *Opera mathematica,* Moguntiae, 1611, vol. I, *Prolegomena,* p. 5: « Cum igitur disciplinae mathematicae veritatem adeo expetant, adament excolantque, ut non solum nihil quod sit falsum, verum etiam nihil quod tantum probabile existat, nihil denique admittant quod certissimis demonstrationibus non confirment, corroborentque, dubium esse non potest quin eis primus locus inter alias scientias omnes sit concedendum'; quoted by E. Gilson, *Discours de la Méthode, Texte et Commentaire,* Paris, 1925, p. 128.

[321] Descartes, Letter to Mersenne, 11 March 1640, *Oeuvres,* vol. III, p. 39 ff.: 'In physics I believe I know nothing unless I not only know how things can be but can also prove that they cannot be otherwise: for having reduced physics to mathematical laws this is now possible, and I believe I can do it in relation to all that little that I believe I know of it, though I have not done this in my Essays because I did not want to give my principles there, and I still cannot see anything which would persuade me to do so in the future'.

APPENDIX
THE ELIMINATION OF GRAVITY

I THE GALILEANS

We are now faced with a problem, a problem which we mentioned at the beginning of this study: if, as we believe we have shown, Galileo did not formulate the principle of inertia, how was it that his successors and followers were able to believe that it could be found in his work? There is another problem: if, as we believe we have also shown, Galileo not only did not conceive but could not have conceived of rectilinear inertial motion, how did it happen, or rather how was it brought about, that this concept, at which even the mind of a Galileo had baulked, could seem to his followers and successors to be simple, self-evident, to be taken for granted?

In our view this latter question provides the answer to the former. For it was precisely because the concept of inertial motion, i.e., of perpetual rectilinear motion, seemed self-evident and clear to Galileo's followers and successors, and subsequently to many historians, that they believed that they saw it formulated and asserted in the master's work. Now it should be said straight away that if this concept could seem self-evident to them, if they were able to overtake Galileo on the road that led to the geometrisation of space and to the mathematisation of reality, and were able almost without noticing it to break the last bond, the bond of gravity, which had kept Galileo shackled to the rock of physics, and to fly off freely into the skies of mathematical entities, they were able to do this only because of Galileo, because of Galileo's example, because of what Galileo had taught them, because of the education they had received from him. Therefore, they were not altogether wrong in attributing to Galileo a discovery which he had not made, and in finding in his work something which, while certainly not there explicitly, was there in embryo.

Let us look at this in more detail; and to do this let us allow the Galileans to speak for themselves.

1. Cavalieri

Cavalieri's *Speccio Ustorio* was published in 1632, the same year as the *Dialogue*. But what a difference in style! If one were to date it on internal evidence one would believe that it had been written twenty years later. Galileo's book, as we have seen, was polemical and combative. Cavalieri's book is simply a scientific work. One has the feeling that for Cavalieri Galileo's great battle belongs to the past, and that Galileo's victory is so complete that it is no longer even a subject for discussion. The central philosophical problem—Plato or Aristotle, mathematics or perceptual experience—is already resolved. It is taken for granted that physics is mathematics, and the move from the purely geometrical study of curves and conic sections to that of their 'effects' in physical reality is an untroubled one. It is scarcely noticeable. Thus the investigation of motion, of the motion of fall, of projectile motion, is conceived from the outset as a mathematical investigation; the bodies which Cavalieri now sets in motion are mathematical bodies.

They are, of course, 'heavy bodies'. Cavalieri does of course speak of the 'internal gravity' of bodies. But this internal gravity, which one cannot but accept, is no longer thought of as something inseparable from the physical body. Even though Cavalieri still refers to gravity as 'internal', it is, for him, completely externalised: and, precisely as a result of this, all distinction between 'natural' and 'violent' motion is definitively absent from his thought. But let us listen to Cavalieri himself.[1]

'While, in relation to heavy bodies, there is a great variety of possible considerations which could be made, all fine and all interesting, yet we do not attempt anything more than to determine the nature of the path which is followed by the heavy body, first when it is moved by internal gravity, then when it is projected, and finally when it is moved by both of these at the same time, in order to find out whether the conic sections are involved here and, if this is so, to see which ones.

'Now I say that if we consider the motion of the heavy body which results from internal gravity alone, and regardless of the manner of its action, the body will always be led straight towards the universal centre of heavy things,[2] that is towards the centre of the earth, and all heavy bodies are universally guided towards this centre. . . .

'Next I say that if one considers a body which is projected towards some given place, and if there is no other motive force which pulls it in another direction, then it will go to the selected place if it is projected towards it in a straight line and is moved by the force impressed on it solely in a straight line; and that it is not reasonable that the moving body will move off this straight line, because there is no other motive force to deflect it from it. Thus, for example, a cannon-ball leaving the mouth of the cannon, if it has no other [motive force] than that impressed on it by being fired, will go from the place where it is fired straight towards a point located on a line which is an extension of the cannon's barrel; but because there is another motor, namely the internal gravity of the ball, it follows that it will be forced to deviate from this straight line to go towards the centre of the earth.'

Thus if its internal gravity did not pull the cannon-ball towards the centre of the earth, its motion would take place in a straight line. It seems at first sight that there is nothing new, nothing of note, in Cavalieri's assertion. For it had always been said that violent motions are rectilinear. It had even been believed that they are like this in reality, and that a cannon-ball, on leaving the mouth of the cannon, moves first in a straight line. Had it not been precisely one of Galileo's great discoveries, that a projectile's trajectory is curved from the very first instant that it is projected? Cavalieri knows this perfectly well. But we should be careful. This in no way prevents him from imagining a cannon-ball *without gravity*, not acted upon by the force of gravity, and moving *solely* under the influence of the cannon's fire. This supposition has, in his eyes, nothing absurd about it, nor is it impossible. Internal gravity acts on the ball just like any other force, with the single exception that it acts uniformly, that it is an *unchanging force*, and one can abstract from it just as from any other.

So Cavalieri continues:[3] 'I say furthermore that not only will the projectile go in a straight line towards its target, but that in equal times it will not cover more than equal distances on this line, given that the moving body is indifferent to the direction of motion, and given also that the medium offers it no resistance, because there would be nothing to cause either retardation or acceleration.' If we recall the struggle of Galileo's thought to overcome the conception of impressed force which equates it with the cause of the motion, if we recall the long and laborious arguments by which he attempted to persuade us of the uniformity of the motion of a heavy body on *a horizontal plane*, then we will appreciate, and will realise the merit of, the conciseness of Cavalieri's argument, the argument of a man who has long since got used to the idea of motion as a real entity which persists by itself and remains constant as long as nothing destroys or changes it; then we will understand that, since *gravity has become for him a separable force*, this uniform motion will continue *along a straight line* and not in a circle. 'Thus a heavy body moved [only] by internal gravity will go nowhere but towards the centre of the earth, whereas one which is moved by a force impressed on it can advance in any direction whatever.'[4] Here again the difference from Galileo, and to be honest the advance beyond Galileo, is obvious. To get his bodies to move 'in any direction whatever' Galileo, even when he wrote the *Discourses*, needed to have them supported by imaginary planes so as thereby to counteract the unavoidable action of gravity. There is nothing like this in Cavalieri: in order to eliminate the action of 'internal' gravity it is enough simply to abstract from it: and to investigate the concrete motion of a cannon-ball it is enough to have both forces, that of gravity and that of projection, act on it simultaneously and to calculate the outcome simply by aggregating the 'partial' effects of each of them taken separately, because it is *self-evident* that these two forces, i.e., these two motions, have no influence on each other.

So, to continue:[5] 'If there are two motive forces on the projectile, gravity and the impressed force, then each of them separately would move the body in a straight line, as has been said, but combined together they do not make it move in a straight line except in the two following cases: first, if the body is projected by the impressed force perpendicular to the horizon; second, if not only the impressed force but also gravity moved the body uniformly, because the distances moved towards the centre of the earth in equal times . . . would always be equal, as would also be the distances covered by the body in these same times along the line of projection; thus the body would always be on a straight line. But if one of the two is not uniform then the body moved by both gravity and by the impressed force would not move in a straight line, but on a curve, the quality and condition of which would depend on the respective uniformity and non-uniformity of the combined motions. For gravity is certainly acting on the body which, being impelled by the projector is led to move in some direction or other, for example on a line at an angle to the horizon, but this gravity does nothing else than draw the body away from this straight line, and has nothing to do with the other motion, except in as much as the body moves away from the centre of the earth while the gravity impresses on it a tendency towards its centre, just as towards any other place: so the motion [produced by gravity] is indifferent to that given to the body by the projector, and in the absence of resistance from the medium this latter would be uniform'.

It is scarcely necessary to point out that the motion of projection and that of fall are treated identically; and this identity of treatment goes so far as to use the same term for them both. It is clear that for Cavalieri all motions are of the same nature and that the distinction between 'violent' and 'natural' is simply a matter of terminology. In fact he says this explicitly:[6] 'It remains for us to consider the motion towards the centre of the earth of the body moved by internal gravity, which is said to be natural motion, and the motion away from it, resulting from an impulsion conferred on the body, which is called violent motion. The body starting from rest and moving towards the centre moves with continuous acceleration to the extent that it moves toward the centre, or rather to the extent that it moves away from its point of departure,[7] and the violent motion, that away from the centre, has a continuous retardation'. Now while the philosophers had always been familiar with this fact, it was only Galileo who had, in his *Dialogue*, determined the precise rate of acceleration and deceleration, and it was this in turn which enabled Cavalieri to prove, using in his proof a method of calculation which he had himself invented, that the trajectory of any body projected in any direction whatever is a conic section, and even that it is a parabola.

The modern reader will probably be disappointed. Perhaps he will even accuse us of being the victim of the same optical illusion that we have sometimes diagnosed in certain historians of Galileo. He might say to us that if Cavalieri was really at home with the principle of inertia then he would have stated it as such, as a fundamental law of nature, as a fundamental axiom of mechanics, as did Descartes and Newton. He would not have limited himself to stating in passing something in which *we* can see an expression of the principle of inertia, but which nobody, *not even Galileo*, was able to recognise. This is certainly possible. It is possible that Cavalieri did not himself understand the significance of his formulation; in fact he does not say that motion, once started, will persist *indefinitely*; and it is true that Galileo read the *Speccio Ustorio* but derived no benefit from it, and that in his *Discourses* he stated the principle of the persistence of motion with the qualifications with which we are now familiar. It is possible that he saw in it only the effect, or an example, of that thorough-going mathematisation that he had himself outlined in the *Dialogue*. It is also possible that Cavalieri himself saw no more in it than this.

Objectively Cavalieri's formulation contains the principle of inertia. But does it do so subjectively? This can be doubted. It must be doubted even.

Now this very fact, the fact that Cavalieri leaves us uncertain as to his real thought, the fact that, in any case, he was unable to give to the principle of inertia the status and significance which it deserved, makes clear for us the role and importance of the work of Descartes. For what we have just said in relation to Cavalieri could also be said, though with certain qualifications, of Torricelli.

2. Torricelli

For Torricelli also does not state the principle of inertia *as a principle*. He formulates it, as it were, in passing, just as Cavalieri did, in the context of his investigation of projectiles. 'Let the body be projected from point *a*', he says, 'in any direction AB, inclined to the horizon. *It is clear that without the pull of gravity the moving body would proceed with a uniform, rectilinear motion along the line of direction AB*'[8]

It is fascinating to observe the change in scientific attitude since Galileo, and even since Cavalieri. '*It is clear that . . .*'; this is all that is necessary, in Torricelli's judgement, in introducing the principle of inertia. But, as in the case of Cavalieri, it might be asked whether this really is the principle of inertia. After all, Galileo knew perfectly well that if bodies were not drawn downwards by gravity they would move, indefinitely even, in a straight line. But he also knew that this does not happen and never could happen. Torricelli also knows this. Thus, he continues: 'But internal gravity acting internally, the body begins immediately to deviate from the direction of projection and, the extent of this deviation continuously increasing, it follows a certain curve'. Here again Torricelli's manner might be admired: it is unnecessary to spend time demonstrating the independence of the motions; for Torricelli's readers, students of Galileo, this is just as self-evident as the conservation of these motions. But, here again, it must be asked whether there is anything different in Torricelli from what we have already found in Galileo. Does the phrase '*it is clear*' mean anything more than a reference to a case which is not only unreal, but physically impossible? Or, to put it another way, in the face of this physical impossibility will Torricelli simply come to a halt, as his master Galileo had done, or will he go beyond it as had done Cavalieri? In fact he does neither one nor the other. But, having deeply thought about the structure of physical science, about the conditions of application of geometry to physics, about the very essence of the 'method of resolution', or to give it at last its true name, the '$\delta\iota\alpha\iota\rho\eta\sigma\iota\varsigma$', that he saw at work in Galileo and Kepler, Torricelli came to understand the physical impossibility of the motions studied by rational mechanics, but claimed for geometry the right to extend to the limit its analysis of the real, i.e., to extend it to the unreal, and even to the impossible.

Like all the Galileans, like Galileo himself, Torricelli was an

Archimedean.[9] 'Among all the works concerning the mathematical disciplines', he says, 'it seems that the discoveries of Archimedes, which amaze the mind by the miracle of their subtlety, can claim the first place.' Now, while the mathematical genius of Archimedes is accepted by everybody, the science founded by him, i.e., mechanics—we might remain faithful to Torricelli's intention if not to his exact words in saying, mathematical physics—has been attacked as being based on two false presuppositions.[10] Archimedes, in fact, accepts or presupposes as true two things which are obviously false, namely, 'first, *that gravity is attributed to surfaces, whereas they do not in fact have it*; and second, *that the threads by which weights are hung on a balance are equidistant, whereas in reality they must meet at the centre of the earth*'. As for me, Torricelli goes on, 'in my opinion either none of these propositions is false, or all the other principles of geometry are as false as they are, and in the same way. For it is false that a circle has a centre, that a sphere has a surface, and a cone solidity. I am speaking of abstract figures, those which are studied in geometry, and not of physical and concrete figures. We must therefore admit that the centre of the circle, the sphere's surface, the cone's solidity, and other things of this kind, which are not subject to dispute, have no existence other than that which they have by definition and by the intellect. Thus gravity is in geometrical figures in exactly the same way as are centres, perimeters, surfaces, solidity, etc.' It can be seen clearly that for Torricelli mechanics is simply a part of geometry. It is not a matter of investigating the phenomena of the physical world, the motions of real bodies, acted upon by real forces; it is not a matter of explaining free fall, or gravity. Gravity, within Torricelli's science, is not a 'quality' or a 'power' of 'heavy' bodies; it is a magnitude, or to use Torricelli's own term, a dimension. No doubt it is a different kind of dimension from those of length, breadth and thickness. But the geometer treats it exactly as he treats the others, without being concerned about the physical possibility of the objects which he studies. Therefore, there is nothing to prevent him from withholding gravity from a 'body', and attributing it to a surface or a line. We are not in the physical world; we have been carried over into the Archimedean world of realised geometry, and the 'bodies' of this world are neither more nor less real than are its lines without breadth or its surfaces without thickness. Reasoning in mechanics is no different in nature from that in geometry. As in the latter so also in mechanics one is free to define its objects and to confer being on them by definition. One can even 'by mechanical reasoning give birth to [geometrical] figures by means of new definitions'.[11] Thus, for example, one can define the square 'as a quadrilateral of which the particular points, since it is equiangular and equilateral, have the 'moment' to move towards some part of the world along lines parallel to each other'.[12] Which means, unless we are mistaken, that it is impossible to separate mechanics from geometry, because the idea of motion is an idea used by geometry in its definitions.[13] 'This should be enough to eliminate all reason for doubting the value and the truth of Archimedean science among those who do not accept his mechanics in the spirit in which it ought to be accepted.'[14]

This is all by way of answer to the first of the objections against Archimedes, i.e., that he attributed weight to geometrical figures. As for the second:[15] 'I now come to the second presupposition which has been judged to be false. It is a very common objection, and even the most serious people have charged that Archimedes presupposed *something false when he asserted that the threads from which weights are hung on the arms of a balance are parallel to each other, while in reality they must meet each other at the centre of the earth.* As for me (I say this with no disrespect for these illustrious men), I believe that the foundations of Mechanics must be viewed in quite a different way. I agree that if physical magnitudes hang freely on the balance then the material supporting threads will be convergent, because each of them is drawn to the centre of the earth. Nevertheless, if this same balance, even though corporeal, were considered to be not on the earth's surface but in the highest regions beyond the sun's sphere, then the threads, while still drawn to the centre of the earth, would be very much less convergent to each other, would be quasi-equidistant. Let us imagine a mechanical balance transported beyond the starry balance [i.e., the constellation of that name] in the firmament, to an infinite distance. It will be understood by everybody that the suspension threads would no longer be convergent, but would be exactly parallel. Thus, when I consider a balance which weighs geometrical figures I do not think of it as being between the covers of a book in which I can see a picture of it; and I do not believe that the point towards which these magnitudes tend is the centre of the earth; but I imagine the balance to be infinitely distant from the point towards which heavy bodies on it tend'.

The separation from physical reality, the geometrisation of space, the identification of physical space with geometrical space which had been, according to Torricelli, incompletely achieved by Archimedes, 'is now accomplished. Physics = mechanics; mechanics = geometry. So Torricelli has no hesitation in transporting his 'corporeal' balance into 'imaginary' space beyond the sphere of the stars, to a *really infinite* distance. Geometrical space *is* infinite; and now at last the space of mechanics, and hence of physics—regardless of the actual dimensions of the real Universe—have become infinite in their turn. Torricelli's 'abstract' space is a more attractive version of Bruno's infinite Universe. But let us listen to Torricelli:[16] 'If, after it is placed there [at an infinite distance] and after certain of its geometrical relations have been deduced, we then, in imagination, bring back this balance to our own regions, I admit that the equidistance of the suspension threads will be destroyed; but the ratios of the figures which have been deduced will not thereby be destroyed. There is a particular advantage for the Geometer in carrying out all his operations, with the aid of this abstraction, by means of the intellect. Who, then, would wish to prevent me from freely considering figures hanging on a balance imagined to be at an infinite distance beyond the confines of the world? Or who would prohibit the consideration of a balance located on the surface of the earth of which, however, the abstract magnitudes [weights] no longer tend toward the earth's middle point, but towards the centre of the Dog star or of the Pole star?'

There is, in fact, no reason for limiting the freedom of the study of mechanics-geometry once we have been advised that the balance located on the earth's surface is mathematical and not real, and that abstract weights-magnitudes are attached to it. 'The triangles and the parabolas and even the spheres and cylinders of the Geometer, being in themselves indifferent to motion, no more tend towards the centre of the earth than towards the centre of Saturn. Therefore all advantage is lost if these figures are seen as tending solely towards the centre of the earth.'[17] The operation described by Torricelli consists, in fact, in the replacement of the real, physical body by an 'abstract', mathematical body (which implies the transformation of natural gravity into a freely variable 'magnitude' or dimension) and in the reinsertion of this 'body' into the spatial framework of the real world. To limit the possible direction of gravity, to confine it, or rather to reconfine it, to the centre of the earth, would be to throw away everything that had been gained from the operation. So Torricelli continues:[18] 'Why then should I not be permitted to consider the points of any figure whatever as being endowed with a virtue such that they all tend with equal moments along parallel lines towards any place in the world?' For this 'motive force' or 'virtue' is no more than a dimension or a magnitude that one can *add* to the parts as one wishes; there is no need to be able to put them there. 'If we assume these facts to be true, in the same way that the properties which are attributed to figures in and by definitions are true, then similarly all the theorems will be true which will be deduced from them, by arguments in Mechanics, by those who perform this abstraction, and they will not at all be proved by the use of false propositions'—and this because, as Torricelli has just explained to us, the basic propositions, the assumptions, do not refer to perceptible, physical reality in the established sense of the term, but to an abstract, mathematical 'reality' which replaces it.

'Therefore', Torricelli goes on to say,[19] 'the foundation of Mechanics, namely the proposition that the threads of the balance are parallel, could be said to be false if the magnitudes attached to the balance were real, physical things, tending towards the centre of the earth. But it will not be false when the magnitudes (whether they be abstract or concrete) tend neither to the centre of the earth nor to any point close to the balance, but towards some infinitely distant point.'

3. Gassendi

As E. Wohlwill has quite rightly pointed out, Gassendi's work was heavily influenced by that of Galileo, far more heavily than Gassendi himself admitted.[20] Gassendi's merit was nonetheless very great. He had a deep *understanding* of Galileo. By this we mean that he understood, and clarified, the *ontology* which constituted the infrastructure of the new science. Moreover, thanks to Democritus and, strangely, thanks to Kepler, he succeeded in ridding himself of the last obstacles deriving from tradition and from common sense which had hindered the progress of Galileo's thought, and thereby won for himself the everlasting glory of having been the first to *publish*—if not the first to state—a correct formulation of the principle of inertia. It is, therefore, extremely instructive to study his thought. Moreover it seems to us that it entirely confirms the explanation of Galileo's failure which we have elaborated above.

Unlike Cavalieri and Torricelli, Gassendi was not a mathematician.[21] It was the physical aspect, and even the physical mechanism, of the phenomena investigated by Galileo which interested him and which he wished to understand. He was not wrong in this, as we shall see below; being able to explain gravity enabled him to abstract from it.

Furthermore, Gassendi did not have the imperious attitude towards experiment that Galileo had. Thus it is with the account of an experiment that he begins; it is the famous experiment of the fall of a cannon-ball dropped from the top of the mast of a moving ship.[22] As was mentioned above[23] he had performed this experiment himself, and he uses it now so as to deduce from it the two fundamental principles of the new science; the principles of the relativity and interdependence of motions.

The experiment refutes traditional doctrine. The stone falls at the foot of the mast. Gassendi explains at length to his correspondent how it comes about that the stone which, as a result of the combination of the motions by which it is impelled,[24] and regardless of whether it is thrown from below upwards or from above downwards, in fact describes a complex motion, namely a parabola,[25] appears to us to move in a straight line. The reason is that only relative motion is observable. Now we ourselves are carried along by the moving ship. Therefore,[26] 'it is not surprising that to all of us who were in the galley, the motion seemed to be vertical; for only the stone's downwards motion was observable by us; for the forwards motion was not observable because it was common to us and the stone'.

Gassendi is well aware that a traditionalist would not be convinced, or at least would not be satisfied, with this explanation. For it does not matter to him whether the horizontal motion is observable or not. It exists, and for Gassendi's explanation to work it must be able to combine with the motion of fall or of projection without the two motions being able to mutually impede each other. It can be agreed that this is possible for two violent motions. But how could a violent motion combine, without producing an impediment, with the natural motion? Gassendi replies by denying the validity of the traditional distinction. Not that he is completely opposed to using the two terms. In his opinion they are useful in referring to the difference between those motions which take place spontaneously, or at least without opposition, and those which are opposed by the nature of the moving body: 'Thus, the trajectory of a sphere through the air is violent . . . whereas its rotation on a plane is natural, since nothing opposes it'.[27]

But if one wishes to give this distinction some deeper meaning, the consequences would be quite different from those which are taught by traditional physics. For,[28] 'it seems that there is no motion, except the very first, which could not be considered violent: for there is none which takes place except by the impulsion of one thing by another; this is why Aristotle looked for an external motor even for falling things'. One could, of course, invoke the well known saying: 'Nothing violent can be perpetual'. Now, this saying does not seem self-evident in the slightest to Gassendi; in his opinion it is without foundation, and a perpetual violence is not at all an absurdity.[29] But let this pass. Let us suppose that the saying is correct. It would follow that, conversely, everything natural must be perpetual, and from this it would follow that the motion of fall could never be considered a natural motion, if only because it is not uniform, for 'it is clear that the origin of perpetuity is uniformity, and the origin of cessation is non-uniformity; for only that which neither increases nor decreases can persist: and nothing can infinitely increase or decrease by the force of nature. Therefore if one were to seek among these composite things a motion which is natural to the maximum, it would be clearly seen to be in the heavens; for this motion is above all others uniform and perpetual, being chosen by the creator to be circular in form; this form having neither beginning nor end can be uniform and perpetual'.[30]

Thus it is the circularity of celestial motions which explains their uniformity and hence their perpetuity: their circularity by itself. Therefore circular motions on the earth, and horizontal motion in particular, will share this uniformity, eternity and naturalness.

So Gassendi continues:[31] 'I do not rehearse here how the stone [dropped] from the top of the mast while the ship is in motion apparently falls only vertically, whereas it really follows the oblique line which we have described: [I will say] only that the stone does not move of itself, but is moved by the force impressed on it by the hand, which derives from the translation of the hand by the ship which it follows together with the mast. Therefore it comes to the same thing, whether the hand drops the stone from the top of the mast or whether the stone falls from the top of the mast having been thrown upwards from its base. This is why the downward motion of the stone can be called violent, as much as the upward motion. It might be said that the oblique motion, being mixed or composed of the vertical and the horizontal [motions], . . . could be considered violent, but not the vertical [motion], this being natural. In fact it is obvious that [the motions of] the stone when thrown upwards and when nevertheless advancing obliquely, are both violent, since the cause is external in both cases, and is the very force of the ship and the hand's own force. But it is not so obvious that its motion when dropped downwards, and when it nevertheless advances obliquely, is also violent in respect of both [components]: for while the cause of the horizontal [motion] is still external, namely the impulsion or force of the ship, the cause of the vertical [motion] is no longer the hand's own force. This is why it seems necessary that this motion of the stone is caused by an internal principle, and that its motion is therefore not violent but natural. However, it seems worth considering first of all that if one of these two motions which make up the oblique [motion], namely the vertical and the horizontal, were to be considered natural, it would be the horizontal rather than the vertical. For, since the projectile is one part of a whole which is moving horizontally, i.e., in a circle, it also moves in a circle, in imitation of this whole, and it therefore moves naturally and quite uniformly. Thus the vertical motion increases or decreases, while the horizontal on the other hand always flows evenly and continues without variation. This might seem less surprising if it were a matter of the earth's motion (assuming it to be moving around on its axis); for then one could say that the stone moved uniformly because it spontaneously follows the motion of the whole, whether it be attached to it or separated from it. But, no doubt, what is surprising is that it is impressed by the ship's passage, or by something else, or by the hand itself, because the stone's relations with these things or their motions is not the same. From all this it can be concluded that horizontal motion, from whatever cause it arises, is by its nature perpetual, unless a cause intervenes to deflect the moving body and perturb its motion'. To be convinced of this it is enough to think of a body in motion, there being no cause of deviation or perturbation: an example would be a perfect sphere, perfectly smooth, made of some homogeneous substance, and placed on the horizontal, i.e., on 'the ambit of the earth'. 'Imagine that a motion is impressed on it, however weak: it will be understood that this motion will never stop, but that the sphere, having completed one circuit will complete a second and will again cover the whole circuit, and having done this it will do it again and again, and will continue to do so indefinitely.'

Gassendi explains that a perfect sphere rolling on a horizontal surface always remains in the same position in relation to this surface: when one half of it gets lower the other gets higher to the same degree—an argument which we know to be derived from Nicholas of Cusa. Moreover this sphere, like any object moving on a horizontal plane, i.e., on a spherical surface, that of the earth, is in a special situation in relation to the earth, or more accurately in relation to its centre:[32] 'Moreover there is no reason at all why it would ever either slow down or accelerate, because its distance from the centre of the earth never either increases nor decreases; nor is there any reason why it would stop, as there would be if there were some unevenness on the surface'.

The situation is now a Galilean one; heavy bodies—and for Gassendi, just as for Galileo himself, all bodies are 'heavy' or 'have gravity'—once set in motion, conserve the motion conferred on them and move with a constant, uniform, and hence perpetual motion, given that they move in a circle 'around a centre', or more accurately around the centre of the earth or around that which is the centre for heavy bodies in general.

It is at this point that reflection on Kepler's work, and also of course on that of Gilbert, enables Gassendi to take a step, a decisive step, forward. For, to the question 'What then is gravity?', he does not confine himself, as Galileo had had to do, to replying 'It is a name whereby we refer to something of which the nature is unknown to us'. For Gassendi specifies both its positive and, more importantly, its negative nature: gravity is a force like any other; it is an attraction, something analogous to magnetic force.

It could, of course, be objected that Gassendi's step forward here is deceptive, that there is no great gain in substituting the *name* attraction for the *name* gravity, or even that there is no gain at all since the nature of the thing in question, which is designated by these 'names', is still unknown. From a certain point of view, and in particular from that of Galileo, this objection is perfectly justified. It is clear that Gassendi is as completely ignorant of this thing's nature as are Gilbert, Kepler and ourselves, and that the analogies he uses in order to give us an idea of this thing and to account for its action (various different analogies, which are moreover incompatible; little strings, little chains, hooks, the action of particles, etc.) are quite unable to do the job that Gassendi assigns to them. However, the very fact of giving an explanation, even though it is simply a matter of words, proves to be of decisive importance.

Attraction is a force like any other, and this means that it is an *external* force: and ultimately it comes down to impact, thrust or pressure. For Gassendi, just as for Descartes, there is no other way in which material forces act except by contact. No material force can act at a distance. A body can only act where it is; and no body can create motion, it can only transmit it. Gassendi says this quite clearly: all motion is caused by impulsion, and 'when I say impulsion I make no exception of attraction; for to attract is nothing other than to impel towards oneself by means of a curved instrument'.[33] Thus gravity loses its mystery, or its special ontological status;[34] and the motion produced by gravity thereby loses its exceptional character. 'Gravity, which is in the parts of the earth themselves, and in all terrestrial bodies, is not so much an internal force as a force impressed by the attraction of the earth.' How? The example of the magnet can help us to understand.[35] 'Take a small strip of iron weighing a few ounces, and hold it in the hand. If, now, a very strong magnet is placed beneath the hand, a weight not of ounces but of pounds will be felt. It will be agreed that this weight is not so much inside the iron as impressed on it by the attraction of the magnet underneath the hand, and similarly in the case of the weight or gravity of a stone or of any other terrestrial body, it will be understood that this gravity does not so much come from this body in itself as from the attraction of the earth.'

Now if this is so, if a body's gravity is only an effect of an external force, then one can easily abstract from it without having to change one's conception of what a body is considered to be in itself. Or one could say rather that since gravity is only an external effect it must therefore be excluded from our conception of the body considered in itself. We can, therefore, imagine a non-heavy body, i.e., a body not under the influence of that action of the earth which produces gravity in it. This is not even particularly difficult. For all action implies direct or indirect contact. So let us eliminate this contact, or if this is not enough, eliminate the very cause of the action.[36]

'Think of a stone in the space which we imagine to extend beyond this world, and in which God could create other worlds: would this stone set off for the earth from this place where it was formed? Would it not rather, once placed there, remain motionless, *having so to speak neither up nor down towards which it might tend, or from which it might recede*? Let us go further, and imagine that not only the earth but even the whole world is reduced to nothing, and that space is empty as it was before God created the world. Then clearly there would be no centre, and all spaces would be the same; thus the stone would not be carried here, but would remain motionless in its place. Restore the world, and the earth in it, and what would happen to the stone? Would it tend towards the earth? For this to happen the earth would have to be perceived by the stone . . .'[37] In fact it is the earth which is to attract the stone, and for this to happen some action from the earth would have to reach the stone. Therefore there would have to be some contact between them. For '. . . if some of the space of the air which surrounds us were to be made completely empty by God, and if nothing, neither from the earth nor from elsewhere, arrived to refill it, then would a stone located in this space tend towards the earth or towards its centre? Certainly no more than would the stone in the spaces beyond the world; because for this stone, which would have no communication with the earth nor with anything else in the world, it would be as if the world, and the earth, or its centre, were not, and as if nothing existed'.[38]

Thus, gravity is not only a phenomenon which is external to, and not an essential constituent of, the physical body; it is even an effect which can easily enough be eliminated (in imagination if not in reality): for in order to withdraw the action of gravity from any body one need only place it far enough away,[39] or place it in empty space.

'You ask then what would happen to this stone, which I have assumed can be conceived to be in empty space, if it were to be impelled by some force or other and hence disturbed from rest. I reply that it is probable that it would move with a uniform and endless motion: and it would move slowly or rapidly depending on whether a large or small *impetus* had been impressed on it. The proof of this I derive from the uniformity of horizontal motion, discussed above: this can be seen to come to an end only as a result of being combined with vertical motion, and it follows that since there would be no combination with vertical [motion] in empty space, the motion, in whatever direction it were to take place, would be like horizontal [motion] and would neither accelerate nor slow down, and thus would never stop.'[40]

We can see then that for the Democritean Gassendi there is nothing easier than to imagine a Universe, or at least a space, which is empty and without limit; and, liberated by Gilbert and Kepler from the obsession with gravity, nothing is easier for him than to imagine in the empty space a real body in perpetual rectilinear motion, never either accelerating or slowing down its motion.

But one objection remains. Had not Kepler stated that the body, by nature inert, has a natural tendency to rest, and a natural powerlessness to move? Yes, he had, but he was mistaken. Without naming Kepler Gassendi explains:[41] 'I add that stones and other bodies which are called heavy do not have that resistance to motion that is usually attributed to them. You will see that if a very large weight is hung on a wire it is very easy to displace it and to make it move backwards and forwards'. But this is not all. Gassendi has studied Galileo thoroughly and he knows that the motion of a pendulum is isochronal. So he adds:[42] 'Do you not see that the motion, once impressed on the suspended stone, is most constantly retained, i.e., that it performs all its swings not only in equal times but also through equal arcs? Now all this can only tend to make us understand that motion impressed on a body in empty space, where nothing attracts nor offers any resistance, will be uniform and perpetual. From this we can infer that all motion impressed on a stone is, in itself, of this nature. Thus if you throw a stone in any direction whatever, and if you assume that at the very moment when it leaves the hand everything except the stone is reduced to nothing by divine force, then its motion would continue perpetually and with the same direction that the hand had given it. If it does not do so it is clearly because of the combination with the vertical motion which intervenes as a result of the earth's attraction, which deflects it from its course (and which does not stop until it has brought it to the earth), just as iron filings close to a magnet do not move in a straight line but are deflected towards the magnet'. Thus if bodies fall and if their trajectories are curved, this is because they are acted upon by external influences. *De jure* and in itself all motion must be rectilinear, and all motion must be conserved for ever.[43]

One last objection: is not motion something positive, something more than rest? Is not a force necessary to produce motion, and to conserve it? When a body is thrown a force is exerted. Is it not this force, this *impetus*, that is impressed on the body, and must it not be used up? Not at all, for there is no need for a *force* to be impressed on a body in order for motion to take place. The motor does not do this. 'Nothing over and above *the motion* is impressed on the body by the motor. It is the motion which it has when it is in contact with the motor which is impressed on it and this motion would persist eternally if it were not diminished by some opposing motion.'[44] Thus motion is conserved by itself.

'I add', Gassendi goes on,[45] 'that one can thereby understand what our reaction should be to the difficulty usually raised concerning the force impressed on projectiles. It is asked what does this force do in the moving body? How is it impressed on it? How does it persist and how does it disappear? Now it is usually taken to be an active force, which moves the stone. It can be seen, however, that the active force, which is the cause of the projection, is in the projector itself and not at all in the thing projected, which is merely passive. That which is in the projectile is motion, which though it is sometimes called force, *impetus,* etc. (as we have ourselves called it, since we have retained familiar terminology as far as possible so as to be more easily understood), is in reality nothing over and above the motion itself. Certainly, one and the same motion, according to Aristotle, is action and at the same time passion; action in so far as it is in the motor, passion in so far as it is in the moving body. For in the motor it is an active force by means of which the body is moved; and when the body moves the active force is not to be found in it, for it is necessarily only in the motor; there is in it only a passive force and this, it is said, is brought to the act. Now there is nothing to prevent it happening that the received motion should persist if the motor becomes separated, or even if it vanishes altogether. For the motor is not required to transmit a force to the body, over and above the motion, a force which would then produce the motion. It is enough that it produces in the body a motion which can persist without it. Motion can do this; it is its natural property, given that it has an enduring subject and that nothing occurs to oppose it: it persists without the continuous action of its cause.'

Certainly we are still a long way from the lucidity and the metaphysical depth of Descartes. But we are also a long way from the hesitations of Galileo and the errors of Kepler. The deliberate elimination of the idea of *impetus*, the possession of a theory of gravity, and the definitive geometrisation of space enable Gassendi to break through the barriers which had held back these two great intellects.

II DESCARTES

1. Le Monde

Let us now turn to Descartes, to the Descartes of the period after 1630. Let us take a look at his *Le Monde*. As one goes from Galileo to Descartes, from the *Dialogue Concerning the Two Chief World Systems* to *The World*, one has a strange feeling, a feeling that could be described, inadequately no doubt, by saying that we experience an abrupt change of intellectual atmosphere.

The era of battle, the era of struggle, has been left far behind. Descartes is not in the slightest concerned to wage war on the endlessly rehearsed, continuously reworked, arguments of the partisans of geocentric astronomy. Copernicanism is calmly and innocently displayed and developed in his work as the only possible conception. Discussion is no longer necessary.

Nor is he interested in criticising Aristotelian physics, in analysing its foundations, its weaknesses and contradictions: it is enough to make occasional fun of its primary matter or of the imaginary space of the philosophers.[46] For Descartes, traditional physics is not only dead, but buried as well. One does not bother with it any more. What needs to be done, what Descartes confidently sets out to do, is to replace it. It is to establish and develop a new, and true, physics, to offer a new picture of the world, which is to say, in particular, a new concept of matter and a new concept of motion.

It is a matter of constructing, or reconstructing, the world, and of doing this *a priori*, going from causes to effects and not from effects to causes.

Nothing illustrates better Descartes', lack of concern for the traditional theories than the fact that he gives *The World* the form of a fiction: he is not undertaking to describe our own world, he explains, but a quite different one, a world created by God, infinitely remote from ours, somewhere in imaginary space; and created, one might say, with the means to hand. Therefore Descartes does not undertake to explain the laws of our world;

quite the contrary, he intends to deduce the laws of that other world, the laws which God imposes on nature and by means of which he will create all the variety and multiplicity of objects which will be found in that other world.[47]

We said that this is a fiction, a joke. So it is, because in fact it is our own world that Descartes is planning to reconstruct. But this device reveals an attitude which is typical of Descartes; for he does not actually ask questions about our own world. He does not ask, as Galileo had done, what nature's way of acting actually is. What Descartes asks is somewhat different: it is, one could say, what way of acting *must* nature adopt? The laws of nature are laws which nature must obey, rules to which it cannot but conform. For it is these rules themselves which make it what it is.

The Cartesian universe, as is known only too well, is built with very meagre materials. Matter and motion; or rather (since Cartesian matter, homogeneous and uniform, is nothing but extension) extension and motion; or better still, since Cartesian extension is strictly geometrical, space and motion. The Cartesian universe is the realisation of geometry.

The supreme law of the Cartesian universe is the law of persistence. That which is, remains. That which God has created is kept in being by him. The two realities of the Cartesian universe, space and motion, once created, remain forever. Space does not change; this is obvious. But neither does motion which, so to speak, carves them out of the homogeneous mass of neither increase nor decrease. It remains constant. This implies that, in the Cartesian world, motion has its own reality. It is created by God, created even before things. For it is as a result of motion that things exist. It is motion which, so to speak, carves them out of the homogeneous mass of extension or space. Thus things could not exist unless there were, and even previously had been, motion in the world.

But this is metaphysics, and Descartes does not want to engage in this at the moment. He considers his world, as it were, at a later stage in its evolution. There are already things; and there is already motion in these things. This ought to be enough. So he says:[48] 'I do not stop to search out the cause of their motions: for it is enough if I think that they began to move as soon as the World came into being. Given this, I discover by my reason that it is impossible that their motions ever cease, or even that they change except by change of subject. That is to say that the virtue or power which a body has of moving itself can pass in whole or in part into another, and thereby be no longer in the first, but it cannot no longer be in the world. I say that I am satisfied by reason that this is so; but I have not yet had occasion to tell these reasons to you. You can imagine, however, if you wish, as do the majority of Scholars, that there is some Prime Mover which, turning around the world with unimaginable speed, is the origin and source of all other motions to be found in it'. However, the 'Prime Mover', transposed into Descartes' new world, will perform there a very different role from that which it performed in Aristotle's world. It might well be, if one wishes to see it like this, the source and origin of all the motions in this world. But its function is limited to this. It is no longer needed once the motion has been produced. Because, and this is the basic difference, the prime mover does not have to sustain motion. Motion is sustained and preserved all by itself, without a 'motor', and this we know to be completely at odds with Aristotelian ontology. Motion can pass from one subject to another; it 'changes' subjects. Because of motion bodies have the virtue or power to move by themselves.[49]

What is this strange entity? What is its ontological status? It is clearly not the motion of the 'philosophers'. For what exactly is the motion of the philosophers? 'The philosophers also assume many motions, which they think can occur without any body changing place, such as those which they call *Motus ad formam, motus ad calorem, motus ad quantitatem* (motion as to form, motion as to heat, motion as to quantity), and a thousand others. As for me, I know of none other than that which is easier to conceive than the lines of the Geometers, namely that whereby bodies pass from one place to another and successively occupy all the spaces between them.'[50] It might seem as if, in contrast to the philosophers who assume many kinds of motions, Descartes assumes only one, namely that which the philosophers call 'local'. But we should not be deceived by appearances. In fact even as regards local motion the philosophers confess that its nature is 'very little understood; and in order to render it intelligible in any way they have not yet known how to explain it any more clearly than in these terms: *Motus est actus entis in potentia, prout in potentia est*, which as far as I am concerned are so obscure that I am forced to leave them here in the original because I would not know how to interpret them. (In fact these words: motion is the actualisation of that which exists potentially, in so far as it exists potentially, are no clearer for being translated). But, on the contrary, the nature of that motion of which I speak here is so easy to understand that even the Geometers, who are among all men those who are most practised in conceiving very distinctly those things which they consider, have judged the nature of motion to be simpler and more intelligible than that of their surfaces and their lines, as is clear from the fact that they have explained the line by the motion of a point, and the surface by that of a line'.[50]

Thus Cartesian motion, this motion which is the simplest thing and the easiest to understand, a completely intelligible essence which, in the order of reason as in the order of things, comes *before* all other material essences, even before the forms of space, this motion is that of the geometers. This should be remembered, for as we will soon see it is very significant.

But let us take a closer look. The philosophers, as we have just seen, wrongly distinguish several kinds of motion and at the same time fail to grasp the nature of that one kind that Descartes accepts as real. They believe that motion is in essence a transition from one state to another, that it is a process. For this very reason they deny it the degree of being that they attribute to qualities and to modes. But on the other hand they see in motion the actualisation of a potential, the transition from nothingness to being, and for this reason they attribute to it a degree of being, or of reality, greater than that attributed to its opposite, immobility, the absence of motion.

Thus[51] 'they attribute to the least of these motions a more real and substantial being than they do to rest, which they say is no more than privation. As for me, I conceive that rest is just as much a quality, to be attributed to matter when it remains in one place, as is motion, which is attributed to it when it changes place'. Thus Cartesian motion is in no way a process, but rather a quality or a *state*: and the identification which Descartes explicitly makes between the ontological status of motion and that of rest (a point which is of the very greatest importance, and to which we will return later[52]) is enough to explain why, in the new world built by Descartes, the persistence and the indefinite continuation of motion has no more need of a cause than that of rest had in the old world.

Descartes goes on, however (we apologise for quoting these passages, which everyone knows, or ought to know, at such length; but it is never a waste of time to re-read Descartes, and there is no end to the interpretation of such passages as these, being as they are so rich and so packed with such a wealth of content): 'finally, the motion of which they speak is so bizarre in nature that, while everything else has perfection as its end, and seeks only to preserve itself, their motion has no other end or aim than rest, and contrary to all the laws of nature, seeks to destroy itself.[53] But, in contrast, that motion which I assume obeys the same laws of nature which universally bring about all the arrangements and all the qualities which are in matter, as well as those which the Scholars call *Modos et entia rationis cum fundamento in re* (the modes and entities of reason with foundation in things) and *Qualitates reales* (their real qualities), in which I frankly confess I can discover no more reality than in the others'.

Motion, like rest, is a *state*. As such it obeys the general laws of nature, that is to say the laws of persistence and conservation which have been laid down for it by God.

Thus[54] 'without going any further into these metaphysical considerations, I set down here two or three of the main rules in accordance with which we must believe that God makes nature act. . .

'The first is, that each part of matter continues always to be in the same state as long as it is not forced to change it by coming into contact with others. Which is to say that: if it has a certain size, it will never become smaller unless it is divided by others; if it is round or square this shape will never change unless others force it to do so; if it is at rest in some particular place, it will never leave this place unless it is expelled from it by others; and if it has once begun to move it will continue always, with equal force, until it is stopped or slowed down by others'. So, every change needs a cause. Moreover, for Descartes, who has eliminated from nature all the forms (qualities or forces) of traditional physics, every change needs an external cause (quite right, Aristotle would say, all motion needs a motor). Thus no body can change or be modified by itself, spontaneously: nor can it spontaneously modify its state. In particular it cannot *set itself in motion.*[55] But once in motion it remains in motion. It cannot stop by itself: for motion is no longer a change. Certainly the moving body changes place; but is this still a change, in the Cartesian world?[56]

'There is nobody who does not believe that this same rule is obeyed in the ancient world as far as size, shape, rest and a thousand other similar things are concerned: but the philosophers have made an exception of motion, which is, however, the thing which I most expressly wish to include. But do not think because of this that I intend to contradict them: the motion which they speak of is so utterly different from that which I have in mind that it could easily happen that what is true of the one should not be so of the other.'[57]

We can only repeat what we have already said above:[58] 'We know that Descartes was right. His motion-state, the motion of classical physics, has nothing in common with the motion-process of Aristotelian and Scholastic physics. This is the reason why it is of their essence that they obey completely different laws. While in Aristotle's well-ordered Cosmos motion-process *self-evidently* needs a cause to sustain it, in Descartes' world of extension motion-state *self-evidently* persists by itself, and continues indefinitely in a straight line into the infinity of the completely geometrised space which Cartesian philosophy has opened up before it.'

But we should slow down a bit. We have as yet far from exhausted the specific essence of Cartesian motion.

Motion, we said, is a state. But it is also, and above all, a quantity. There is a determinate quantity of motion in the world. Each body in motion also has a certain, completely determinate, quantity of it. Now in all action, i.e., in every 'passage' of motion from one body to another, or to use Cartesian terminology, each time that motion changes subject, it obeys the following rule:[59] 'When one body impels another it cannot give to it any motion without at the same time losing an equal amount of its own; nor take any from it without its own increasing by the same amount. This rule, taken together with the previous one, agrees very well with all those observations in which we see that a body begins or ceases to move because it is driven or stopped by some other body. For having accepted this rule we are relieved of the difficulty which the Scholars find themselves in when they wish to account for the fact that a stone continues to move for some time after having left the hand of the person who throws it. For we will be asked, rather, why it does not continue to move for ever. But it is easy to give the reason for this. For who could deny that the air in which it moves offers some resistance?' From which it follows that the old question: *a quo moveantur projecta*?, which kept the Scholars so busy and on which so much ink has been used, is now definitively, and very simply, answered: *a motu*, or *a seipso*, or *a nihilo*, since the continuation of the motion of *projecta* is implied in the very fact of their motion. This solution shows that this famous problem was quite simply not a real problem; or was a wrongly formulated question. From which it directly follows that if the external resistance (of the air, etc.) is set aside, the moving body, conserving its motion, will never stop and will never even slow down.

It should be noticed, however, that resisting the motion of a moving body means receiving, or absorbing, some motion. For the body does not stop nor even slow down unless it can give up all or part of its motion to another. Motion, i.e., the quantity of motion, in the world is constant. Thus:[60] 'if one fails to explain the effect of resistance according to our second rule, and thinks that the more a body can resist the greater its capacity to stop the motion of others, as one might think at first glance, then one will once again have considerable difficulty in explaining why this stone's motion is more absorbed on contact with a soft body of only moderate resistance than on contact with a harder body which offers greater resistance. And also why, as soon as it has exerted a little effort against the latter, it immediately turns around as if to retrace its steps, rather than stopping or interrupting its motion on its account. Whereas, if we accept this rule, then there is no difficulty at all in this: for it shows us that a body's motion is not slowed down by contact with another to the extent that this latter resists motion, but only to the extent that its resistance is overcome and that in submitting to the motion the latter itself takes on the power to move which is lost by the other'.

This is very profound. It is also very ingenious. The Cartesian conception makes it possible to explain the phenomenon of the resistance to motion of a stationary body, a phenomenon which Kepler had found very puzzling and which, Kepler having misunderstood it, led him to formulate his concept of *inertia*, internal and essential to matter.[61] The body as such does not resist motion at all; it absorbs it and takes it from whatever impels it. This conception also enables Descartes to explain, as we shall see below, the phenomenon of a body's rebound after impact; and this in a physics in which there is no room for elasticity.[62]

But let us return to the passage which we have just quoted. Descartes appears here to give a justification for his conception by an appeal to experience. But we should not be deceived by this. Descartes knows perfectly well that experience, at least everyday experience, the raw data of experience, is of no use to us in laying down the true foundations of physics. Quite the contrary. Experience shows us bodies which, far from continuing to move indefinitely, stop almost as soon as they are thrown: it can only feed our prejudices. It is not experience but reason whereby we discover the truth, for[63] 'while everything which our senses have experienced in the real world obviously appears to be contrary to what is contained in these two rules, yet the reason which has taught them to me seems to me so strong that I cannot but think myself obliged to assume them of the new [world] which I am describing to you. For what more solid, more unshakeable, foundation could be found on which to base the truth, if one could choose any at will, than the very strength and immutability of God?'

As is well known, only the immutability of God can serve, for Descartes, as the metaphysical foundation of the conservation laws. So he continues:[64] 'Now these two rules obviously follow from this fact alone, that God is immutable, and that always acting in the same way he always produces the same effect. For supposing that he put a certain quantity of motion into matter overall at the first instant when he created it, then it must be admitted that this quantity is always conserved by him, or else we must give up our belief that he always acts in the same way. And also supposing that at this first instant the various particles of matter among which these motions are unequally distributed, began either to keep them or to transfer them from one to another depending on their various degrees of strength, then it must necessarily be thought that he always continues to make them do this; and this is what these two rules contain'.

Motion, then, is conserved. But which motion? Beeckmann, from whom Descartes had learned this fundamental law[65] (which he did not at that time believe had to depend on God's immutability, any more than Beeckmann did) certainly accepted the conservation of rectilinear motion. But also that of circular motion.[66] Moreover Beeckmann, and Descartes followed him in this, accepted the law of the conservation of motion only for motion *in a vacuum*. Now, when he wrote *Le Monde* Descartes no longer accepted the existence, or even the possibility, of a vacuum, the only environment in which rectilinear motion is possible: and yet he now restricts the conservation law to rectilinear motion. So we have the strange situation in which Descartes formulates the principle of inertia at the very moment when the newly acquired foundations of his physics make its realisation strictly impossible. Descartes, moreover, is well aware of this. Thus, he tells us, it is not the actual or real motion of bodies that is involved, but their 'action' or 'inclination' to motion.

He says:[67] 'I will add for the third [rule] that when a body moves, although its motion most often follows a curve, and though it can never follow a curve which is not in some degree circular, as was said above,[68] nevertheless each of its individual parts always tends to continue its own [motion] in a straight line: and therefore their action, i.e., their inclination to move, differs from their motion'. What is this 'action' or 'inclination' to move that Descartes pronounces to be different from the body's actual motion? Could it by any chance be an internal force, an *impetus*? Not at all. The 'action' or 'inclination' to move is nothing other than the motion itself; the motion which persists and which is maintained, and which passes from one subject to another. It is the state of motion, which Descartes quite correctly distinguishes from the motion which is actually achieved, accomplished, which includes an actually effected translation, a change of location or place. It is, then, the state of motion; a state which persists but which, on the other hand, exists at an instant, and which Descartes had earlier called 'point of motion' or 'moment'.[69] It is this motion at a point (the differential) which is always directed in a straight line.

'For example, if a wheel is turned on its axle, whereas all its parts go round because being joined to one another they cannot move otherwise, nevertheless their inclination is to go straight, as would be seen clearly if one of them happened to become detached from the others; for as soon as it is freed its motion ceases to be circular, and continues in a straight line.'[70] Remember the protracted arguments used by Galileo to show that the centrifugal force is tangential,[71] and compare them with this simple observation which Descartes considers sufficient. 'Similarly if a stone is whirled around in a sling not only does it go in a straight line as soon as it leaves it, but also, during the whole time that it is in it, it presses against the middle of the sling and stretches the cord, showing thereby . . . that it only goes round by constraint.'[72] Once again we must apologise for stressing this point and for giving the quotation, but is this not necessary in order to convey the distance between this and the precisely contemporaneous work of Galileo? The millenary privilege of circular motion had never up to this point been so firmly, and yet simply, rejected.

'This rule rests on the same foundation as the other two, and depends only on the fact that God conserves each thing by a continuous action, and consequently that he does not conserve it as it might have been some time before, but exactly as it is at the very instant at which he conserves it. Now, it is the case that of all motions *only rectilinear motion is completely simple and such that its whole nature may be grasped in an instant*. For, to understand it, it is enough to think of a body actually moving in a particular direction, and this is so at any given instant during the time that it is moving. Whereas, to understand circular motion or any other possible kind, it is necessary to consider at least two of its instants, or rather two of its parts, and the relation between them.'[73]

Let us pause here for a moment. The passage just quoted seems to us to be of fundamental importance. It enables us, we believe, to understand why Descartes succeeded where Galileo had failed: in other words, why Descartes managed to formulate the principle of inertia, which, as we have seen, Galileo did not and could not do.

Ultimately, no doubt, the reason lies in the fact that Descartes' thought was so radical, that he actually carried out the programme that had been merely suggested in *The Assayer*, that he reduced the real to the mathematical, to the geometrical, and excluded from the constitution of the physical body *everything* over and above its essential constitution; everything, and this means above all, gravity. As we have already pointed out, Galileo asked: what are in fact the processes of nature? Descartes asks: how *must* its constitution be, how *must* it act? Galileo, who was a physicist as much as, if not more than, he was a geometer, did not go beyond the phenomena; he accepted the authority of the real. Descartes, mathematician above all else, denies the authority of the phenomena. Thus Galileo tells us[74] that it is not his concern to know *if* God might have been able to create an infinite world; it is enough to know that he did not *in fact* make it so. Descartes, on the other hand, explains that God *could not* have made it other than infinite, simply because finite space is an absurdity.

Basically this was all that was needed. Galilean bodies, which, as we have seen, are *heavy* bodies [*des graves*], cannot move in a straight line in just any direction whatsoever. They are necessarily, and naturally, drawn 'downwards'. Moreover, their motion cannot in fact continue indefinitely: the *real* finiteness of the world prevents this. There is nothing like this in Descartes. His bodies, which are Euclidean and not Archimedean, are not drawn or attracted anywhere. Nor do they have any intrinsic tendency or quality. They have no relations, except spatial relations, with neighbouring bodies; they neither attract them nor do they tend towards them.

Therefore, once their motion has started they can continue to move indefinitely *in the same direction.*

Of course they cannot do this in fact. Of course motion never actually takes place in a straight line (except by mechanical means). But this is of no importance at all for Descartes. For him, he has just told us, the entire motion is contained *in an instant.* In this he is both beneficiary and victim of what we have called thorough-going geometrisation, for he forgets the essential relation between motion and time, whereas Galileo, who kept this relation in mind, was thereby rewarded with the solution to the problem of fall.[75]

It could of course be said that in Galileo also motion is concentrated in an instant, and that having been able to formulate the idea of moment, instantaneous speed, element (or differential) of motion, is even one of his great achievements: we ourselves have stated that this idea is identical with the Cartesian idea of 'moment'.[76] It could be added that Descartes does not say that in being realised motion can do without time, or that it can be realised in a single instant. On the contrary, he explicitly denies this, and 'so that the Philosophers, or rather the Sophists, do not take this opportunity to practise their superfluous subtleties', he asks us to note that he does not say that 'rectilinear motion can take place in an instant; but only that all that is required to produce it is in the bodies at each specifiable instant during the motion, whereas this is not so of everything required to produce circular motion'.[77] But, for all that, it is nonetheless true that Cartesian motion, on Descartes' own account of it, has only an indirect relation with time, and that the Cartesian conception of it differs precisely in this respect from that of Galileo.

That motion which Descartes tells us is the only one he knows, *the motion of the geometers,* is above all a translation, i.e., a change of place: and this is what it always remained. But Galilean motion, or perhaps one should say motion as it was seen by Galileo, is above all a speed. Every translation, of course, takes place at a certain speed, and all speed implies translation: thus the ultimate elements of which motion is constituted are equivalent in Galileo and Descartes. However, speed and translation are not the same thing, and the fact of emphasising one or the other of these aspects of motion is not without consequence. For what we have just said is not in fact the case; it is not true that every translation implies a speed; this is certainly true of every real translation, but not of geometrical translation.

The 'motion' of a point which makes a line, the 'motion' of a line which makes a plane, 'motions' such as these have no speed. Having no speed they do not take place in time. Now it is on the model of these non-temporal 'motions' that Descartes fashions his idea of motion, an idea which he asserts is simple and easy to understand, the simplest, easiest, and clearest of our ideas. Of course. For what makes the idea of motion so obscure is precisely its connection with that of time. It is easily understandable that those philosophers who investigated *temporal* motion were able to define it only in a very obscure manner: and that Descartes, having eliminated time from motion, and having substituted *being* for *becoming*, could find that there was no trace of this obscurity left in it.

But is it possible to talk of geometrical motion? Is non-temporal motion still motion at all? In other words what is left of motion when time has been 'eliminated' from it? Is anything left?

To eliminate time, is this not to freeze motion? Certainly; it is to freeze it, or to spatialise it. Thus what is left of motion when its temporal character has been suppressed is precisely whatever in it is immobile: position, direction, trajectory, functional relations. The thorough-going geometrisation to which Descartes yields undoes the work of time—a moving picture of motionless eternity—and gives us a motionless and finished picture of the essential incompleteness of motion. But it also enables Descartes to understand the infinity of motion at an instant.

To substitute the trajectory for the motion is very serious, and even very hazardous. It can sometimes give rise to error.[78] However, on other occasions it leads to the truth. For example, it is in fact very difficult to know which of circular and rectilinear motion is the simpler; but it is very easy to see that the straight line is simpler than the circle;[79] that the circle, like any curve, is a straight line that has been bent; and therefore that motion which follows a straight line, and which has the same direction at each point, is simpler than that which describes a circle and which must change direction at each point. So there is no need for a lengthy discussion to understand that 'for example, if a stone moves in a sling, and follows the circle AB, and if you were to consider it exactly as it is at the instant that it arrives at point A, you would certainly find that it has an action to move, for it does not stop there, and to move in a particular direction, namely towards *c*, for its action at this instant is determined towards that point; but you would not be able to find in the stone that which makes its motion circular. So much so that if we imagine that it then begins to leave the sling, and assume that God continues to preserve it as it is at this instant, it is certain that he would not preserve it with an inclination to go in a circle following the line AB, but with an inclination to go straight towards the point *c*.'[80]

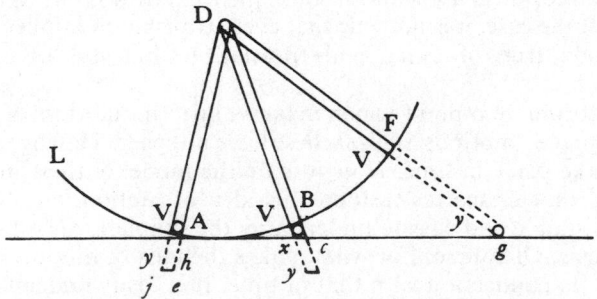

Thus for Descartes it is thorough-going geometrisation which smooths the way for the victory of the straight line over the circle. It is a victory of which, curiously enough, he makes a point of making God the foundation:[81] 'Thus, according to this rule, we must say that God alone is the Author of all motions in the world, in so far as they exist, and in so far as they are straight. But it is the various arrangements of matter which make them irregular and curved. As the Theologians teach us, God is also the Author of all our actions, in so far as they take place and in so far as there is some goodness in them; but that it is the various dispositions of our wills which can make them evil.'

2. Les Principes

In relation to the problem we are concerned with, the discovery and formulation of the principle of inertia, the *Principles* does not contribute much that is new. In fact its contribution is not always progressive, except in relation to the construction. Thus the infrastructure and the epistemological and metaphysical foundation of the physics are explicitly and systematically developed and are in their place at the beginning of the book. The exposition is now clearer, more economical, more precise and more detailed; in other words, more scholarly. The impulsive nonchalance of *The World* has vanished. This is easily understood; the *Principles* is a new model, and it is not aimed at the same audience. The earlier book had been aimed at the cultivated public; the later is a text-book for scholarly use.

Moreover, since 1630 Descartes had become a more important figure, his position in the world had changed a great deal. He was no longer unknown as he had been earlier. He was now the famous, the important philosopher, admired by some, opposed by others. He was the leader of a school. Now this inevitably implied a change in tone. Finally, and this is very important, Descartes, older now, had become more cautious: too cautious even, for some people's liking. In the light of what had happened to Galileo, and of incidents that had affected Descartes himself, he felt it necessary to take precautions; which he did, though they were rather clumsy. For while the Copernicanism which had been so openly on display in *The World* had disappeared from the *Principles*, or rather had been hidden behind an odd and peculiar theory of motion, on the other hand the infinity of the world was still there, explicitly asserted.

'We know, furthermore, that this world, or the extended matter which makes up the universe, has no limits; because wherever we might imagine them to be we can always imagine further space indefinitely extended beyond them, and because we do not simply imagine this, but conceive that this further space is in fact as we imagine it: so that it contains indefinitely extended corporeal substance, . . . for the idea of the extension that we conceive any particular space to have is the true idea that we should have of corporeal substance.'[82]

The fundamental laws of Nature in the *Principles* are the same as in *The World*. The two versions differ only in the order of their exposition and in the greater emphasis, in the *Principles*, on their metaphysical foundation.

In the final analysis, the order adopted in the *Principles* (in which Descartes reverses the order of the second and third rules) is more logical than that adopted in *The World*. The laws of Nature are now arranged in order of increasing specificity. Thus the first rule states the law of the conservation of motion, the second specifies that it is rectilinear motion that is involved, and finally the third gives the laws governing the transmission of motion.

As in *The World* the first law, or rule, of Nature rests on the general principle of conservation:[83] 'From the fact that God is not subject to change and that he always acts in the same way, we can know certain rules, which I call the laws of nature, and which are the secondary causes . . . of the various motions observable in all bodies: and this means that they are of great interest here. The first is that each individual thing . . . remains in the same state as far as possible, and that its state is never changed except by contact with others. Thus we see every day that if some particular bit of matter is square . . . it always remains square unless something external occurs which changes its shape: and that, if it is at rest, it does not start to move by itself. But once it has started to move, we have no reason at all to think that it need ever stop moving with the same force . . ., so long as it does not come into contact with anything which slows down or stops its motion. So we must conclude that once a body has started to move it continues to move thereafter and that it never stops by itself'. As in *The World*, Descartes explains that the opposite belief, i.e., the belief that a moving body can spontaneously come to rest, is only a prejudice based on misunderstood experience, and that it 'obviously contravenes the laws of nature; for rest is the opposite of motion, and nothing has a natural tendency to change to its opposite, or to its self-destruction'.[84]

As in *The World* Descartes believes that this first law is, on the contrary, confirmed by everyday experience when this is properly understood: and that the problem *a quo moveantur projecta* is thereby solved:[85] 'We see the proof of this first rule every day in projectiles. For there is no other reason at all why they should continue . . . to move once they have left the projector's hand, except that, according to the laws of nature, every body in motion continues to move until it is stopped by some other body . . . And it is obvious that the air and other fluid substances in which we see things move, gradually decrease the speed of their motions'.

The second law, which is also deduced from the immutability of God, is similarly confirmed by experience:[86] 'The second law which I observe in nature is that each particular particle of matter never tends to continue to move along a curve, but along a straight line, although many such particles are often forced to deviate, because they meet others on their path, and because . . ., when a body moves it always gives rise to a circle or ring in the matter which moves together with it. This rule, like the previous one, depends on the fact that God is immutable, and that he conserves the motion in matter by a very simple operation; for he does not conserve it as it might have been some while beforehand, but exactly as it is at the very instant that he conserves it. And although it is true that motion does not take place in an instant, nevertheless it is obvious that the motion of any moving body is determined [at each instant] to move in a straight line and not in a circle . . .; for when a stone A is whirled round in a sling EA along the circular path ABF, at the instant when it is at point A, it is determined to move in some particular direction, namely towards C, along the straight line AC, assuming that this is the tangent. But it could not be imagined that it was determined to move in a circle, because although it has come from L towards A along a curve, we do not think that there is any part of this curvature in the stone once it is at point A:[87] and this is confirmed by experience, because the stone goes straight towards C when it leaves the sling, and in no way tends to move towards B. This shows quite clearly that every body which moves in a circle has a continuous tendency to depart from that circle. We can even feel this with the hand while we are whirling the stone around in the sling'.

It is clear that the formulation and the deduction of the first two laws of nature are no different from those given in *The World*: it is simply a little more economical, duller, lacking in richness. But the statement of the third law does contribute some new details concerning the transmission and exchange of motion, and lays down concrete rules (almost all false, however):[88] 'The third law to be observed in nature is that if a body in motion meets another, and if the former has less strength to continue moving in a straight line than the latter has the strength to resist it, then it loses this determination, without losing any of its motion: and that if it has greater strength, it moves the other body with it, and loses as much of its motion as it gives to the other. Thus a hard body, which we have impelled against a larger body, which is hard and solid, rebounds in the direction from which it came and loses none of its motion; but if the body which it meets is soft, it stops straight away because it transfers its motion to it . . .'

It is well known that the concrete rules given by Descartes for the transmission of motion are almost all false; but as we have often said, the mistakes of a Descartes are as interesting and instructive as his discoveries. We will, therefore, return to this, and will look for an explanation for Descartes' error, which has, in our view, been more often denounced than explained, if indeed it ever has been completely explained at all.[89]

But for the moment we are faced with another question, namely what is this motion the laws of which are given by Descartes? For as we have pointed out the concept of motion to be found in the *Principles* is not exactly the same as that in *The World*. The latter is based on a purely geometrical idea, whereas the *Principles* tries to give a physical definition of motion, based on the principle of the relativity of motion. For this, and for other reasons, the Cartesian definition is no longer as violently opposed as it once was to the Scholastic definition, but is in fact closely related to it.[90] 'Motion (i.e., from one place to another, for I conceive no other, and furthermore do not think that it is necessary to assume any other in nature), motion then, as it is usually understood is nothing other than *the action whereby a body passes from one place to another.* And just as we have pointed out that one and the same thing can simultaneously change place and not change it at all,[91] similarly we can say that it simultaneously both moves and does not move at all . . .[92]

'But if, instead of remaining with what has no other foundation than ordinary usage, we wish to know what motion truly is, we would say, so as to attribute to it a determinate nature, that *it is the translation of one part of matter or of a body from the vicinity of those which directly touch it, which we take to be at rest, to the vicinity of some others.* . . And I say that it is the translation and not the force or the action which translates it, so as to show that motion is always in the moving body and not in that which moves it.'[93] And 'since it is not a matter here of the action which does the moving or stops the motion . . . it is obvious that this translation is not outside the moving body; but that a body is in a different condition when it is translated than when it is not; so that motion and rest are only two different *modes* of a body . . .'[94] Moreover Descartes states 'that motion, correctly understood, only refers to those bodies which are contiguous with the one which is said to be moving', and even that it refers, among these contiguous bodies, to those 'which we take to be at rest', since 'in itself, translation is reciprocal'.[95]

P. Mouy, one of the most recent and most penetrating among historians of Cartesian physics, sums up these passages, in which the strictest idea of the relativity of motion is asserted and developed, by saying: 'Therefore motion is *not a being*, but a 'mode' of the translated body; it is entirely relative and purely kinetic; there is no force hidden behind it'.[96] But if it is not a being, but simply a 'mode', how then is this motion conserved in the world? Descartes does give us an answer to this, an answer which is, compared with *The World*, much more detailed.

'Having examined the nature of motion we must now consider its cause, and since this can be taken in two ways we will start with the primary and most universal cause, the general cause of all the motions in the world. Later we will consider the other cause, . . . that whereby the parts of matter come to have motion when they did not have it previously. As for the *primary* cause, it seems evident to me that it is none other than God, who *by his omnipotence* created matter together with motion and rest, and who now conserves in the universe by his ordinary concourse as much motion and rest as he placed there when he created it. For, although motion is only a *mode* of the moving body, nevertheless it does have a definite quantity . . . which never increases nor decreases . . . though there is sometimes more and sometimes less of it in the several particular parts of matter. This is why, when a part of matter moves twice as fast as another part, and when this latter is twice as large as the former, we must reckon that there is as much motion in the smaller as in the larger part; and that whenever the motion of one part diminishes, that of some other part increases in proportion.'[97]

Therefore, it is indeed this reciprocal, relative, and purely kinetic motion which, created by God and sustained by him in the world, is perpetually conserved there with the same quantity. Descartes tells us this quite explicitly. However . . . P. Mouy, who has thoroughly studied the laws of impact put forward by Descartes, quite correctly points out that the 'way in which motion is treated [in relation to these laws] completely contradicts the fundamental relativity attributed to it by Descartes'.[98] Now we cannot explain Descartes' error simply by the fact that Descartes 'wanted to conform to experience, but used ideas which were far too simple for this purpose',[99] but only, and above all, by the fact that Descartes himself never took seriously the relativistic idea of motion just expounded at such length, and never used it as the basis for deduction. In fact it is not only with the laws of impact that the kinetic relativity of motion is incompatible. It is so even with the law of the conservation of motion, if this is taken as Descartes explicitly wished to take it, as the conservation of the *quantity* of motion. For clearly if one attributes (and reciprocity and kinetic relativity legitimate this) *the same speed* to a large body and to a small one when these are approaching each other, or moving away from each other, one will obtain vastly different *quantities of motion.* Now, it is impossible to believe that Descartes remained unaware of such blatant contradictions, or that he overlooked them.

The ultra-relativism of his idea of motion was not original with Descartes. It is our opinion that he only adopted it so as to be able to reconcile Copernican astronomy, or more simply the mobility of the earth, which was manifestly implied by his physics,[100] with the official doctrine of the Church. It was an initiative which succeeded only in making Cartesian mechanics self-contradictory and obscure. Now while Cartesian mechanics may be false, it is not in itself self-contradictory, and the (admittedly incorrect) laws of impact formulated by Descartes do follow quite logically from his own concept of motion, a concept of which he had given a very clear account in *The World.* We can clarify this concept by reference to these laws.

Let us go back, then, to *The World.* It will be recalled that in this book Descartes explicitly assimilated, or even identified, the ontological status of motion with that of rest. This enabled us to understand straight away why Cartesian motion, in contrast with that of Aristotle, could persist in the absence of a motor or cause. But any assimilation or identification can be construed in either of the two opposite directions. Thus we have assimilated motion to rest; now we must, conversely, assimilate rest to motion. Since, for Descartes, rest has as much *reality* as motion, it must therefore no longer be seen as a merely negative state, as an absence of motion, as infinitely slow motion, etc., etc.; but as a state which has reality, a positive power to act and react. So it is no longer enough to say that a body at rest has a quantity of motion equal to zero. It must be said, over and above this, that it has a certain quantity of *rest.*[101] It is precisely because of their 'quantity of rest' that bodies resist and oppose being set in motion.

In Cartesian physics motion is the principle of separation. Conversely, rest is the principle of connection and adhesion. In fact it is the only principle of adhesion in this physics. Two particles which are contiguous, or even which are merely at rest in relation to each other, are interconnected by virtue of this fact alone, in such a way that 'some force, however small, is needed to separate them; for once they are in this position they will never take it upon themselves to redistribute themselves'.[102] The unity of a body, and even its hardness, are constituted by the fact that its parts are at rest in relation to each other, 'for what other glue or cement could be imagined there, to make them hold more firmly one to another?'[103]

Thus rest is a positive force or power. The *Principles* say this explicitly.[104] But what is its magnitude, or in Cartesian terminology, its quantity? It is clear, for Descartes at least, that for a given body it is precisely equal to the quantity of motion of a body *of equal dimensions* which is in motion in relation to the body at rest. Therefore, the quantity of rest is, so to speak, a variable magnitude, which one might say is a function of the speed of the moving body. This is a necessary consequence of the *physical*, i.e., *dynamic* relativity of motion. It necessarily follows that for each pair of bodies, of which one is at rest and the other moving, the ratio of the powers of rest and motion is the same as that of their sizes. Thus when Descartes tells us that when a small body hits a large one it can never, regardless of its speed, make the large one move[105] (Galileo, we know, believed on the contrary that whatever the size of a body at rest, a body which hits it, however small it may be, always confers some motion on it), it is not at all his desire to conform to experience which leads him to make this mistake (when a ball is thrown against a wall it rebounds whereas the wall does not visibly move): he knows perfectly well that the case which he is investigating is *never* given in experience; but he is drawing a necessary conclusion from his concept of motion. From that of rest also.

A strange and peculiar concept, certainly; and unfortunate even, one might say, for it leads Descartes into error and brings Cartesian physics to an impasse. Yet, in spite of error, the greatness of Cartesian physics remains intact. For the Cartesian concept of motion is the logically inevitable consequence of his original (but enormously fruitful!) sin: thorough-going geometrisation. Only at the price of self-contradiction, an infinitely more serious sin for a philosopher, could Descartes have avoided falling into error.

Geometer's motion is not, as we have seen, real motion: and the 'bodies' which have this motion are not real either. Strictly speaking they are no more at rest than they are 'in motion'. Ultimately this is the reason why Descartes' God, in creating his world, i.e., in bestowing real being on Euclidean space, was obliged to create as much rest in it as he had created motion.

NOTES

NOTES

APPENDIX

THE ELIMINATION OF GRAVITY

[1] B. Cavalieri, *Lo Speccio Ustorio overo Trattato Delle Settioni Coniche et alcuni loro mirabili effetti intorno al Lume, Caldo, Freddo, Suono e Moto ancora*, Bologna, presso Clemente Ferroni, 1632, chap. XXXIX, p. 153: « Del movimento de' corpi gravi. Benche intorno à' corpi gravi diversissime' cose si potessero considerare, tutte belle, et tutte curiose, però non cercaremo altro, se non che forte di linea sia quella, per la quale si move esso grave, mercè prima dell'interna gravità, poi del proiciente, e finalmente dell'uno et'dell'altro accoppiati insieme, per vedere, se vi havessero che fare le Settioni Coniche, et quali siano quando ciò sia vero.

Dico adunque, se noi consideraremo il moto del grave fatto per la sola interna gravità, in qualcunque modo poi ella si operi, che quello sarà sempre indrizzato verso il centro universale delle cose gravi, ciò è verso il centro della terra, et universalmente conspirare tutti i gravi à questo centro, poiche si veggono in tutti i luoghi della superficie terrestre scendere non impediti a perpendicolo sopra l'Orizonte . . .

Dico piu oltre, che considerato il mobile che da un proiciente viene spinto verso alcuna parte, se non havesse altra virtù motrice, che lo cacciasse verso un'altre banda, andarebbe nel luogo segnato dal proiciente per dritta linea, mercè della virtù impressali pur per dritta linea, dalla quale drittura non è ragionevole, che il mobile si discosti, mentre non vi è altra virtù motrice, che ne lo rimova, e ciò quando fra li duoi termini non sia impedimento ; come per essempio, una palla d'Artiglieria uscita dalla bocca del pezzo, se non havesse altro, che la virtù impressali dal fuoco, andarebbe à dare di punto in bianco nel segno posto à drittura della canna, ma perche vi è un altro motore, che è l'interna gravità di essa palla, quindi avvienne, che da tal drittura sia quella sforzata deviare, accostandosi al centro della terra. »

[2] 'The universal centre of heavy things' stands in for Aristotle's 'centre of the world'.

[3] *Ibid.*, p. 155: « Dico ancora, che quel proietto non solo andarebbe per dritta linea nel segno opposto, ma che in tempi eguali passarebbe pur spatij eguali della medesima linea, mentre que i mobile fosse a tal moto indifferente; e mentre ancora il mezzo non li facesse qualche resistenza, poiche non ci farebbe causa di ritardarsi, ne di accelerarsi. »

[4] *Ibid.*: « si che il grave, mercè della interna gravità, non anderà se non verso il centro della terra, ma quello, mercè della virtù impressali, potrà incaminarsi verso ogni banda. »

[5] *Ibid.*: « Essendo due adunque nel proioetto le virtù motrici, l'una la gravità, l'altra la virtu impressa, ciascuna di loro separatamente farebbe ben caminare il mobile per linea retta, come si è detto, ma accopiare insieme non la faranno andare per linea retta, se non in questi due casi, nel primo, quando dallo virtù impressa sia spinto il grave per la perpendicolare all' Orizonte ; il secondo, quando non solo la virtù impressa ma anco la gravità mova il grave uniformemente, perche gli accostamenti fatti in tempi eguali al centro della terra, partendosi da una retta linea, sariano sempre eguali, come anco li spatii decorsi ne' medesimi tempi dell' istessa linea, per la quale viene spinto esso grave ; e perciò il mobile farebbe sempre nella medesima linea retta. Ma quando uno de' duoi non fosse uniforme, allhora non caminarebbe il mobile spinto dalla gravità, e dalla virtù impressa, altrimente per linea retta, ma si bene per una curva, la cui qualità e conditione dipenderebbe dalla detta uniformità, e difformità di moto accoppiate insieme. Hora nel grave, che, spiccandosi dal proiciente, viene indrizzato verso qual si sia parte, per essempio, mosso per una linea elevata sopra l'Orizonte, vi è bene la gravità, che opera, ma quella non fà altro, che ritirare il mobile dalla drittura della sudetta linea elevata, non

havendo che far niente con l'altro moto, se non per quanto viene il grave allontanato dal centro della terra, astraendo adunque nel grave la inclinatione al centro di quella, come anco ad altro luogo, egli resta indifferente al moto conferitoli dal proiciente, e perciò se non vi fosse l'impedimento dell'ambiente, quello sarebbe uniforme : ragionevolmente adunque si potrà supporre, che i gravi spinti dal proiciente verso qualunque parte, mercè della virtù impressa, caminino uniformemente, non havendo risguardo all'impedimento dell'aria, che per esser tenuissima, e fluidissima, per qualche notabile spatio, può esser, che i permetta la sudetta uniformità. »

6   *Ibid.*, p. 157: « Resta hora, che facciamo riflessione all'accostamento del grave, fatto al centro della terra mercè dell' interna gravita, che vien detto moto naturale, e al discostamento da quello, per l'impulso conferitoli, che si chiama moto violento ; che il grave, che si parte dalla quiete, e si move al centro, si vada sempre velocitando, quanto più si accosta al centro, o per dir meglio, quanto più si allon tana dal suo principio, e che il violento, o dal centro si vada sempre ritardando, ciò è stato saputo da tutti i Filosofi ancora, ma con qual proportione s'acceleri il moto naturale, et si ritardi il violento, ce lo insegna nouvamente e singolarmente il Sig. Galileo ne' suoi Dialogi alla p. 217, dicendo esser l'incremento della velocita, secondo il progresso de' numeri dispari continuati dall'unita. »

7   Notice the persistence of this formulation! Cf. above, Part II, p. 69.

8   Evangelistae Torricellii, *Opera geometrica*, Florence, 1644, *De Motu Projectorum*, Book II, p. 156. Emphasis added.

9   Torricelli, *Opera Geometrica, Proemium:* « Inter omnia opera ad Mathematicas disciplinas pertinentia, iure optimo Principem sibi locum vindicare videntur Archimedis inventa; quae quidem ipso subtilitatis miraculo terrent animos ».

10   Torricelli, *Opera Geometrica*, De Dimensione Parabolae, *Proemium*, p. 8: « Veniamus ad objectiones quae circa artis fundamenta versantur. Indignor equidem Lucam Valerium, vere nostri saeculi Archimedem, cum optimam causam suscepisset, pessima defensione usum fuisse. Solent ab eruditis culpari figurarum Geometricarum dimensiones, quae Mechanicis fundamentis innixae stabiliuntur, tamquam duplex falsum supponant : alterum *quod superficies gravitatem non habentes habere tamen concipiuntur :* alterum vero, *quod fila quae magnitudines ad libram suspendunt aequidistantia supponuntur, cum tamen in centro terrae concurrere debeant.* Ego vero in ea sum sententia, vel nullam ex his suppositionibus esse falsam, vel reliqua omnia principia Geometriae falsa existere eodem modo. Falsum enim est, quod circulus habeat centrum, sphaera superficiem, conus soliditatem. Loquor de figuris abstractis quales Geometria considerare solet; non autem de physicis et concretis. Necesse igitur erit fateri quod circuli centrum, superficies sphaerae, soliditas coni, et reliqua huiusmodi non controversa, nullam aliam habeant existentiam, praeter illam quam accipiunt per definitionem et per intellectum. Eodem prorsus modo gravitas est in figuris Geometricis, quomodo in iisdem est centrum, perimeter, superficies, soliditas, etc. »

11   *Ibid.*, p. 9: « Laudarem igitur in Mechanicis contemplationibus nova definitione figuras generare ; hoc, aut alio non absimili modo. »

12   *Ibid.:* « Quadratum est quadrilaterum, quod cum aequilaterum, et aequiangulum sit, singula ipsius puncta momentum habent procedendi versus aliquam mundi plagam per lineas inter se parallelas. »

13   The same idea is found in Descartes; see above, p. 253.

14   *Ibid.:* « Huiusmodi enim definitio omnem demeret occasionem dubitandi, illis, qui Mechanica Archimedis opera, secundum ipsius mentem non accipiunt. Sed hucusque dictum sit pro obliteranda primae falsitatis nota, quod figurae Geometricae graves sint. »

15   *Ibid.:* « Venio nunc ad secundum (ut aliqui existimant) falsum. Principio, vulgatissima est etiam apud gravissimos viros obiectio illa, videlicet *Archimedem supposuisse aliquod falsum, dum fila magnitudinum ex libra pendentium consideravit tanquam inter se parallela, cum tamen re vera in ipso terrae centro concurrere debeant.* Ego vero (quod pace clarissimorum virorum dictum sit) crediderim fundamentum Mechanicum longe alia ratione esse considerandum. Concedo si fisicae magnitudines ad libram libere suspendantur, quod fila materialia suspensionum convergentia erunt ; quandoquidem singula ad centrum terrae respiciunt. Verumtamen si eadem libra, licet corporea,

consideretur non in superficie terrae, sed in altissimis regionibus ultra orbem Solis ; tum fila (dummodo adhuc ad terrae centrum respiciant) multo minus convergentia inter se erunt. Sed quasi aequidistantia. Concipiamus iam ipsam libram Mechanicam ultra stellatam libram firmamenti in infinitam distantiam esse provectam, quis non intelligit fila suspensionum iam non amplius convergentia, sed exacte parallela fore ? Quando ego considero libram, figuras Geometricas ponderantem, non concipio illam esse inter cartas librorum in quibus depicta conspicitur ; neque suppono punctum, ad quod magnitudines ipsius tendunt, esse centrum terrae ; sed libram fingo in infinitum remotam esse ab eo puncto, ad quod ipsius gravia contendunt. »

16 *Ibid.,* p. 10: « Si postea ibi conclusero triangulum aliquod triplum esse cuiusdam spatii; retrahatur imaginatione ipsa libra ad nostras regiones ; concedo quod retracta libra destruetur aequidistantia filorum suspensionis, sed non ideo destruetur proportio iam demonstrata figurarum. Peculiare quoddam beneficium habet Geometra, cum ipse abstractionis ope, omnes operationes suas mediante intellectu exequatur. Quis igitur mihi hoc negaverit, si liberat considerare figuras appensas ad libram, quae quidam libra ultra mundi confinium in infinitam distantiam remota supponatur ? Vel quis proibebit considerare libram in superficie terrae constitutam, cuius tamen abstractae magnitudines tendant, non ad medium terrae punctum, sed ad centrum caniculae, sive stellae polaris ? »

17 *Ibid.:* « Triangula et parabolae, immo etiam sphaerae cylindrique Geometrici, cum nullam per se habeant motus differentiam, non magis ad ipsius terrae, quam ad Saturni centrum contendunt. Destruit ergo beneficium suum quisquis figuras illas, tamquam ad unicum terrae centrum tendentes, contemplatur. »

18 *Ibid.:* « Cur denique non licebit mihi considerare puncta cuiuscunque figurae eiusmod virtute praedita, ut singula versus eandem mundi plagam per lineas inter se parallelas aequali momento contendant ? His ita suppositis, quae vera sunt, quemadmodum sunt verae passiones figurarum, quae in definitionibus adhibentur, vera etiam erunt quaecunque Theoremata per Mechanicas rationes ab ipsis abstrahentibus fuerint considerata, neque per falsas positiones demonstrabuntur. »

19 *Ibid.,* p. 11: « Tunc itaque falsum dici poterit fundamentum Mechanicum, nempe fila librae parallela esse, quando magnitudines ad libram appensae fisicae sint, realesque, et ad terrae centrum conspirantes. Non autem falsum erit, quando magnitudines (sive abstractae, sive concretae sint) non ad centrum terrae, neque ad aliud punctum propinquum librae respiciant ; sed ad aliquod punctum infinite distans connitantur. »

20 See E. Wohlwill, 'Die Entdeckung des Beharrungsgesetzes', *Zeitschrift für Völkerpsychologie,* vol. XV, p. 355, note 2.

21 He was so far from being a mathematician that he did not manage to understand Galileo's deduction of the law of fall, and believed that in order to obtain the law that distances are as the squares of the times it was necessary to assume a simultaneous action of attraction and of the medium's reaction. Cf. Petri Gassendi, *De motu impresso a motore translato,* Paris, 1642, cap. XVII, p. 64 ff.; cap. XVIII, p. 69 ff.

22 See *De motu impresso a motore translato,* cap. V, p. 14 ff. The whole of the first part of the work is devoted to the explanation of the transmission of motion from the motor to the moving body, or more accurately, of the fact that a moving body, connected with a moving system, participates in its motion.

23 See above, Part III, note 149.

24 *De motu impresso a motore translato,* cap. VI, p. 22 ff.

25 *Ibid.,* cap. VII, p. 27 ff.

26 *Ibid.,* cap. V, p. 17: « Neque est jam mirum, si omnibus nobis, qui eadem triremi eramus, apparebat motus perpendicularis ; quippe observabilis nobis solum erat motus lapidis deorsum ; nam ille quidem ad anteriora observari non poterat, quoniam erat nobis communis cum lapide . . . »

27 *Ibid.,* cap. IX, p. 35: « Preterea cum motus naturalis, et violenti voces non videantur nobis esse confundendae, ea mihi semper utriusque notio visa est commodissima, ut naturalis appelletur, qui aut sponte, aut sine ulla repugnantia fit : violentus, qui praeter naturam, aut cum aliqua repugnantia . . . Ita trajectio globi per aerem violenta, quia praeter naturam ; volutio supra planum naturalis quia nihil repugnat. »

28 *Ibid.:* « Nullus videtur motus, qui secluto primaevo illo, non possit censeri violentus: quatenus nullus est, qui nisi cum impulsione unius rei in aliam fiat, ex quo effectum est, ut Aristoteles, etiam rerum cadentium quaesierit motorem externum. » The 'very first' or 'primaeval' motion is that of the sphere of the fixed stars.

29 *Ibid.:* « Neque videri absurdum debet, esse continuam aliquam in rebus naturae violetiam ».

30 *Ibid.,* cap. IX, p. 36: « Et sane cum sit commune effatum, Nihil violentum esse perpetuum ; cui est consentaneum, ut quod est naturale perpetuum sit ; constat radicem perpetuitatis esse aequabilitatem, cessationis inaequabilitatem ; quatenus id solum, quod neque invalescit, neque debilitatur, perdurare potest ; nihilque potest naturae vi aut increscere, aut decrescere infinite. Adhaec, si quis requirat motum in hisce rebus compositis, qui sit maxime naturalis, perspicuum videtur eum esse caelestem ; quatenus est prae ceteris aequabilis, atque perpetuus ; delecta ab authore circulari forma, secundur.ι quam, principio, et fine carentem, esse aequabilitas, et perpetuitas posset. »

31 *Ibid.,* cap. X, p. 38 ff.: « Non repeto heic, quemadmodum lapis a vertice mali, dum navis movetur, apparenter solum secundum perpendiculum cadat, reipsa vero oblique per eam, quam descripsimus lineam ; innüo duntaxat lapidem non sponte moveri, quia movetur vi a manu impressa ex translatione manus a navi, cui una cum malo insistit. Atque id quidem seu manus in fastigio mali consistens lapidem dimittat, seu lapis ex radice mali projectus, ubi pervenerit ad summum, postea recidat ; ut proinde intelligas posse vel ex hoc capite motum lapidis decidentis, recidentisve dici violentum. Dices, cum hic obliquus motus mistus, seu compositus fit ex perpendiculari et horizontali ; id quidem, quod est ex horizontali, existimari posse violentum, at quod ex perpendiculari, id saltem esse naturale. Nam quod lapis quidem sursum projectus, et nihilominus oblique incedens, secundum utrumque violentus sit, videri perspicuum : quoniam utriusque causa externa, impellensque est, illius nempe ipsa vis navis, huius vero vis manus propria: at quod deorsum dimissus, et oblique nihilominus incedens, secundum utrumque violentus sit, non posse perinde esse in confesso : quippe horizontalis quidem causa similiter externa, impellensque, vis navis est ; sed perpendicularis causa non est perinde vis propria manus. Quare et necesse videri lapidem eo motu moveri ab interno principio: esseque proinde eum motum non violentum, sed naturalem. Attamen id videtur primum consideratione dignum, si ex duobus his motibus, perpendiculari nempe, et horizontali, qui obliquum illum componunt, alter habendus naturalis sit, illum horizontalem potius, quam perpendicularem esse. Id vero patet; quia cum projectum pars fuerit aliqua totius, quod secundum horizontem, seu circulariter movebatur, ideo ad ejus imitationem movetur circulariter, ac naturaliter proinde, et prorsus equabiliter ; adeo ut, quantumcumque motus perpendicularis increscat semper, aut decrescat ; ipse tamen horizontalis uno semper tenore fluat, invariabiliterque procedat. Ac forte res minus mirabilis esset, de impressione ex motu terrae, si quis vellet ipsam supra axem suum mobilem supponere ; siquidem lapis dici posset moveri uniformiter, ob spontaneam consequutionem, ad uniformem motum totius ; seu cum eo cohaerens, seu abiunctus foret ; Sed mirabile sane est de impressione ex navi, equo, curru, aliave re, aut ex sola manu : quando lapis non habet cum rebus eiuscemodi, motibusve earum parem relationem. Ex quo par est existimare, motum horizontalem, a quacumque causa is fiat, ex sua natura perpetuum fore, nisi causa aliqua intervenerit, quae mobile abducat, motumque exturbet. Id, ut minus absurdum habeas, concipiendum est mobile, quod tantundem sese reducat, quantum abductum fuerit. Huiusmodi autem esse potest exquisitus, et uniformis materiae globus, si volvi ipsum imagineris supra horizontem, seu ambitum terrae, quem aliunde esse exquisite complanatum concipias. Si supponas enim te illi vel leviculum imprimere motum ; intelliges sane hunc motum nunquam cessaturum, sed globum revolutum iri secundum totum ambitum, ac revolutione peracta revolutum iterum iri, et consequentur iterum, et ita continuo perseveraturum. »

32 *Ibid.,* cap. X, p. 40: « Accedit, quod nulla sit causa, quamobrem suum cursum vel retardet unquam, vel acceleret, quatenus nunquam magis, vel minus a centro terrae abscedit, aut ad id accedit : neque cur proinde unquam debeat a motu cessare, quemadmodum fieret, si supponeres aliquam in superficie inaequabilitatem. »

33 *Ibid.,* cap. XVII, p. 68: « Neque vero, cum impulsum dico, attractum non intelligo :

quippe cum attrahere nihil aliud sit, quam recurvato instrumento versum se impellere ; et perspicuum sit lapidem, globumve memoratum tam impelli uno, pluribusve ictibus posse, si quis ipsum antecedendo curvis digitis adigat, quam si subsequendo devexeris propellat. » On the problem of attraction as it was seen at that time, see *La Correspondance du Père Marin Mersenne*, vol. II, p. 234 ff.

34   It is interesting to compare the way in which gravity is exorcised by Cavalieri and Torricelli with that in which this is done by Gassendi: Cavalieri and Torricelli make it a *magnitude* or a *dimension*. Gassendi, following Kepler (cf. above, pp. 144 ff., 146 ff.), makes it a mechanical force.

35   *De motu,* II, cap. VIII, p. 116: « . . . gravitatem, quae est in ipsis partibus Terrae, terrenisve corporibus, non tam esse vim insitam, quam ex attractu Terrae impressam ; idque posse intelligi adjuncto exemplo ipsius magnetis. Accipito enim, et contineto manu laminulam ferri paucarum unciarum. Si supponatur deinde manui magnes aliquis robustissimus, experiere pondus non jam unciarum, sed librarum aliquot esse. Et quia fatebere hoc pondus non tam esse insitum ferro, quam impressum ex attractione magnetis manui supposti ; idcirco ubi agitur de pondere seu gravitate lapidis, alteriusve corporis terreni, intelligi potest ea gravitas non tam convenire huiusmodi corpori ex se, quam ex attractione suppositae Terrae. » The identification, or assimilation, of terrestrial attraction with magnetic attraction was of course the fundamental idea of the work of Gilbert, and was accepted and shared by Galileo. Cf. above, p. 187 ff. As for Kepler, he provided Gassendi with the idea of the attractive chains or bonds. Cf. cap. XV, p. 61 ff.: « Fit denique, ut si duo lapides, duove globi ex eadem materia veluti ex plumbo, unus pusillus alius ingens, simul dimittantur ex eadem altitudine, eodem momento ad Terram perveniant, ac pusillus, tametsi una uncia ponderosior non sit, non minore velocitate, quam ingens, tametsi sit centum, et plurium librarum. Videlicet pluribus quidem chordulis attrahitur ingens, sed plureis etiam particulas attrahendas habet ; adeo ut fiat commensuratio inter vim, ac molem, et ex utraque utrobique tantum sit quantum ad motum sufficit eodem tempore peragendum. Id permirum ; si globi fuerint ex diversa materia, ut alter plumbeus, alter ligneus, vix quicquam tardius attingi Terram ab uno, quam ab alio, hoc est a ligneo, quam a plumbeo ; quoniam pari modo fit commensuratio, dum totidem particulis totidem chordulae destinantur. »

36   Emphasis added. *De motu,* cap. XV, p. 59: « Concipe certe lapidem in spatiis illis imaginarus, quae sunt protensa ultra hunc mundum, et in quibus posset Deus alios mundos condere ; an censeas ipsum illico ubi constitutus illeic fuerit, versus hanc Terram convolaturum, et non potius ubi fuerit semel positus, immotum mansurum, ut puta quasi non habentem neque sursum, neque deorsum, quo tendere, aut unde recedere valeat ? Si censeas fore, ut huc feratur ; imaginare non modo Terram, verum etiam totum mundum esse in nihilum redactum, spatiaque haec esse perinde inania, ac antequam Deus mundum conderet ; tunc saltem, quia centrum non erit, spatiaque omnia erunt similia ; censebis lapidem non huc accessurum, sed in loco illo fixum permansurum. Restituatur mundus, et in ipso Terra, an lapis statim huc contendet ? Si fieri dicas, oportet sane sentiri Terram a lapide, debereque proinde Terram transmittere in ipsum vim quandam, atque adeo corpuscula, quibus sui sensum illi imprimat, seseque restitutam, ac in eodem loco denuo existentem veluti renunciet. Secus enim quomodo capis posse lapidem allici ad Terram ?·»

37   Gassendi is right. Moreover this is the conclusion accepted by Telesio and Patrizzi.

38   Gassendi, *De motu,* cap. XV, p. 60: « . . . fac jam certum aliquod aëris nos ambientis spatium fieri a Deo prorsus inane, adeo ut neque ex Terra, neque aliunde aliquid in ipsum perveniat : an constitutus in eo lapis feretur in Terram, centrumve ipsius ? Certe non magis, quam constitutus in spatiis illis ultra-mundanis ; quia ipsi nihil neque cum Terra, neque cum alia re quacumque mundi ipsius communicanti, perinde erit, ac si Mundus Terraque, aut centrum non esset, nihilque rerum existeret ? »

39   Gassendi like everybody else, and in particular like Gilbert and Kepler, imagined the action of attraction to be *finite*. It required the genius and the boldness of Newton to extend its action to *infinity*.

40   *De motu,* cap. XVI, p. 62 ff.: « Quaeres obiter, quidnam eveniret illi lapidi, quem assumpsi concipi posse in spatiis illis inanibus, si a quiete exturbatus aliqua vi

impelleretur ? Respondeo probabile esse, fore, ut aequabiliter, indesinenterque moveretur ; et lente quidem, celeriterve, prout semel parvus, aut magnus impressus foret impetus. Argumentum vero desumo, ex, aequabilitate illa motus horizontalis iam exposita ; cum ille videatur aliunde non desinere nisi ex admistione motus perpendicularis ; adeo ut, quia in illis spatiis nulla esset perpendicularis admistio, in quamcumque partem foret motus inceptus, horizontalis instar esset, et neque acceleraretur, retardareturve, neque proinde unquam desineret. »

41 *De motu,* cap. XV, p. 60 ff.: « Addo saxa, et caetera corpora, quae dicuntur gravia, non eam habere ad motum resistentiam, quam vulgo concipimus. Vides quippe si ingens moles appendatur funiculo, quam levicula vi fit opus, ut e loco dimoveatur, et, prorsum, retrorsumque eat. Cur maiore ergo opus sit, ut cieri deorsum possit? Nec dicas vero esse majorem, ob motum magis pernicem ; etenim cum primum deorsum contendit, motus illius pernix non est, sed lentissimus potius, causaque dicenda mox est, ob quam deinceps acceleretur. Adnoto interea vim illam quae ex chordularum insensilium singularibus viribus conflatur, et constat, comprobari tantam, quantam superari oportet, ut manus, aut res alia gravitantem rem, velut lapidem, abducat a Terra. Et vides profecto quid fieri videatur, dum lapis tibi ipsum e Terra attollere conanti resistit. Nempe tot illae chordulae suis deflexionibus, et quasi decussationibus, illum implexum detinent ; et, nisi vis major interveniat, quae eas deflexiones, decussationesque promoveat, strictionesque fieri ulterius cogat, nunquam a Terra lapis tolletu. Heinc fit, ut quanto vis externa, seu quae a manu, aliave re extrinsecus imprimitur, pluribus gradibus vim illam chordularum superaverit, tanto lapis efferatur sublimius; quanto paucioribus, tanto humilius. Fit etiam, ut impressa vis initio pollens vehementer pellat, quia nondum refracta est ; deinceps vero segnius, segniusque, quoniam ipsi semper aliqui gradus adimuntur: donec ille solus supersit, quo exaequetur vi chordularum. »

42 De motu, cap. XVI, p. 65 ff.: « An non capis fore ut lapis appensus impressum semel motum constantissime tueatur ; scilicet omneis vibrationes non aequalibus modo temporibus peragens, sed aequalibus etiam arcubus continuo perficiens ? Haec porro omnia alio non tendunt, quam ut intelligamus motum perspatium inane impressum, ubi nihil neque attrahit, neque omnino renititur, aequabilem fore, ac perpetuum : atque exinde colligamus, omnem prorsus motum, qui lapidi imprimitur esse ex se huiusmodi ; adeo ut in quamcumque partem lapidem conjeceris si quo momento a manu emittitur, supponas omnia vi divina, lapide excepto, in nihilum redigi ; eventurum sit, ut lapis motum suum perpetuo, ac in eadem partem, in quam manus ipsum direxerit, moveatur. Nisi iam faciat, causam videri admistionem motus perpendicularis, ob attractionem a terra factam intervenientis, quae divergere illum a tramite faciat (neque cesset, quousque ipsum ad Terram usque perduxerit) ut dum ramenta ferri prope magnetem transiecta non recta pergunt, sed versus magnetem divertuntur ; aut dum universe rei, quae movetur, oblique occurrimus, ipsamque in obliquam deflectimus plagam. »

43 *De motu,* cap. XII, p. 46: « praèter causam impellentem, videtur esse necessarium ad attrahentem recurrere, quae id muneris exsequatur. Ceterum, haec vis quaenam alia sit, quam qui totius globi Telluris propria sit, et magnetica dici possit ? »

44 *De motu,* cap. XIX, p. 75 ff.: « . . . [mobili] a movente nihil imprimi aliud quam motum . . . Imprimi, inquam, qualem movens habet, donec mobile est ipsi conjunctum, et qualis continuandus esset, futurusque perpetuus, nisi a motu aliquo adverso labefactaretur. »

45 *De motu,* cap. XIX, p. 74: « Unum addo; nempe licere ex istis intelligi, quid sentiendum sit de difficultate vulgo excitata circa vim impressam projectilibus. Requiritur quippe quidnam haec vis sit in re mobili ? quamodo in ea imprimatur ? quomodo perduret ? quomodo evanescat ? Enim vero, cum haberi soleat ut vis activa lapidem movens ; videtur tamen vis activa, quae projectionis causa est, esse in ipso projiciente non vero in projecta re, quae mere passive se habet. Id quod in re projecta est, motus est, qui licet interdum nominetur vis, impetus etc. (ut etiam aliquoties a nobis factitatum est, dum, ut facilius intelligamur, familiares voces, quantum possumus, retinemus) non propterea tamen aliud quidpiam est reipsa, quam ipsemet motus. Et sane unus, idemque motus, vel per Aristotelem, actio simul et passio est ; actio prout est a movente, passio, prout in mobili ; quare ut in movente est vis activa, qua moveat, ita in mobili vis passiva, qua

moveatur ; et dum mobile reipsa movetur, non in eo querenda est vis activa, quae in movente solo necessaria fuit, sed passiva solum, quae in eo est, et redacta quidem, ut vocant, ad actum. Neque obstat, quod movens separatum sit, aut interiisse etiam, constante motu accepto, possit ; nam non propterea requiritur, ut aliam, praeter motum, vim a seipso transmiserit, quae motum deinceps efficiat ; sed sufficit ut motum semel in mobili fecerit, qui continuari absque ipso possit. Potest autem ; quoniam est ejus naturae accidens, ut modo subjectum perseverans habeat, neque contrarium quidpiam occurrat ; perseverare absque continua causae suae actione valeat. »

[46] Cf. Descartes, *Le Monde ou Traité de la Lumière, Oeuvres,* vol. XI, pp 32, 33, 35.

[47] *Le Monde,* pp. 33, 34. Cf. *Discours de la Méthode, Oeuvres,* vol. VI, p. 72 ff. Descartes joke was turned round and used against him by Père Daniel in his very amusing *Voyage du Monde de M. Descartes,* Paris, 1690.

[48] *Le Monde,* p. 11 ff.

[49] See above, Part II, p. 91 ff.; cf. Letter to Mersenne of 28 October 1650, *Oeuvres,* vol. III, p. 213: 'He was right in saying that it was a big mistake to accept as a principle that no body moves by itself. For it is certain that it is enough that a body has started to move for it to have in it the power to continue to move; in the same way that it is enough that it has come to rest in some place for it to have the power to remain there'.

[50] *Le Monde,* p. 39.

[51] *Le Monde,* p. 40.

[52] Cf. above, Part II, p. XXX ff., and Part III, p. XXX ff.

[53] *Le Monde,* p.40. See above, p. 17. Descartes misinterprets Scholastic doctrine; it is not *motion* which tends towards rest, it is the *moving body,* which is quite a different matter. But this Cartesian misinterpretation is revealing: Descartes does not really understand the motion of the philosophers.

[54] *Le Monde,* p. 38.

[55] Therefore Descartes' God has to create motion by a special act of will. It is not enough for him to create matter.

[56] *Cf. E. Meyerson,* Identité et Réalité, 3rd edition, p. 123 ff.

[57] *Le Monde,* p. 38.

[58] Above, Part II, p. 91.

[59] *Le Monde,* p. 41.

[60] *Le Monde,* p. 41.

[61] See above, p. 151 ff.

[62] Only rigid bodies are known to Cartesian physics, and this makes impact impossible. Therefore Huyghens, even though he was thoroughly Cartesian, found himself obliged to postulate elasticity, and thereby to betray Descartes. On Huyghens' physics, see the book mentioned above by P. Mouy, *Le développement de la physique cartésienne,* Paris, 1934.

[63] *Le Monde,* p. 43.

[64] *Ibid.*

[65] See-above, Part II, p. 79 ff Cf. *Correspondance du Père Marin Mersenne,* Paris, 1936, vol. II, p. 600 ff.

[66] See above, Part II, note 61.

[67] *Le Monde,* p. 43.

[68] Cf. *Le Monde,* pp. 19, 20.

[69] See above, Part II, notes 71, 83.

[70] *Le Monde,* p. 44.

[71] See above, p. 192 ff.

[72] *Le Monde,* p. 44.

[73] *Le Monde,* p. 44; emphasis added.

[74] Cf. above, Part III, note 112.

[75] Cf. above, Part II, p. 95 ff.

[76] Cf. above, Part II; note 136.

[77] *Le Monde,* p. 45.

[78] Cf. above, Part II, p. 68 ff., p. 82 ff.

[79] This is simple above all for Descartes: the equation for the circle is of one degree higher

than that for the straight line.
80 *Le Monde*, p. 45 ff.
81 *Le Monde*, p. 46.
82 *Principes de Philosophie*, Part II, section 21 (*Oeuvres*, vol. IX, 2, p. 74).
83 *Principes*, II, 37, p. 84.
84 *Principes*, II, 37, p. 85.
85 *Principes*, II, 38, p. 85.
86 *Principes*, II, 39, p. 85.
87 Certainly nobody had ever imagined any curvature in the stone! Descartes *isolates* the stone from the rest of the world and considers the motion at an instant.
88 *Principes*, II, 40, p. 86 ff.
89 Beeckman's role seems to have been negligible. Cf. *Correspondance du Père Marin Mersenne*, vol. II, p. 600 ff.
90 Cf. P. Duhem, *Le mouvement absolu et le mouvement relatif*, Montligeon, 1907, p. 179 ff.
91 Descartes gives the example of a 'man sitting in the stern of a ship which the wind carries
92 *Principes*, II, 24, p. 75.
93 *Principes*, II, 26, p. 76. This is in contrast to the Scholastic doctrine according to which the motion is in the motor as much as or even more than in the moving body.
94 *Principes*, II, 27, p. 77.
95 *Principes*, II, 29, p. 78.
96 See P. Mouy, *Le développement de la physique cartésienne*, Paris, 1934, p. 19.
97 *Principes*, II, 36, p. 83.
98 P. Mouy, *op. cit.*, p. 22.
99 P. Mouy, *op. cit.*, p. 22. The phenomenon of impact is indeeed simple in appearance only (as Mouy rightly says) and it is greatly to the credit of Huyghens (Cf. Mouy, *op. cit.*, p. 192 ff.) that he sorted out its real complexity, and thereby destroyed the Cartesian edifice.
100 On this point see the penetrating remarks by P. Mouy, *op. cit.*, p. 22 ff.
101 The idea of rest as a positive reality, and also that of the quantity of rest, can be found in Chazdaî Crescas; cf. H. A. Wolfson, *Crescas' Critique of Aristotle*, Cambridge (Mass.), 1929, p. 287 ff.
102 *Le Monde*, p. 12.
103 *Le Monde*, p. 13.
104 *Principes*, II, 43, p. 88: ' . . . it must be pointed out that the power with which one body acts on another or with which it resists the action of another, consists only in this, that each thing persists as far as possible in the state that it is already in, in conformity with the first law given above . . . So that if one body is joined to another body it has some power which prevents them from becoming separated: and when they are separated, it has some power which prevents it from becoming joined to it: and also that when it is at rest, it has some power for remaining at rest and for resisting anything which could change it'. Cf. *Principes*, II, 44: motion is not opposite to another motion, but to rest.
105 *Principes*, II, 49, p. 90: 'If a body C is bigger to any degree, however small, than B, and if it is completely at rest . . . then however fast B comes towards it, it will never have the power to move it'.

NAME INDEX